S0-BZA-745

***Modeling and Simulation in Science, Engineering and Technology***


***Series Editor***
Nicola Bellomo
Politecnico di Torino
Italy

***Advisory Editorial Board***

*M. Avellaneda* (Modeling in Economy)
Courant Institute of Mathematical Sciences
New York University
251 Mercer Street
New York, NY 10012, USA
avellaneda@cims.nyu.edu

*K.J. Bathe* (Solid Mechanics)
Department of Mechanical Engineering
Massachusetts Institute of Technology
Cambridge, MA 02139, USA
kjb@mit.edu

*P. Degond* (Semiconductor & Transport Modeling)
Mathématiques pour l'Industrie et la Physique
Université P. Sabatier Toulouse 3
118 Route de Narbonne
31062 Toulouse Cedex, France
degond@mip.ups-tlse.fr

*M.A. Herrero García* (Mathematical Methods)
Departamento de Matematica Aplicada
Universidad Complutense de Madrid
Avenida Complutense s/n
28040 Madrid, Spain
herrero@sunma4.mat.ucm.es

*W. Kliemann* (Stochastic Modeling)
Department of Mathematics
Iowa State University
400 Carver Hall
Ames, IA 50011, USA
kliemann@iastate.edu

*H.G. Othmer* (Mathematical Biology)
Department of Mathematics
University of Minnesota
270A Vincent Hall
Minneapolis, MN 55455, USA
othmer@math.umn.edu

*L. Preziosi* (Industrial Mathematics)
Dipartimento di Matematica
Politecnico di Torino
Corso Duca degli Abruzzi 24
10129 Torino, Italy
preziosi@polito.it

*V. Protopopescu* (Competitive Systems, Epistemology)
CSMD
Oak Ridge National Laboratory
Oak Ridge, TN 37831-6363, USA
vvp@epmnas.epm.ornl.gov

*K.R. Rajagopal* (Multiphase Flows)
Department of Mechanical Engineering
A&M University
College Station, TX 77843, USA
KRajagopal@mengr.tamu.edu

*Y. Sone* (Fluid Dynamics in Engineering Sciences)
Professor Emeritus
Kyoto University
230-133 Iwakura-Nagatani-cho
Sakyo-ku Kyoto 606-0026, Japan
sone@yoshio.mbox.media.kyoto-u.ac.jp

# Advances in Multifield Theories for Continua with Substructure

Gianfranco Capriz
Paolo Maria Mariano
Editors

Birkhäuser
Boston • Basel • Berlin

Gianfranco Capriz
Dipartimento di Matematica
Università di Pisa
56127 Pisa
Italy

Paolo Maria Mariano
Dipartimento di Ingegneria Strutturale
e Geotecnica
Università di Roma "La Sapienza"
00184 Roma
Italy

**Library of Congress Cataloging-in-Publication Data**

Advances in multified theories for continua with substructure / Gianfranco Capriz, Paolo Maria Mariano, editors.
p. cm. – (Modeling and simulation in science, engineering and technology)
ISBN 0-8176-4324-9 (acid-free paper) – ISBN 3-7643-4324-9 (Basel : acid free paper)
1. Continuum mechanics. I. Capriz, G. (Gianfranco) II. Mariano, Paolo Maria, 1966-
III. Series.

QA808.2.A385 2003
531–dc21

2003052406
CIP

AMS Subject Classifications: 74A10, 74A15, 74A30, 74A40, 74A35, 74A45, 74A50, 74A60, 74C15, 74N05, 74N15, 74N20, 74N30, 74R15, 74E25, 74E30, 74H33, 76F25, 76A05

Printed on acid-free paper.
©2004 Birkhäuser Boston

*Birkhäuser*

All rights reserved. This work may not be translated or copied in whole or in part without the written permission of the publisher (Birkhäuser Boston, c/o Springer-Verlag New York, Inc., 175 Fifth Avenue, New York, NY 10010, USA), except for brief excerpts in connection with reviews or scholarly analysis. Use in connection with any form of information storage and retrieval, electronic adaptation, computer software, or by similar or dissimilar methodology now known or hereafter developed is forbidden.
The use of general descriptive names, trade names, trademarks, etc., in this publication, even if the former are not especially identified, is not to be taken as a sign that such names, as understood by the Trade Marks and Merchandise Marks Act, may accordingly be used freely by anyone.

ISBN 0-8176-4324-9 SPIN 10920787
ISBN 3-7643-4324-9

Printed in the United States of America.

9 8 7 6 5 4 3 2 1

Birkhäuser Boston • Basel • Berlin
*A member of BertelsmannSpringer Science+Business Media GmbH*

# Contents

# Preface

To achieve design, implementation, and servicing of complex systems and structures in an efficient and cost-effective way, a deeper knowledge and understanding of the subtle cast and detailed evolution of materials is needed. The analysis in demand borders with the molecular and atomic one, spanning all the way down from classical continua. The study of the behavior of complex materials in sophisticated devices also opens intricate questions about the applicability of primary axioms of continuum mechanics such as the ultimate nature of the material element itself and the possibility of identifying it perfectly. So it is necessary to develop tools that allow us to formulate both theoretical models and methods of numerical approximation for the analysis of material substructures.

Multifield theories in continuum mechanics, which bridge classical materials science and modern continuum mechanics, provide precisely these tools. Multifield theories not only address problems of material substructures, but also encompass well-recognized approaches to the study of soft condensed matter and allow one to model disparate conditions in various states of matter. However, research in multifield theories is vast, and there is little in the way of a comprehensive distillation of the subject from an engineer's perspective. Therefore, the papers in the present volume, which grew out of our experience as editors for an engineering journal,[1] tackle some fundamental questions, suggest solutions of concrete problems, and strive to interpret a host of experimental evidence. In this spirit, each of the authors has contributed original results having in mind their wider applicability.

In Chapter 1, M. Silhavy deals with the energetics of ideal elastic-plastic crystalline materials under large deformations, showing that the convexification of a dissipation function (introduced to describe the set of possible free energies) provides the optimal lower bound of the plastic work. New subtle consequences ensue from the celebrated Il'yushin condition: for example, the existence of a new kind of energy function and the validity of a condition stronger than the common normality condition.

In Chapter 2, G. Parry also investigates crystal plasticity: When defective crystals are described by lattice vectors, one is able to deduce elastic invariants under deformation paths. An unusual but physically significant decomposition of the deformation gradient into elastic and plastic parts is obtained by purely geometrical

[1] G. Capriz and P. M. Mariano, eds., *Multifield Theories*, *Internat. J. Solids Structures*, **38**-6–7 (2001).

arguments, without resorting to the notion of stress; thus, issues in kinematics are discerned from issues associated with interactions. The latter ones are, in fact, power conjugated with the former and are, in this sense, subsequent to them.

Motivated by questions that arise in the mechanics of complex fluids, such as spin glasses, H. Cendra, J. E. Marsden, and T. S. Ratiu introduce, in Chapter 3, in the general context, related Poisson brackets with cocycles. Possible connections with the variational multisymplectic approach are suggested, with implications for further research. Thus the subtle and geometrical nature of mechanics, in particular mechanics of complex bodies, is emphasized.

In Chapter 4, O. Nairmark discusses physical interpretations of a wide number of detailed experimental results on defect-induced transitions and studies their consequences for the macroscopic irreversible mechanisms of metal plasticity and failure. Dynamic phenomena involving solitary waves are characterized by their connection with the essential features of the defect substructure of bodies under experimental examination. Here as well, the data and the qualitative behavior lead to additional theoretical questions.

Even simple substances, such as monoatomic gases, sometimes display complex behavior. When a monoatomic gas is rarefied, the Navier–Stokes–Fourier scheme does not hold satisfactorily. Extended thermodynamics become a necessary tool, and hundreds or thousands of moments of the distribution function enter the picture. In Chapter 5, I. Müller examines this issue, provides an analysis of the scattering of light in a gas, and investigates the shock structure in a shock tube.

When turbulence is present, drag reduction is observed in polymeric fluids as a consequence of their macromolecular substructure. When the polymeric flow becomes semidense, a multifield approach can be used to show that the terms involving spatial gradients (in particular Laplacians), required in the balance equations, are consequences of microstress, that is, of contact interactions between neighboring families of polymeric chains. This matter is discussed by E. De Angelis, C. M. Casciola, P. M. Mariano, and R. Piva in Chapter 6.

The choice of the manifold $M$ of substructural states can be, at times, quite unusual. Nevertheless the material substructure can always be characterized by an abstract, extensive property. The notion of stress still makes sense and can be achieved without the notion of "velocity." The material elements are considered as integral manifolds of a flux vector field. Such a point of view is discussed in Chapter 7 by R. Segev, and the results are illustrated by an application to the case in which the extensive property is the electric charge.

Composites are natural examples of complex materials, and they may be constructed in a number of different ways. For polycrystals, one possible method of analysis is to deduce "effective" elastic constants by homogenization. However, in general, there is no guarantee that the tensors of material properties be symmetric. Nevertheless, exact relations among material constants can be obtained which are preserved by the homogenization procedure. They give us information on the resulting composite, regardless of the complexity of the material texture. This issue is discussed by Y. Grabovsky in Chapter 8.

Shape memory behavior arises in the study of materials undergoing solid-to-solid transformations between austenite and martensite phases. Their exotic behavior is a natural proving ground for the potential of multifield theories. In Chapter 9, D. Bernardini and T. J. Pence consider three morphological descriptors of the substructural state: Two are of a scalar nature, and the third is a second-order tensor. This approach accounts not only for pseudoelasticity and shape memory behavior, but also for the high temperature reorientation of martensitic variants.

Interfaces and junctions are common in complex materials. Often interfaces are characterized by their own surface energy, and an additional energy accrues along junctions. The surface energy must be positive, while, at the junction, there may occur a change of sign. The modeling of sharp interfaces and junctions becomes more intricate when one views them as structured surfaces or lines. Surface microstresses and self-forces, as well as line stresses and self-forces, enter the picture and need to be balanced. The matter is discussed in Chapter 10 by G. Capriz and P. M. Mariano.

This collection of invited papers had its origins at Taormina, Italy, during the 2001 International Congress of the Italian Association for Theoretical and Applied Mechanics (AIMETA); it was there that many of the authors discussed with us the basic organization of this text. We are grateful to the organizers of the Congress and to the Italian National Group of Mathematical Physics for the grants that made our work possible. Furthermore, we express our appreciation to Tom Grasso and his staff at Birkhäuser Boston for providing kind support. Last but not least, we thank Nicola Bellomo for accepting our book into his series Modeling and Simulation in Science, Engineering and Technology.

*Gianfranco Capriz*
*Paolo Maria Mariano*

October 2003

*Advances in Multifield Theories for Continua with Substructure*

# 1. Work Conditions and Energy Functions for Ideal Elastic-Plastic Materials

M. Šilhavý*

**Abstract.** The paper deals with the cyclic second law and Il'yushin's condition for isothermal, ideal, isotropic, elastic-plastic materials at large deformations. The second law is equivalent to the existence of the elastic potential and the nonnegativity of plastic power. The material admits infinitely many free energies: The set of all energy functions is described in terms of a dissipation function. Its convexification provides the optimal lower bound for plastic work; it also figures in the maximal and minimal energies. Il'yushin's condition is equivalent to the existence of the elastic potential and a new condition that is stronger than the normality of the plastic stretching and the convexity of the stress range. Il'yushin's condition is also equivalent to the existence of a new kind of energy function called "the initial and final extended energy functions." Materials of type C are introduced for which the initial extended energy function has additional convexity properties. It can be viewed as a stored energy of a Hencky hyperelastic material associated with the elastic-plastic material.

## 1.1. Introduction

This paper deals with the cyclic second law and Il'yushin's condition for isothermal, ideal, isotropic elastic-plastic materials at large deformations. The ideal nature of the material excludes hardening. Both the isothermal cyclic second law and Il'yushin's condition assert the nonnegativity of the work of external forces on for certain classes of cyclic processes. The former speaks of cycles in the state space; the latter of cycles in the space of total deformations.

Sections 1.3–1.5 analyze the material from the point of view of the general theory of actions and potentials on thermodynamical systems by Coleman and Owen [5]. That theory provides an unbiased derivation of potentials like the free energy; moreover, it enables one to discuss their uniqueness/nonuniqueness and to determine the maximal and minimal potentials. For materials treated here, the plastic flow rule can be "nonassociated," and the restrictions equivalent to the second law are the existence of the elastic potential for the elastic stress function and the nonnegativity

*Mathematical Institute of the AV ČR, Žitná 25, 115 67 Prague, Czech Republic, silhavy@math.cas.cz and Department of Mathematics, University of Pisa, Via F. Buonarroti 2, 56127 Pisa, Italy, silhavy@dm.unipi.it.

of the plastic power in every process. Using that, and defining the free energy as any state function satisfying the dissipation inequality, one finds that the elastic potential is a possible free energy. However, the free energies are highly nonunique here: Infinitely many of them differ mutually by a nonconstant state function. A complete description of the set of all free energy functions is given using a dissipation function that is closely related to the plastic power. The central result of this part says that the optimal lower bound for the plastic work is the convexification of the dissipation function. The maximal/minimal free energies are the elastic potential plus/minus the convexification of the dissipation function at the logarithmic plastic stretching. Analyses of elastic-plastic materials from the standpoint of [5] have been given previously by Coleman and Owen [6], [7] for unidimensional infinitesimal ideal materials, by Lucchesi [14] for three-dimensional infinitesimal materials with hardening, and by Lucchesi and Šilhavý [17] for large deformation three-dimensional infinitesimal materials with hardening.

In the context of infinitesimal deformation theory, Il'yushin's condition is classically known to imply the convexity of the stress range (CSR) (the stress range is the region below the yield surface) and the normality rule (NR) asserting that during loading, the plastic stretching is proportional to the exterior normal to the yield surface (e.g., [25]). The analysis has been extended to large deformations by Lucchesi and Podio-Guidugli [15] (see also [16] and [10]). Their analysis showed that Il'yushin's condition still implies the CSR and NR provided that the stress range is interpreted as the set of elastically reachable Kirchhoff stresses (as opposed to Cauchy stresses; see Section 1.2 for definition); in addition, it implies the existence of the elastic potential (EP). However, their analysis shows that conversely EP, CSR, and NR lead to Il'yushin's condition only for deformation cycles that are small in a precisely defined sense. The main feature of the analysis in Sections 1.6–1.7, and what distinguishes it from the previous work, is the possibility of treating arbitrarily large deformation cycles. Namely, Il'yushin's condition is shown to be equivalent to EP and a new Condition E (Condition (ii) of Theorem 2). Condition E is stronger than NR and CSR. On the basis of Il'yushin's condition for large cycles, new energy functions are constructed satisfying dissipation inequalities stronger than those based on the second law. Their main feature is that they are nonlocal in time (have no localized counterparts). There are two types of the new energy functions; I call these the initial and final (extended) energy functions.

In Sections 1.9–1.10 materials called materials of type C are studied that permit a more explicit analysis. The main requirement is a restricted convexity in the logarithmic deformation on symmetric deformation gradients of fixed determinant. Materials of type C satisfy Il'yushin's condition and the initial extended energy admits a description in terms of convex-conjugated functions in the logarithmic deformation. The initial extended energy can be interpreted as a stored energy of some hyperelastic material; this material is the large deformation analogue of the hyperelastic material in the infinitesimal Hencky theory of plasticity. Its stress relation coincides with the nonlinear elastic response of the plastic material for deformation gradients from the original elastic range; if the deformation is outside the elastic range, then the stress is determined by some nonlinear projection onto the elastic range (in the space of the logarithmic deformations; see Section 1.9 for a precise description). Unexpectedly, the extended energy grows logarithmically at large deformations, in contrast to the infinitesimal Hencky theory, where the growth is linear (cf. Témam [31]). As an illustration, an elastic-plastic material of type C is considered for which the extended energy and all other derived objects can be calculated explicitly.

The above description makes it clear that the logarithmic deformation and logarithmic convexity/concavity emerge naturally from the analysis. The reason is that for large deformations theory, the processes of minimal dissipation are those for which the logarithm of plastic deformation, and not the plastic deformation itself, is linear in time.

## 1.2. Ideal elastic-plastic materials

Let Lin denote the set of all second-order tensors on a three-dimensional space; we use the scalar product $\mathbf{A} \cdot \mathbf{B} = \mathrm{tr}(\mathbf{A}\mathbf{B}^T)$ on Lin and the euclidean norm $|\mathbf{A}| := \sqrt{\mathbf{A} \cdot \mathbf{A}}$. The deformation gradients are interpreted as the elements of the set $\mathrm{Lin}^+$ of all second-order tensors with positive determinant. $\mathrm{Sym}^+$ is the set of all positive definite symmetric tensors.

For the considerations of the paper, it is convenient to describe the response in terms of states and processes (cf. [20], [5], [6], [7], [28], [16], [29], [2], [3], [24]). The *state space* $\Sigma$ is

$$\Sigma = \{\sigma = (\mathbf{E}, \mathbf{P}) : \mathbf{E} \in \mathcal{E}, \mathbf{P} \in \mathrm{Unim}\} = \mathcal{E} \times \mathrm{Unim},$$

where $\mathcal{E} \subset \mathrm{Lin}^+$ and Unim is the set of all tensors with determinant 1. The set $\mathcal{E}$ is closely related to the elastic range (see below). The *states* $\sigma$ are pairs $(\mathbf{E}, \mathbf{P})$ where $\mathbf{E} \in \mathcal{E}$ is the *elastic deformation* and $\mathbf{P}$ is an unimodular *plastic deformation.* The *total deformation* and the *Kirchhoff stress* of $\sigma = (\mathbf{E}, \mathbf{P})$ are

$$\hat{\mathbf{F}}(\sigma) := \mathbf{E}\mathbf{P}, \quad \hat{\mathbf{T}}(\sigma) := \bar{\mathbf{T}}(\mathbf{E}), \tag{1}$$

where $\bar{\mathbf{T}} : \mathcal{E} \to \mathrm{Sym}$ is a given function and Sym is the set of all symmetric tensors. Equation (1) expresses the multiplicative decomposition of the deformation gradient into the elastic and plastic parts [13]. The Kirchhoff stress $\mathbf{T} := (\det \mathbf{F})\mathbf{T}_C$ is more convenient than the Cauchy stress $\mathbf{T}_C$ here. Let $\bar{\mathbf{S}} : \mathcal{E} \to \mathrm{Lin}_0$ denote the mapping defined by

$$\bar{\mathbf{S}}(\mathbf{E}) = \mathbf{E}^T\bar{\mathbf{T}}(\mathbf{E})\mathbf{E}^{-T} - \tfrac{1}{3}(\mathrm{tr}\bar{\mathbf{T}}(\mathbf{E}))\mathbf{1},$$

where $\mathrm{Lin}_0$ is the set of all traceless tensors. The definition is motivated by the expression for the plastic work (4) below. The *stress range* $\mathcal{S}$ is defined by

$$\mathcal{S} := \bar{\mathbf{S}}(\mathcal{E}).$$

To specify the *class of processes* $\Pi$, we assume that the plastic deformation changes only when $\mathbf{E}$ is on the boundary $\partial\mathcal{E}$ of $\mathcal{E}$ and that the direction of the plastic stretching is determined by a prescribed function $\bar{\mathbf{M}}$ mapping $\partial\mathcal{E}$ into the set of all traceless symmetric tensors $\mathrm{Sym}_0$, normalized by $|\bar{\mathbf{M}}(\mathbf{E})| = 1$, $\mathbf{E} \in \partial\mathcal{E}$. Thus we assume the flow rule of the form

$$\mathbf{D}^p(t) = \alpha(t)\bar{\mathbf{M}}(\mathbf{E}(t)), \tag{2}$$

where the *plastic stretching* $\mathbf{D}^p(t)$ is the symmetric part of $\mathbf{L}^p(t) = \dot{\mathbf{P}}(t)\mathbf{P}(t)^{-1}$, and $\alpha(t) > 0$ is a coefficient of proportionality. Formally, $\Pi$ consists of all functions $\pi = (\mathbf{E}(\cdot), \mathbf{P}(\cdot))$ mapping closed intervals of the type $[0, d_\pi]$, $d_\pi > 0$ into $\Sigma$ that are continuous and piecewise continuously differentiable and that satisfy the following condition: For every $t \in [0, d_\pi]$ for which $\mathbf{D}^p(t) \neq \mathbf{0}$, one has $\mathbf{E}(t) \in \partial\mathcal{E}$ and (2) holds for some $\alpha(t) > 0$. For convenience it is assumed that every process starts at time $t = 0$. The number $d_\pi$ can be different for different processes and it is called the *duration of the process*. The states $\pi^{\mathrm{i}} = \pi(0)$ and $\pi^{\mathrm{f}} = \pi(d_\pi)$ are called the *initial* and *final states* of the process $\pi$. We also write $\pi^{\mathrm{i}} = (\mathbf{E}^{\mathrm{i}}, \mathbf{P}^{\mathrm{i}})$, $\pi^{\mathrm{f}} = (\mathbf{E}^{\mathrm{f}}, \mathbf{P}^{\mathrm{f}})$, and $\mathbf{F}^{\mathrm{i}} = \hat{\mathbf{F}}(\pi^{\mathrm{i}})$, $\mathbf{F}^{\mathrm{f}} = \hat{\mathbf{F}}(\pi^{\mathrm{f}})$. The considerations in the paper require the construction of new processes from given ones via the operation of continuation. If $\pi_1, \pi_2 \in \Pi$ are two processes with $\pi_1^{\mathrm{f}} = \pi_2^{\mathrm{i}}$ then the *continuation* $\pi_1 * \pi_2$ *of* $\pi_1$ *with* $\pi_2$, is

$$\pi_1 * \pi_2(t) = \begin{cases} \pi_1(t) & \text{if } t \in [0, d_{\pi_1}], \\ \pi_2(t - d_{\pi_1}) & \text{if } t \in [d_{\pi_1}, d_{\pi_1} + d_{\pi_2}]. \end{cases}$$

For every $\delta > 0$ let

$$\mathcal{E}_\delta = \{\mathbf{A} \in \mathcal{E} : \det \mathbf{A} = \delta\}. \tag{3}$$

A set $M \subset \mathrm{Lin}$ is said to be objective or isotropic if

$$\mathbf{RA} \in M, \quad \mathbf{RAR}^T \in M,$$

respectively, for every $\mathbf{A} \in M$ and $\mathbf{R} \in \mathrm{Rot}$. Here Rot is the proper orthogonal group. A function $f : M \to \mathbb{R}$, $M \subset \mathrm{Lin}$ is said to be objective or isotropic if $M$ is objective or isotropic, respectively, and

$$f(\mathbf{RA}) = f(\mathbf{A}), \quad f(\mathbf{RAR}^T) = f(\mathbf{A}),$$

respectively, for every $\mathbf{A} \in M$ and $\mathbf{R} \in \mathrm{Rot}$. A function $\mathbf{G} : M \to \mathrm{Lin}$ is said to be objective or isotropic if $M$ is objective or isotropic, respectively, and

$$\mathbf{G}(\mathbf{RA}) = \mathbf{RG}(\mathbf{A})\mathbf{R}^T, \quad \mathbf{G}(\mathbf{RAR}^T) = \mathbf{RG}(\mathbf{A})\mathbf{R}^T,$$

respectively, for every $\mathbf{A} \in M$ and $\mathbf{R} \in \mathrm{Rot}$. $\mathbf{G}$ is said to be scalar-objective if $M$ is objective and

$$\mathbf{G}(\mathbf{RA}) = \mathbf{G}(\mathbf{A})$$

for every $\mathbf{A} \in M$ and $\mathbf{R} \in \mathrm{Rot}$.

**Definition 1.** *The objects $\mathcal{E}$, $\bar{\mathbf{T}}$, $\bar{\mathbf{M}}$ are said to determine an ideal elastic-plastic material if they satisfy the following conditions:*

(i) *$\mathcal{E}$ is a closure of its interior, any two points of $\mathcal{E}$ can be connected by a piecewise smooth curve in $\mathcal{E}$, and $\bar{\mathbf{T}}$ is continuous.*

(ii) *For every $\mathbf{C} \in \mathrm{Sym}_0$, $|\mathbf{C}| = 1$, we have $\bar{\mathbf{M}}(\mathbf{E}) = \mathbf{C}$ for at least one $\mathbf{E} \in \partial\mathcal{E}$.*

(iii) *$\mathcal{S}$ is bounded and $\bar{\mathbf{S}}$ maps the interior and the boundary of $\mathcal{E}$ onto the interior and boundary of $\mathcal{S}$, respectively.*
(iv) *For each $\delta > 0$, the set $\mathcal{E}_\delta$ is nonempty.*
(v) *The set $\mathcal{E}$ and the function $\bar{\mathbf{T}}$ are objective and isotropic and the function $\bar{\mathbf{M}}$ is scalar-objective and isotropic.*

The quadruple $\mathcal{M} = (\Sigma, \Pi, \hat{\mathbf{F}}, \hat{\mathbf{T}})$ is called the *ideal elastic-plastic material* or, briefly, the *material*. The material just constructed is a special case of a material with elastic range (cf. [25], [22], [23], [27], [15], [16]). For processes starting at $\sigma = (\mathbf{E}, \mathbf{P})$ and remaining in the elastic range $\mathcal{E}(\sigma) := \mathcal{E}\mathbf{P}$ the material behaves like an elastic material with the elastic response

$$\hat{\mathbf{T}}^*(\sigma, \mathbf{F}) = \bar{\mathbf{T}}(\mathbf{F}\mathbf{P}^{-1}), \quad \mathbf{F} \in \mathcal{E}(\sigma).$$

Item (v) of Definition 1 expresses the objectivity and isotropy of the material. The objectivity and isotropy of $\mathcal{E}$ and $\bar{\mathbf{T}}$ need not be commented; however, note that they imply that $\bar{\mathbf{S}}$ is scalar-objective and isotropic; moreover, its values are symmetric tensors. The scalar-objective nature of $\bar{\mathbf{M}}$ is consistent with the following transformation rules under a change of frame and change of reference configuration: If $\pi = (\mathbf{E}(\cdot), \mathbf{P}(\cdot))$ is a process, $\mathbf{Q}$ a piecewise continuously differentiable function on $[0, d_\pi]$ with values in Rot, and $\mathbf{R}$ a fixed element of Rot, then

$$\mathbf{F} \mapsto \mathbf{QF} \Rightarrow \mathbf{E} \mapsto \mathbf{QE}, \quad \mathbf{P} \mapsto \mathbf{P},$$
$$\mathbf{F} \mapsto \mathbf{FR} \Rightarrow \mathbf{E} \mapsto \mathbf{E}, \quad \mathbf{P} \mapsto \mathbf{PR}.$$

Other transformation laws are possible since the plastic deformation is determined only to within a rotation in isotropic materials [27].

## 1.3. The cyclic second law: Its first consequences

The *work* of external forces on the material in the process $\pi = (\mathbf{E}(\cdot), \mathbf{P}(\cdot))$ is

$$w(\pi) = \int_0^d \mathbf{T} \cdot \mathbf{L}\, dt,$$

where $\mathbf{L} = \dot{\mathbf{F}}\mathbf{F}^{-1}, \mathbf{F} = \mathbf{EP}, d = d_\pi$, and $\mathbf{T} = \bar{\mathbf{T}}(\mathbf{E})$ is the time evolution of the Kirchhoff stress in the process. Using the isotropy and symmetry of $\bar{\mathbf{T}}$ we find that

$$w(\pi) = w^E(\pi) + w^P(\pi),$$

where

$$w^E(\pi) = \int_0^d \mathbf{T} \cdot \dot{\mathbf{E}}\mathbf{E}^{-1}\, dt, \quad w^P(\pi) = \int_0^d \mathbf{S} \cdot \mathbf{D}^p\, dt \tag{4}$$

are the *elastic* and *plastic works* in $\pi$, respectively, where $\mathbf{S} = \bar{\mathbf{S}}(\mathbf{E})$ is the time-evolution of the traceless part of the Kirchhoff stress. A process $\pi$ is said to be a $\sigma$-cycle if $\pi^{\mathrm{i}} = \pi^{\mathrm{f}}$. For isothermal materials the second law reduces to the following assertion [5].

**The cyclic second law.** For every $\sigma$-cycle $\pi$, $w(\pi) \geq 0$.

A function $p : \mathcal{E} \to \mathbb{R}$ is said to be an *elastic potential* for $\bar{\mathbf{T}}$ if

$$p(\mathbf{E}^{\mathrm{f}}) - p(\mathbf{E}^{\mathrm{i}}) = w^E(\pi) \tag{5}$$

for every process. It follows that if $p$ exists, it is objective and isotropic and continuously differentiable in the interior of $\mathcal{E}$ with the *stress relation* $\bar{\mathbf{T}}(\mathbf{E}) = \mathrm{D}p(\mathbf{E})\mathbf{E}^T$ prevailing there.

**Proposition 1.** *The material satisfies the cyclic second law if and only if* $\bar{\mathbf{T}}$ *has an elastic potential and*

$$\bar{\mathbf{S}}(\mathbf{E}) \cdot \bar{\mathbf{M}}(\mathbf{E}) \geq 0 \tag{6}$$

*for every* $\mathbf{E} \in \partial\mathcal{E}$.

Inequality (6) says that the plastic power, and hence also the plastic work, is nonnegative: $\mathbf{S} \cdot \mathbf{D}^p = \alpha(t)\mathbf{S} \cdot \mathbf{M} \geq 0$. Let us emphasize that this consequence does not hold for general, nonideal, elastic-plastic materials.

*Proof.* Suppose that the material satisfies the cyclic second law. To prove the existence of the potential, consider the elastic process $\pi = (\mathbf{E}(\cdot), 1)$, where $\mathbf{E}(\cdot) : [0, 1] \to \mathcal{E}$ is a path with values in $\mathcal{E}$. If $\mathbf{E}(0) = \mathbf{E}(1)$ then the cyclic second law asserts $w(\pi) \geq 0$. Applying the same to the time reversal $\bar{\pi}$ we obtain $w(\bar{\pi}) = -w(\pi) \geq 0$ and thus $w(\pi) = 0$. Since $w(\pi) = w^E(\pi)$, the vector field $\bar{\mathbf{T}}$ is path independent on $\mathcal{E}$; hence it has an elastic potential. Proof of (6): Let $\mathbf{E} \in \partial\mathcal{E}$ and set $\mathbf{D} := \bar{\mathbf{M}}(\mathbf{E})$. Consider an orthonormal basis of eigenvectors of $\mathbf{D}$ so that $\mathbf{D} = \mathrm{diag}(d_1, d_2, d_3)$. Let $\mathbb{P}^3$ be the group of all 3 by 3 permutation matrices $Z$ which we identify with orthogonal tensors $\mathbf{Z}$ in our basis. Enumerate the elements of $\mathbb{P}^3$ arbitrarily to obtain a sequence $\mathbf{Z}_\alpha$, $\alpha = 1, \ldots, n \equiv 3!$. Since

$$\mathbf{Z}_\alpha \mathbf{D} \mathbf{Z}_\alpha^T = \mathrm{diag}(d_{\sigma_\alpha(1)}, d_{\sigma_\alpha(2)}, d_{\sigma_\alpha(3)}),$$

where $\sigma_\alpha$ is the permutation corresponding to $\mathbf{Z}_\alpha$, the family $\{\mathbf{Z}_\alpha \mathbf{D} \mathbf{Z}_\alpha^T, \alpha = 1, \ldots, n\}$ is commutative. Using $\mathrm{tr}\mathbf{D} = 0$, one finds that

$$\sum_{\alpha=1}^{n} \mathbf{Z}_\alpha \mathbf{D} \mathbf{Z}_\alpha^T = \mathbf{0}. \tag{7}$$

Define inductively the processes $\pi_\alpha = (\mathbf{E}_\alpha(\cdot), \mathbf{P}_\alpha(\cdot))$ of duration 1 by

$$\mathbf{E}_1(t) = \mathbf{Z}_1 \mathbf{E} \mathbf{Z}_1^T, \quad \mathbf{P}_1(t) = \mathbf{Z}_1 e^{\mathbf{D}t} \mathbf{Z}_1^T,$$

and for $\alpha > 1$,

$$\mathbf{E}_\alpha(t) = \mathbf{Z}_\alpha \mathbf{E} \mathbf{Z}_\alpha^T, \quad \mathbf{P}_\alpha(t) = \mathbf{Z}_\alpha e^{\mathbf{D}t} \mathbf{Z}_\alpha^T \mathbf{P}_{\alpha-1}^{\mathrm{f}},$$

$t \in [0, 1]$. These are really processes and one finds that $\mathbf{D}_\alpha^p(t) = \mathbf{Z}_\alpha \mathbf{D} \mathbf{Z}_\alpha^T$ and

$$w^P(\pi_\alpha) = \bar{\mathbf{S}}(\mathbf{E}) \cdot \bar{\mathbf{M}}(\mathbf{E}). \tag{8}$$

Let $\rho_\alpha = (\mathbf{U}_\alpha(\cdot), \mathbf{P}_\alpha^{\mathrm{f}}))$, $\alpha = 1, \ldots, n$, be elastic processes such that

$$\mathbf{U}_\alpha^{\mathrm{i}} = \mathbf{Z}_\alpha \mathbf{E} \mathbf{Z}_\alpha^T, \quad \mathbf{U}_\alpha^{\mathrm{f}} = \mathbf{Z}_{\alpha+1} \mathbf{E} \mathbf{Z}_{\alpha+1}^T,$$

if $i < n$, and

$$\mathbf{U}_n^{\mathrm{i}} = \mathbf{Z}_n\mathbf{E}\mathbf{Z}_n^T, \quad \mathbf{U}_n^{\mathrm{f}} = \mathbf{Z}_1\mathbf{E}\mathbf{Z}_1^T.$$

One finds that the process $\pi = (\mathbf{E}(\cdot), \mathbf{P}(\cdot)) := \pi_1 * \rho_1 * \pi_2 * \rho_2 \cdots \pi_n * \rho_n$ can be constructed from

$$\mathbf{E}^{\mathrm{i}} = \mathbf{Z}_1\mathbf{E}\mathbf{Z}_1, \quad \mathbf{E}^{\mathrm{f}} = \mathbf{Z}_1\mathbf{E}\mathbf{Z}_1,$$

and

$$\mathbf{P}^{\mathrm{i}} = \mathbf{1}, \quad \mathbf{P}^{\mathrm{f}} = \exp[\sum_{\alpha=1}^{n} \mathbf{Z}_\alpha\mathbf{D}\mathbf{Z}_\alpha^T] = \mathbf{1}.$$

Here we use (7) and the commutativity of the family $\{\mathbf{Z}_\alpha\mathbf{D}\mathbf{Z}_\alpha^T, \alpha = 1, \ldots, n\}$. As also $\mathbf{E}^{\mathrm{i}} = \mathbf{E}^{\mathrm{f}}$, the process $\pi$ is cyclic. Since $w^E(\pi) = 0$ by (5), $w(\pi) \geq 0$ reduces, by (8), to

$$w^P(\pi) = n\bar{\mathbf{S}}(\mathbf{E}) \cdot \bar{\mathbf{M}}(\mathbf{E}) \geq 0$$

and (6) follows. Conversely, if the two conditions of the theorem hold and $\pi$ is a cyclic process then $w^E(\pi) = 0$ by the existence of the potential and $w^P(\pi) \geq 0$ by (6). Hence the cyclic second law holds. □

## 1.4. The free energy functions

Any function $\psi : \Sigma \to \mathbb{R}$ satisfying

$$\psi(\pi^{\mathrm{f}}) - \psi(\pi^{\mathrm{i}}) \leq w(\pi) \tag{9}$$

for every process $\pi$ is referred to as the *free energy function.* Using the nonnegativity of the plastic power, it is shown below that the elastic potential is one example of the free energy function, and a description is given of all free energies. The specific features of the material imply that any free energy function splits into a sum of the reversible elastic potential and a residual energy function that depends only on the plastic deformation.

**Proposition 2.** *Let the material satisfy the cyclic second law and let $p$ be its elastic potential. A function $\psi : \Sigma \to \mathbb{R}$ is a free energy function if and only if*

$$\psi(\sigma) = p(\mathbf{E}) + r(\mathbf{P}) \tag{10}$$

*for every $\sigma = (\mathbf{E}, \mathbf{P}) \in \Sigma$, where $r : \mathrm{Unim} \to \mathbb{R}$ satisfies*

$$r(\mathbf{P}^{\mathrm{f}}) - r(\mathbf{P}^{\mathrm{i}}) \leq w^P(\pi) \tag{11}$$

*for every process $\pi = (\mathbf{E}(\cdot), \mathbf{P}(\cdot)) \in \Pi$. In particular, the function $\psi : \Sigma \to \mathbb{R}$ given by*

$$\psi(\sigma) = p(\mathbf{E}), \tag{12}$$

*$\sigma = (\mathbf{E}, \mathbf{P}) \in \Sigma$ is a free energy function.*

Let us emphasize that a general (nonideal) elastic-plastic material need not have a free energy independent of plastic deformation. Every function $r$ satisfying (11) in every process is called the *residual energy function.* The problem of describing all free energy functions reduces to that of describing all residual energy functions.

*Proof.* By Proposition 1, $\bar{\mathbf{T}}$ has an elastic potential $p$ and the plastic work is nonnegative in every process. If $\psi$ is a free energy function and $\mathbf{P} \in \mathrm{Unim}$, then for every path $\mathbf{E}(\cdot) : [0, 1] \to \mathcal{E}$ the process $\pi = (\mathbf{E}(\cdot), \mathbf{P})$ is elastic and as $w(\pi) = w^E(\pi)$ is given by (5), the dissipation inequality (9) reads

$$\psi(\mathbf{E}^{\mathrm{f}}, \mathbf{P}) - \psi(\mathbf{E}^{\mathrm{i}}, \mathbf{P}) \geq p(\mathbf{E}^{\mathrm{f}}) - p(\mathbf{E}^{\mathrm{i}}). \tag{13}$$

Replacing $\pi$ by its time-reversal $\bar{\pi}$ we obtain the opposite inequality and hence the equality must hold in (13). It follows that for each $\mathbf{P} \in \mathrm{Unim}$, $\psi(\cdot, \mathbf{P})$ differs by a constant $r(\mathbf{P})$ from $p$ which leads to (10). Using (10), one sees that (9) reduces to (11). Reversing the direction of the arguments, one finds that every function $\psi$ of the form (10) is a free energy function. Since the plastic work is nonnegative, the function $r \equiv 0$ satisfies (11) and hence the $\psi$ given by (12) is a free energy function. □

To simplify notation, if $\mathbf{A}, \mathbf{B} \in \mathrm{Lin}$, we write

$$\mathbf{A} \sim \mathbf{B}$$

to mean that $|\mathbf{A}| = 1$, $\mathbf{B} \neq \mathbf{0}$, and $\mathbf{A} = \mathbf{B}/|\mathbf{B}|$. Define the *dissipation function* $m : \mathrm{Sym}_0 \to \mathbb{R}$ by

$$m(\mathbf{D}) := \inf \left\{ \bar{\mathbf{S}}(\mathbf{E}) \cdot \mathbf{D} : \mathbf{E} \in \partial\mathcal{E}, \bar{\mathbf{M}}(\mathbf{E}) \sim \mathbf{D} \right\}$$

if $\mathbf{D} \neq \mathbf{0}$ and $m(\mathbf{0}) := 0$. Note that the infimum is taken over a nonempty set by Definition 1(ii). If the material satisfies the cyclic second law, then $m$ is nonegative by Proposition 1.

If $r$ is a function on an open subset of Unim and $\mathbf{P}$ is in the domain of $r$, then $r$ is said to have a total differential at $\mathbf{P}$ if the function $\mathbf{A} \mapsto r(e^{\mathbf{A}}\mathbf{P})$, $\mathbf{A} \in \mathrm{Lin}_0$ has a Fréchet derivative at $\mathbf{0}$ in the sense of [8, Part I, Chapter VIII]. The differential (derivative) $\mathrm{D}r(\mathbf{P})$ is then defined as the unique element satisfying $\mathrm{tr}(\mathrm{D}r(\mathbf{P})\mathbf{P}^T) = 0$ such that

$$\left. \frac{d}{dt} r(\gamma(t)) \right|_{t=0} = \mathrm{D}r(\mathbf{P}) \cdot \dot{\gamma}(0)$$

for every smooth curve in Unim with $\gamma(0) = \mathbf{P}$. Note that Unim is a regular eight-dimensional surface in the nine-dimensional space Lin and that the unit normal and the tangent space to Unim at $\mathbf{A}$ are $\mathbf{A}^{-T}/|\mathbf{A}^{-T}|$ and $\left\{\mathbf{M} \in \mathrm{Lin} : \mathrm{tr}(\mathbf{M}\mathbf{A}^{-1}) = 0\right\}$. The Haar measure ([8, Part II, Chapter XIV and Part IV, Chapter XIX]) and the surface measures on Unim are mutually absolutely continuous and their images in any local coordinate chart on Unim are absolutely continuous with respect to the eight-dimensional Lebesgue measure in that coordinate chart. Hence there is a well-defined notion of almost everywhere. If $r$ is a locally Lipschitz continuous function on Unim, Rademacher's theorem ([19]) asserts the existence of the total differential for almost every $\mathbf{P} \in \mathrm{Unim}$. Finally note that if $\mathbf{P}_1, \mathbf{P}_2 \in \mathrm{Unim}$ then there exist a unique pair $\mathbf{R} \in \mathrm{Rot}$, $\mathbf{D} \in \mathrm{Sym}_0$ such that $\mathbf{P}_2 = \mathbf{R}e^{\mathbf{D}}\mathbf{P}_1$. This follows from the polar decomposition $\mathbf{P}_2\mathbf{P}_1^{-1} = \mathbf{R}\mathbf{U}$ by writing $\mathbf{U} = e^{\mathbf{D}}$, $\mathbf{D} \in \mathrm{Sym}_0$.

**Proposition 3.** *If the material satisfies the cyclic second law and* $r : \mathrm{Unim} \to \mathbb{R}$ *is a function then the following three conditions are equivalent:*

(i) *$r$ is a residual energy;*

(ii) *for every* $\mathbf{P}_1, \mathbf{P}_2 \in \mathrm{Unim}$ *one has*

$$r(\mathbf{P}_2) - r(\mathbf{P}_1) \le m(\mathbf{D}), \tag{14}$$

*where* $\mathbf{D} \in \mathrm{Sym}_0$ *is determined by the condition that* $\mathbf{P}_2 = \mathbf{R}e^{\mathbf{D}}\mathbf{P}_1$ *for some* $\mathbf{R} \in \mathrm{Rot}$;

(iii) *$r$ is locally Lipschitz continuous, objective, and for almost every* $\mathbf{P} \in \mathrm{Unim}$,

$$\mathrm{D}r(\mathbf{P})\mathbf{P}^T \cdot \mathbf{D} \le m(\mathbf{D}) \quad \textit{for all} \quad \mathbf{D} \in \mathrm{Sym}_0. \tag{15}$$

*Proof.* (i) ⇒ (ii): Let (i) hold and let $\mathbf{P}_1, \mathbf{P}_1, \mathbf{D}, \mathbf{R}$ be as in (ii). Let us prove that

$$r(\mathbf{P}_2) - r(\mathbf{P}_1) \le \begin{cases} 0 & \text{if } \mathbf{D} = \mathbf{0}, \\ \bar{\mathbf{S}}(\mathbf{E}) \cdot \mathbf{D} & \text{if } \mathbf{D} \ne \mathbf{0}, \end{cases} \tag{16}$$

where $\mathbf{E} \in \partial\mathcal{E}$ is any element such that

$$\bar{\mathbf{M}}(\mathbf{E}) \sim \mathbf{D}. \tag{17}$$

Let $\mathbf{E}$ be arbitrary if $\mathbf{D} = \mathbf{0}$ or such that $\mathbf{E} \in \partial\mathcal{E}$ and (17) holds if $\mathbf{D} \ne 0$. Let $\mathbf{Q} : [0, 1] \to \mathrm{Rot}$ be any continuously differentiable function such that $\mathbf{Q}(0) = \mathbf{1}$ and $\mathbf{Q}(1) = \mathbf{R}$. Define $\pi = (\mathbf{E}(\cdot), \mathbf{P}(\cdot))$ by

$$\mathbf{E}(t) = \mathbf{Q}(t)\mathbf{E}\mathbf{Q}(t)^T, \quad \mathbf{P}(t) = \mathbf{Q}(t)e^{\mathbf{D}t}\mathbf{P}_1, \quad t \in [0, 1].$$

This is a process, $\mathbf{D}^p(t) = \mathbf{Q}(t)\mathbf{D}\mathbf{Q}(t)^T$, $\mathbf{P}^{\mathrm{f}} = \mathbf{P}_2$, $\mathbf{P}^{\mathrm{i}} = \mathbf{P}_1$, and $w^P(\pi) = \bar{\mathbf{S}}(\mathbf{E}) \cdot \mathbf{D}$. Thus the residual dissipation inequality gives (16). The definition of $m(\mathbf{D})$ then gives (14).

(ii) ⇒ (iii): Assume that (ii) holds. Let us first prove that $r$ is locally Lipschitz continuous. Recall that it is assumed (Definition 1(iii)) $\mathcal{S}$ is bounded. Set

$$c_1 = \max\{|\mathbf{S}| : \mathbf{S} \in \mathcal{S}\}.$$

By Definition 1(ii), for every $\mathbf{D} \in \mathrm{Sym}_0$, $\mathbf{D} \ne \mathbf{0}$, there exists an $\mathbf{E} \in \partial\mathcal{E}$ such that (17) holds. We then have

$$\bar{\mathbf{S}}(\mathbf{E}) \cdot \mathbf{D} \le c_1|\mathbf{D}|.$$

Thus for any $\mathbf{P}_1, \mathbf{P}_2 \in \mathrm{Unim}$ we have

$$r(\mathbf{P}_2) - r(\mathbf{P}_1) \le c_1|\mathbf{D}|, \tag{18}$$

where $\mathbf{D}$ satisfies $\mathbf{P}_2 = \mathbf{R}e^{\mathbf{D}}\mathbf{P}_1$. Since this holds for any $\mathbf{P}_1, \mathbf{P}_2 \in \mathrm{Unim}$, we can interchange the roles of $\mathbf{P}_1, \mathbf{P}_2$. One finds that $\mathbf{D}$ changes to $\bar{\mathbf{D}} = -\mathbf{R}\mathbf{D}\mathbf{R}^T$ and (18) for the interchanged pair provides $r(\mathbf{P}_1) - r(\mathbf{P}_2) \le c_1|\mathbf{D}|$; hence

$$|r(\mathbf{P}_2) - r(\mathbf{P}_1)| \le c_1|\mathbf{D}|. \tag{19}$$

This inequality obviously implies that $r$ is locally Lipschitz continuous. To prove that $r$ is objective, it suffices to note that if $\mathbf{P} \in \text{Unim}$, $\mathbf{Q} \in \text{Rot}$, then for $\mathbf{P}_1 := \mathbf{P}$, $\mathbf{P}_2 = \mathbf{QP}$, (19) reduces to $r(\mathbf{QP}) = r(\mathbf{P})$. We finally prove (15). Let $\mathbf{P} \in \text{Unim}$ be such that Dr exists. The application of (14) to $\mathbf{P}_2 = e^{\mathbf{D}t}\mathbf{P}$, $\mathbf{P}_1 = \mathbf{P}$, $t > 0$, provides

$$r(e^{\mathbf{D}t}\mathbf{P}) - r(\mathbf{P}) \le m(\mathbf{D}t) = tm(\mathbf{D}).$$

Dividing by $t$, letting $t$ tend to 0, and using the assumed existence of the total differential of $r$ at $\mathbf{P}$ gives (15).

(iii) $\Rightarrow$ (i): Let $r$ satisfy Condition (iii). Note first that a standard consequence of the objectivity asserts that $\text{Dr}(\mathbf{P})\mathbf{P}^T$ is symmetric for every $\mathbf{P} \in \text{Unim}$ for which the total differential exists. Let us now prove (i); i.e., let us prove that the residual dissipation inequality holds for each process $\pi = (\mathbf{E}(\cdot), \mathbf{P}(\cdot))$. We may assume that $\mathbf{P}(\cdot)$ is continuously differentiable; otherwise we divide the process into subintervals where $\mathbf{P}(\cdot)$ is continuously differentiable and apply the forthcoming considerations to each such piece of the process. We want to integrate (15) along the path $\mathbf{P}(\cdot)$. Since the differential Dr exists and satisfies (15) only for almost every (a.e.) $\mathbf{P} \in \text{Unim}$, it may happen that $\text{Dr}(\mathbf{P}(t))$ exists for no $t \in [0, d_\pi]$. Nevertheless, assume first that $\text{Dr}(\mathbf{P}(t))$ exists for a.e. time $t$ and complete the proof in that case first. At the end of the proof we shall employ some limiting procedure that reduces the general case to the above one. The function $s := r(\mathbf{P}(\cdot))$ is Lipschitz continuous and since $\dot{s}(t)$ is given by the chain rule for a.e. time by our assumption and $\text{Dr}(\mathbf{P}(t))\mathbf{P}(t)^T$ is symmetric,

$$\dot{s}(t) = \text{Dr}(\mathbf{P}(t))\mathbf{P}(t)^T \cdot \mathbf{D}^p(t) \le m(\mathbf{D}^p(t)) \le \bar{\mathbf{S}}(\mathbf{E}(t)) \cdot \mathbf{D}^p(t) \tag{20}$$

for a.e. $t \in [0, d_\pi]$; here $\mathbf{D}^p(\cdot)$ is the plastic stretching of $\pi$. We have also used equation (15) and the definition of $m(\mathbf{D})$. The integration of (20) gives the residual dissipation inequality.

To complete the proof in the general case, show that we may perturb $\mathbf{P}(\cdot)$ so that Dr exists for $L$-a.e. $t \in [0, d_\pi]$ during the perturbed process. Here $L$ is the Lebesgue measure on $\mathbb{R}$. We seek the perturbed path in the form $\mathbf{P}(\cdot)\mathbf{O}$ where $\mathbf{O} \in \text{Unim}$ is sufficiently close to $\mathbf{1}$. Using Fubini's theorem on the product space $\text{Unim} \times \mathbb{R}$ with the measure $\nu := \mu \otimes L$, where $\mu$ is the Haar measure on Unim, one can prove that for $\mu$-a.e. $\mathbf{O} \in \text{Unim}$ the differential $\text{Dr}(\mathbf{P}(t)\mathbf{O})$ exists and satisfies (15) for $L$-a.e. $t \in [0, d_\pi]$. Thus the above part of the proof can be applied to $\mu$-a.e. process $\mathbf{P}(\cdot)\mathbf{O}$ which gives

$$r(\mathbf{P}^{\mathrm{f}}\mathbf{O}) - r(\mathbf{P}^{\mathrm{i}}\mathbf{O}) \le w^P(\pi).$$

Letting $\mathbf{O} \to \mathbf{1}$ and using the continuity of $r$ proves the residual dissipation inequality in the general case. □

## 1.5. A lower bound for plastic work and the extremal energies

In this section we calculate a lower bound for the plastic work in processes of fixed initial and final plastic deformation and determine the maximal and minimal residual energies vanishing at a given point. Both these results are stated in terms of the convex hull $m^{**}$ of the dissipation function $m$, i.e., the largest convex function on $\mathrm{Sym}_0$ not exceeding $m$.

The following fact will be useful (see [29, Propositions 18.2.4 and 18.2.5]).

**Proposition 4.** *Let $f : U \to \mathbb{R}$, $U = \mathrm{Sym}, \mathrm{Sym}_0$ be an isotropic function. Then $f$ is convex if and only if its restriction $f_\Delta$ to diagonal arguments (relative to some fixed orthonormal basis) is convex.*

**Theorem 1.** *Let the material satisfy the second law, let $\mathbf{D} \in \mathrm{Sym}_0$, and let $\mathcal{C}(\mathbf{D})$ be the set of all processes $\pi = (\mathbf{E}(\cdot), \mathbf{P}(\cdot))$ satisfying $\mathbf{P}^{\mathrm{f}} = \mathbf{R}e^{\mathbf{D}}\mathbf{P}^{\mathrm{i}}$ for some $\mathbf{R} \in \mathrm{Rot}$. Then*

$$m^{**}(\mathbf{D}) = \inf\left\{w^P(\pi) : \pi \in \mathcal{C}(\mathbf{D})\right\}. \tag{21}$$

*Proof.* Define $s : \mathrm{Unim} \to \mathbb{R}$ by

$$s(\mathbf{P}) = m^{**}(\mathbf{D}), \qquad \mathbf{P} \in \mathrm{Unim},$$

where we write $\mathbf{P} = \mathbf{R}e^{\mathbf{D}}$, $\mathbf{R} \in \mathrm{Rot}$, $\mathbf{D} \in \mathrm{Sym}_0$ and prove that $s$ is a residual energy function. By the construction, $s$ is objective. The convex function $m^{**}$ is locally Lipschitz continuous and hence it has the total differential $\mathrm{D}m^{**}(\mathbf{A})$ for a.e. $\mathbf{A} \in \mathrm{Sym}_0$ with respect to the Lebesgue measure on $\mathrm{Sym}_0$. If $\mathbf{A} \in \mathrm{Sym}_0$ and $\mathbf{P} \in \mathrm{Unim}$ are related by $\mathbf{P} = \mathbf{R}e^{\mathbf{A}}$ for some $\mathbf{R} \in \mathrm{Rot}$, one finds that $m^{**}$ has a differential at $\mathbf{A}$ if and only if $s$ has a differential at $\mathbf{P}$. Since $m$ is isotropic, nonnegative, and positively homogeneous of degree 1, so also is $m^{**}$. Using the isotropy, and working in the basis of eigenvectors of $\mathbf{A}$ (see, e.g., the corresponding considerations in [1] or [29]), one derives the formula

$$\mathrm{D}s(\mathbf{P})\mathbf{P}^T \cdot \mathbf{D} = \mathrm{D}m^{**}(\mathbf{A}) \cdot \mathbf{D} \tag{22}$$

for each $\mathbf{D} \in \mathrm{Sym}_0$. By the convexity and homogeneity of $m^{**}$,

$$Dm^{**}(\mathbf{A}) \cdot \mathbf{D} \le m^{**}(\mathbf{D}) \le m(\mathbf{D})$$

for every $\mathbf{A} \in \mathrm{Sym}_0$, $\mathbf{A} \neq \mathbf{0}$ for which the total differential of $m^{**}$ exists. A combination with (22) provides (15). This proves that $s$ is a residual energy function. Let us now prove that for every process $\pi = (\mathbf{E}(\cdot), \mathbf{P}(\cdot)) \in \mathcal{C}(\mathbf{D})$,

$$w^P(\pi) \ge m^{**}(\mathbf{D}). \tag{23}$$

Since $\pi' := (\mathbf{E}(\cdot), \mathbf{P}(\cdot)(\mathbf{P}^{\mathrm{i}})^{-1})$ is also a process, $w^P(\pi) = w^P(\pi')$, and $s$ is a residual energy,

$$w^P(\pi') \ge s(\mathbf{P}^{\mathrm{f}}(\mathbf{P}^{\mathrm{i}})^{-1}) - s(\mathbf{1}) = m^{**}(\mathbf{D}),$$

which implies (23). This shows that we have the inequality sign $\le$ in (21).

The rest of the proof is devoted to showing the opposite inequality in (21). Thus we seek to prove that for every $\epsilon > 0$ there exists a process $\pi \in \mathcal{C}(\mathbf{D})$ such that

$$w^P(\pi) \le m^{**}(\mathbf{D}) + \epsilon. \tag{24}$$

Let $\{\mathbf{e}_i\}$ be any orthonormal basis and let $\Delta \equiv \Delta(\mathbf{e}_i)$ be the space of all $\mathbf{A} \in \mathrm{Sym}_0$ represented by diagonal matrices in $\{\mathbf{e}_i\}$ so that all elements of $\Delta$ commute. Let $m_\Delta$

and $m^{**}_\Delta$ be the restrictions of $m$ and $m^{**}$ to $\Delta$. Let us show that $m^{**}_\Delta = (m_\Delta)^{**}$ where the last symbol denotes the convex hull of $m_\Delta$ on $\Delta$. Since $m^{**}_\Delta$ is convex, we have

$$m^{**}_\Delta \le (m_\Delta)^{**} \quad \text{on} \quad \Delta. \tag{25}$$

Let us extend the function $(m_\Delta)^{**}$ from $\Delta$ to a function $g : \mathrm{Sym}_0 \to \mathbb{R}$ by isotropy, i.e., by

$$g(\mathbf{B}) = (m_\Delta)^{**}(\mathbf{QBQ}^T), \quad \mathbf{B} \in \mathrm{Sym}_0,$$

where $\mathbf{Q} = \mathbf{Q}(\mathbf{B}) \in \mathrm{Rot}$ is chosen so as to satisfy $\mathbf{QBQ}^T \in \Delta$. Such a $\mathbf{Q}$ exists by the spectral decomposition theorem. Then, first, it is easily seen that $g$ is well defined, i.e., independent of the choice of $\mathbf{Q}$, and second, by Proposition 4, $g$ is convex as a consequence of the convexity of $(m_\Delta)^{**}$. The construction gives that $g \le m$ on $\mathrm{Sym}_0$ and thus since $m^{**}$ is the maximal convex function not exceeding $m$, we have $g \le m^{**}$ on $\mathrm{Sym}_0$ and in particular $(m_\Delta)^{**} \le m^{**}_\Delta$ on $\Delta$. Thus combining with (25) we have $(m_\Delta)^{**} = m^{**}_\Delta$. Let now $\mathbf{D} \in \mathrm{Sym}_0$ and let $\{\mathbf{e}_i\}$ be any basis in which $\mathbf{D}$ is diagonal. The above considerations show that

$$m^{**}(\mathbf{D}) = m^{**}_\Delta(\mathbf{D}) = (m_\Delta)^{**}(\mathbf{D}) \tag{26}$$

and applying the familiar construction of the convex hull [26] to $m_\Delta$, we obtain that

$$(m_\Delta)^{**}(\mathbf{D}) = \inf\left\{\sum_{i=1}^{3} \lambda_i m(\mathbf{A}_i) : \mathbf{A}_i \in \Delta, \lambda_i > 0, \sum_{i=1}^{3} \lambda_i = 1, \sum_{i=1}^{3} \lambda_i \mathbf{A}_i = \mathbf{D}\right\}, \tag{27}$$

which also gives the value of $m^{**}(\mathbf{D})$ by (26). Here the limit 3 in the sums is related to the dimension 2 of $\Delta(\mathbf{e}_i)$ through the Carathéodory theorem. It is noted, and this is the main conclusion of the above considerations, that all the elements $\mathbf{A}_i$ as in (27) mutually commute and commute also with $\mathbf{D}$. Let $\epsilon > 0$ be given. By (27), there exist sequences $\lambda_i > 0$, $\mathbf{A}_i \in \mathrm{Sym}_0$, $i = 1, 2, 3$, such that

$$m^{**}(\mathbf{D}) + \frac{\epsilon}{2} \ge \sum_{i=1}^{3} \lambda_i m(\mathbf{A}_i), \quad \sum_{i=1}^{3} \lambda_i = 1, \quad \sum_{i=1}^{3} \lambda_i \mathbf{A}_i = \mathbf{D}, \tag{28}$$

and $\mathbf{A}_i, \mathbf{D}$ mutually commute. We can also assume that $\mathbf{A}_i \ne \mathbf{0}$ for $i = 1, 2, 3$. By the definition of $m$, for each $i = 1, 2, 3$ there exists a $\mathbf{E}_i \in \partial\mathcal{E}$ such that

$$\bar{\mathbf{M}}(\mathbf{E}_i) \sim \mathbf{A}_i \quad \text{and} \quad m(\mathbf{A}_i) + \frac{\epsilon}{2} \ge \bar{\mathbf{S}}(\mathbf{E}_i) \cdot \mathbf{A}_i, \quad i = 1, 2, 3. \tag{29}$$

The desired process is constructed in the form $\pi = \pi_1 * \rho_1 * \pi_2 * \rho_2 * \pi_3$ as follows. We take $\pi_i$, $i = 1, 2, 3$, as processes of duration $\lambda_i$, where $\pi_i = (\mathbf{E}_i, \mathbf{P}_i(\cdot))$ and

$$\begin{aligned}
\mathbf{P}_1(t) &= e^{\mathbf{A}_1 t}, & t &\in [0, \lambda_1],\\
\mathbf{P}_2(t) &= e^{\mathbf{A}_2 t} e^{\lambda_1 \mathbf{A}_1}, & t &\in [0, \lambda_2],\\
\mathbf{P}_3(t) &= e^{\mathbf{A}_3 t} e^{\lambda_2 \mathbf{A}_2} e^{\lambda_1 \mathbf{A}_1}, & t &\in [0, \lambda_3].
\end{aligned}$$

We further take $\rho_1, \rho_2$ as elastic processes of the form $\rho_1 = (\tilde{\mathbf{E}}_1(\cdot), e^{\lambda_1 \mathbf{A}_1})$, $\rho_2 = (\tilde{\mathbf{E}}_1(\cdot), e^{\lambda_1 \mathbf{A}_1 + \lambda_2 \mathbf{A}_2})$ such that $\mathbf{E}_1 = \tilde{\mathbf{E}}_1^{\mathrm{i}}$, $\mathbf{E}_2 = \tilde{\mathbf{E}}_1^{\mathrm{f}}$, $\mathbf{E}_2 = \tilde{\mathbf{E}}_2^{\mathrm{i}}$, $\mathbf{E}_3 = \tilde{\mathbf{E}}_2^{\mathrm{f}}$. Then the process $\pi = \pi_1 * \rho_1 * \pi_2 * \rho_2 * \pi_3$ can be constructed and

$$\mathbf{P}^{\mathrm{i}} = \mathbf{1}, \quad \mathbf{P}^{\mathrm{f}} = e^{\lambda_1 \mathbf{A}_1 + \lambda_2 \mathbf{A}_2 + \lambda_3 \mathbf{A}_3} = e^{\mathbf{D}},$$

where we have used the commutativity of the $\mathbf{A}_i$ and $(28)_3$. Thus $\pi \in \mathcal{C}(\mathbf{D})$. Furthermore,

$$w^P(\pi) = \sum_{i=1}^{3} w^P(\pi_i) = \sum_{i=1}^{3} \lambda_i \bar{\mathbf{S}}(\mathbf{E}_i) \cdot \mathbf{A}_i \le \sum_{i=1}^{3} \lambda_i m(\mathbf{A}_i) + \frac{\epsilon}{2} \le m^{**}(\mathbf{D}) + \epsilon$$

by $(29)_2$ and (24). □

A residual energy function $r$ is said to be maximal or minimal at $\mathbf{H} \in$ Unim if $r(\mathbf{H}) = 0$ and $r \ge \bar{r}$ or $r \le \bar{r}$, respectively, for any residual energy function $\bar{r}$ such that $\bar{r}(\mathbf{H}) = 0$. Coleman and Owen [5] show that the set of all free energy functions that vanish at a given state is convex and has the largest and the smallest elements. By Proposition 2(ii) the problem of describing the extremal free energy functions is equivalent to that of describing the extremal residual energy functions.

**Proposition 5.** *Let the material satisfy the second law and define* $s, t$ : Unim $\to \mathbb{R}$ *by*

$$s(\mathbf{P}) = m^{**}(\mathbf{D}), \quad t(\mathbf{P}) = -m^{**}(-\mathbf{D}),$$

$\mathbf{P} \in$ Unim, *where we write* $\mathbf{P} = \mathbf{R}e^{\mathbf{D}}$, $\mathbf{R} \in$ Rot, $\mathbf{D} \in \mathrm{Sym}_0$. *Then* $s, t$ *are the maximal and minimal residual energies at* $\mathbf{1}$. *Moreover they satisfy*

$$s(\mathbf{OP}) \le s(\mathbf{O}) + s(\mathbf{P}), \quad t(\mathbf{OP}) \ge t(\mathbf{O}) + t(\mathbf{P}) \tag{30}$$

*for any* $\mathbf{O}, \mathbf{P} \in$ Unim.

The extremal residual energies at a general $\mathbf{H} \in$ Unim are

$$s_{\mathbf{H}}(\mathbf{P}) = s(\mathbf{PH}), \quad t_{\mathbf{H}}(\mathbf{P}) = t(\mathbf{PH}),$$

$\mathbf{P} \in$ Unim.

*Proof.* By the proof of Theorem 1, $s$ is a residual energy; the proof that $t$ is a residual energy is similar. To prove that $s$ is maximal at $\mathbf{1}$, let $r$ be any residual energy vanishing at $\mathbf{1}$, let $\pi = (\mathbf{E}(\cdot), \mathbf{P}(\cdot))$ be any process in $\mathcal{C}(\mathbf{D})$, and let $\pi' = (\mathbf{E}(\cdot), \mathbf{P}(\cdot)(\mathbf{P}^{\mathrm{i}})^{-1})$. The residual dissipation inequality for $\pi'$ reads

$$r(\mathbf{P}^{\mathrm{f}}(\mathbf{P}^{\mathrm{i}})^{-1}) = r(e^{\mathbf{D}}) \le w^P(\pi') = w^P(\pi).$$

Taking the infimum over all processes $\pi \in \mathcal{C}(\mathbf{D})$ and using Theorem 1 we obtain

$$r(e^{\mathbf{D}}) \le s(e^{\mathbf{D}}).$$

Combining with the objectivity of both $r, s$ this gives $r \le s$. To prove (30), consider only $s$. For any function $r$ on Unim and any $\mathbf{H} \in$ Unim let $r_{\mathbf{H}}$ denote the shifted function given by $r_{\mathbf{H}}(\mathbf{P}) = r(\mathbf{PH})$. It follows immediately from, e.g., Condition (iii) of Proposition 3 that $r$ is a residual energy function if and only if $r_{\mathbf{H}}$ is a residual energy function. Note that for any $\mathbf{P} \in \mathrm{Unim}$, the function $s_{\mathbf{P}}(\cdot) - s(\mathbf{P})$ is a residual energy vanishing at $\mathbf{1}$. Hence $s_{\mathbf{P}}(\cdot) - s(\mathbf{P}) \le s(\cdot)$ and $(30)_1$ follows by inserting $\mathbf{O}$. The rest is immediate. □

## 1.6. Il'yushin's condition: Its first consequences

A process $\pi = (\mathbf{V}(\cdot), \mathbf{P}(\cdot))$ is said to be an $\mathbf{F}$-cycle if $\hat{\mathbf{F}}(\pi^{\mathrm{f}}) = \hat{\mathbf{F}}(\pi^{\mathrm{i}})$. Every $\sigma$-cycle (see Section 1.3) is an $\mathbf{F}$-cycle.

**Il'yushin's condition.** For every $\mathbf{F}$-cycle $\pi$, $w(\pi) \ge 0$.

Theorem 2, below, describes the consequences of Il'yushin's condition on the constitutive objects, which include the normality rule.

**Definition 2.** *The material is said to obey the normality rule if for every* $\mathbf{E} \in \partial\mathcal{E}$,

$$\bar{\mathbf{M}}(\mathbf{E}) \in \mathrm{N}_{\bar{\mathbf{S}}(\mathbf{E})}\mathcal{S}, \tag{31}$$

*where* $\mathrm{N}_{\bar{\mathbf{S}}(\mathbf{E})}\mathcal{S}$ *denotes the normal cone to* $\mathcal{S}$ *at* $\bar{\mathbf{S}}(\mathbf{E})$, *defined by*

$$\mathrm{N}_{\mathbf{S}}\mathcal{S} := \left\{\mathbf{D} \in \mathrm{Sym}_0 : (\mathbf{Z} - \mathbf{S}) \cdot \mathbf{D} \le 0 \textit{ for every } \mathbf{Z} \in \mathcal{S}\right\},$$

$\mathbf{S} \in \mathrm{Sym}_0$.

The following lemma gathers some consequences of the normality rule.

**Lemma 1.** *If the material obeys the normality rule then*

(i) *the stress range* $\mathcal{S}$ *is convex;*

(ii) *for every* $\mathbf{D} \in \mathrm{Sym}_0$,

$$m(\mathbf{D}) = \sup\{\mathbf{S} \cdot \mathbf{D} : \mathbf{S} \in \mathcal{S}\} \tag{32}$$

*and in particular,*

$$m(\mathbf{D}) = \mathbf{S} \cdot \mathbf{D} \tag{33}$$

*for any* $\mathbf{S} \in \mathcal{S}$ *such that* $\mathbf{D} \in \mathrm{N}_{\mathbf{S}}\mathcal{S}$;

(iii) *m is isotropic, convex, and positively homogeneous of degree* 1;

(iv) *the subdifferential of m is given by*

$$\partial m(\mathbf{D}) = \begin{cases} \{\mathbf{S} \in \mathrm{Sym}_0 : \mathbf{D} \in \mathrm{N}_{\mathbf{S}}\mathcal{S}\} & \textit{if } \mathbf{D} \ne \mathbf{0}, \\ \mathcal{S} & \textit{if } \mathbf{D} = \mathbf{0}; \end{cases} \tag{34}$$

(v) *the convex conjugate* $m^*$ *of m is given by*

$$m^*(\mathbf{S}) = \begin{cases} 0 & \textit{if } \mathbf{S} \in \mathcal{S}, \\ \infty & \textit{otherwise}. \end{cases} \tag{35}$$

*Proof.* (i): Let us derive (i) from the following assertion, which is easy to prove: *If $M \subset \mathbb{R}^d$ is a closed set with nonempty interior such that for each $x \in \partial M$ we have $\mathrm{N}_x M \neq \{0\}$, then $M$ is convex.* Let us verify that $\mathcal{S}$ satisfies the hypotheses of the assertion. Clearly, $\mathcal{S}$ is closed since it is an image of the closed set $\mathcal{E}$ under the continuous mapping $\bar{\mathbf{S}}$. Further, $\mathcal{S}$ has nonempty interior since $\mathcal{E}$ has nonempty interior by Definition 1(i) and (iv) and $\bar{\mathbf{S}}$ maps the interior of $\mathcal{E}$ onto the interior of $\mathcal{S}$ by Definition 1(iii). Finally, for each $\mathbf{S} \in \partial\mathcal{S}$ we have $\mathrm{N}_{\mathbf{S}}\mathcal{S} \neq \{\mathbf{0}\}$ since $\bar{\mathbf{S}}$ maps $\partial\mathcal{E}$ onto $\partial\mathcal{S}$ by Definition 1(iii) and for each $\mathbf{E} \in \partial\mathcal{E}$ we have (31). (ii): Let $\mathbf{D} \neq \mathbf{0}$ and $\mathbf{E} \in \partial\mathcal{E}$ be any point such that $\bar{\mathbf{M}}(\mathbf{E}) \sim \mathbf{D}$ so that $\mathbf{D} \in \mathrm{N}_{\bar{\mathbf{S}}(\mathbf{E})}\mathcal{S}$ by the normality rule which means that $(\mathbf{S} - \bar{\mathbf{S}}(\mathbf{E})) \cdot \mathbf{D} \leq 0$, i.e.,

$$\mathbf{S} \cdot \mathbf{D} \leq \bar{\mathbf{S}}(\mathbf{E}) \cdot \mathbf{D}$$

for any $\mathbf{S} \in \mathcal{E}$. Fixing $\mathbf{S}$ and taking the infimum of the right-hand side over all $\mathbf{E} \in \partial\mathcal{E}$ such that $\bar{\mathbf{M}}(\mathbf{E}) \sim \mathbf{D}$ we obtain

$$\mathbf{S} \cdot \mathbf{D} \leq m(\mathbf{D})$$

with the equality holding if $\mathbf{S} = \bar{\mathbf{S}}(\mathbf{E})$, where $\mathbf{E}$ is any element as above. Once (ii) has been established, (iii)–(v) follow from the standard duality theory for homogeneous degree 1 convex functions [9], [26]. □

**Theorem 2.** *The material satisfies Il'yushin's condition if and only if the following two conditions are satisfied:*

(i) $\bar{\mathbf{T}}$ *has an elastic potential* $p$;

(ii) *if* $\mathbf{D} \in \mathrm{Sym}_0$ *and* $\mathbf{E}, \mathbf{E}e^{-\mathbf{D}} \in \mathcal{E}$ *then*

$$p(\mathbf{E}e^{-\mathbf{D}}) \geq p(\mathbf{E}) - m(\mathbf{D}). \tag{36}$$

*Moreover, if* (i) *and* (ii) *hold then the material obeys the normality rule.*

Item (ii) is called Condition E in the subsequent discussion.

*Proof.* Assume that the material satisfies Il'yushin's condition. Condition (i) follows from the reversibility of elastic processes via the path-independence argument as in the proof of Proposition 2(i). (ii): Let $\mathbf{E}_0 \in \partial\mathcal{E}$ be such that

$$\bar{\mathbf{M}}(\mathbf{E}_0) \sim \mathbf{D}. \tag{37}$$

Consider a process $\pi = \pi_1 * \pi_0 * \pi_2$, where $\pi_1$ is any elastic process connecting $(\mathbf{E}, \mathbf{1})$ with $(\mathbf{E}_0, \mathbf{1})$, $\pi_2$ any elastic process connecting $(\mathbf{E}_0, e^{\mathbf{D}})$ with $(\mathbf{E}e^{-\mathbf{D}}, \mathbf{R}e^{\mathbf{D}})$, and $\pi_0$ a process of duration 1 with

$$\mathbf{E}_0(t) = \mathbf{E}_0, \quad \mathbf{P}_0(t) = e^{\mathbf{D}t}, \quad t \in [0, 1]. \tag{38}$$

Then $\pi_0$ is really a process by (37) and the process $\pi_1 * \pi_0 * \pi_2$ is an $\mathbf{F}$-cycle. Il'yushin's condition says

$$w(\pi) = w^E(\pi) + w^P(\pi) \geq 0,$$

where $w^E(\pi) = p(\mathbf{E}e^{-\mathbf{D}}) - p(\mathbf{E})$, $w^P(\pi) = w^P(\pi_0) = \bar{\mathbf{S}}(\mathbf{E}_0) \cdot \mathbf{D}$. The last three relations yield $p(\mathbf{E}e^{-\mathbf{D}}) \geq p(\mathbf{E}) - \bar{\mathbf{S}}(\mathbf{E}_0) \cdot \mathbf{D}$. Since $\mathbf{E}_0$ is arbitrary subject to condition

(37), the definition of $m$ gives (36). This completes the proof of (ii). Assume that (i), (ii) hold and prove the normality rule. Let $\mathbf{G} \in \partial\mathcal{E}$. The goal is to prove that

$$(\mathbf{S} - \bar{\mathbf{S}}(\mathbf{G})) \cdot \bar{\mathbf{M}}(\mathbf{G}) \leq 0 \tag{39}$$

for every $\mathbf{S} \in \mathcal{S}$. Assume first that $\mathbf{S}$ is in the interior of $\mathcal{S}$. Use Definition 1(iii) to find that there exists an interior point $\mathbf{E}$ of $\mathcal{E}$ such that $\mathbf{S} = \bar{\mathbf{S}}(\mathbf{E})$. Set $\mathbf{D} = \bar{\mathbf{M}}(\mathbf{G})$ and note that for all $t > 0$ sufficiently small we have $\mathbf{E}e^{-t\mathbf{D}} \in \mathcal{E}$ since $\mathbf{E}$ is in the interior of $\mathcal{E}$. The application of (36) and the use of $m(t\mathbf{D}) \leq t\bar{\mathbf{S}}(\mathbf{E}_0) \cdot \mathbf{D}$, which follows from the definition of $m$, provide

$$p(\mathbf{E}e^{-t\mathbf{D}}) \geq p(\mathbf{E}) - t\bar{\mathbf{S}}(\mathbf{G}) \cdot \mathbf{D}.$$

Dividing by $t$, letting $t$ tend to 0, and using the stress relation, we obtain $\bar{\mathbf{S}}(\mathbf{E}) \cdot \mathbf{D} \leq \bar{\mathbf{S}}(\mathbf{G}) \cdot \mathbf{D}$ and thus eventually (39). Since $\mathcal{E}$ is the closure of its interior, the limit gives (39) for each $\mathbf{S} = \bar{\mathbf{S}}(\mathbf{E})$, where $\mathbf{E}$ is a boundary point of $\mathcal{E}$, and since every boundary point of $\mathcal{S}$ is of the form $\mathbf{S} = \bar{\mathbf{S}}(\mathbf{E})$, where $\mathbf{E}$ is a boundary point of $\mathcal{E}$ (see Definition 1), inequality (39) holds for all $\mathbf{S} \in \mathcal{S}$. Assume conversely that (i) and (ii) hold and prove Il'yushin's condition. Let $\pi = (\mathbf{E}(\cdot), \mathbf{P}(\cdot))$ be an $\mathbf{F}$-cycle. Since (i), (ii) imply the normality rule, Lemma 1(iii) says that $m$ is convex and thus $m = m^{**}$. Then by Theorem 1,

$$w(\pi) = w^E(\pi) + w^P(\pi) \geq p(\mathbf{E}^{\mathrm{f}}) - p(\mathbf{E}^{\mathrm{i}}) + m(\mathbf{D}), \tag{40}$$

where $\mathbf{D}$ is such that $\mathbf{P}^{\mathrm{f}} = \mathbf{R}e^{\mathbf{D}}\mathbf{P}^{\mathrm{i}}$ for some $\mathbf{R} \in \mathrm{Rot}$. Combining with $\mathbf{E}^{\mathrm{f}}\mathbf{P}^{\mathrm{f}} = \mathbf{E}^{\mathrm{i}}\mathbf{P}^{\mathrm{i}}$ we obtain $\mathbf{E}^{\mathrm{f}} = \mathbf{E}^{\mathrm{i}}e^{-\mathbf{D}}\mathbf{R}^T$ and (40) reads $w(\pi) \geq p(\mathbf{E}^{\mathrm{i}}e^{-\mathbf{D}}) - p(\mathbf{E}^{\mathrm{i}}) + m(\mathbf{D}) \geq 0$, where the last inequality is Condition (ii). □

**Remark 1.** *Suppose that the material obeys the normality rule and that* $\bar{\mathbf{T}}$ *has an elastic potential* $p$. *Then we have the following:*

(i) *Condition* E *holds for all pairs* $\mathbf{E}, \mathbf{D}$ *as in that condition with* $|\mathbf{D}|$ *sufficiently small and* $\mathbf{E}$ *in the interior of* $\mathcal{E}$.

(ii) *If* $\mathcal{E}$ *is logarithmically convex in the sense that for every pair of* $\mathbf{E}, \mathbf{D}$ *of tensors with* $\mathbf{D} \in \mathrm{Sym}_0$, $\mathbf{E}, \mathbf{E}e^{-\mathbf{D}} \in \mathcal{E}$ *one also has* $\mathbf{E}e^{-\mathbf{D}t} \in \mathcal{E}$ *for every* $t \in [0, 1]$, *then* E *holds.*

The logarithmic convexity of Item (ii) seems to be hard to verify. The following section gives other sufficient conditions to guarantee E.

*Proof.* (i): Since $\mathbf{E}$ is an interior point of $\mathcal{E}$, if $|\mathbf{D}|$ is small enough, then $\mathbf{E}e^{-\mathbf{D}t} \in \mathcal{E}$ for all $t \in [0, 1]$. Writing $\mathbf{H}(t) := \mathbf{E}e^{-\mathbf{D}t}$, one obtains $dp(\mathbf{H}(t))/dt = -\bar{\mathbf{S}}(\mathbf{H}(t)) \cdot \mathbf{D}$. Lemma 1(ii) gives $\bar{\mathbf{S}}(\mathbf{H}(t)) \cdot \mathbf{D} \leq m(\mathbf{D})$; thus $dp(\mathbf{H}(t))/dt \geq -m(\mathbf{D})$ and the integration over $[0, 1]$ yields (36). (ii): If $\mathcal{E}$ is logarithmically convex, then in the notation of the proof of (i), one has $\mathbf{H}(t) \in \mathcal{E}$ for all $t \in [0, 1]$. The proof is then identical with that of (i). □

## 1.7. The extended energy functions

This section shows that Il'yushin's condition leads to energy functions that satisfy the dissipation inequalities stronger than those arising from the second law.

**Theorem 3.** *Suppose that the material satisfies Il'yushin's condition. Let* $\mathbf{F} \in \mathrm{Lin}^+$ *and define*

$$e(\mathbf{F}) = \inf \left\{ p(\mathbf{F}e^{-\mathbf{D}}) + m(\mathbf{D}) : \mathbf{D} \in \mathrm{Sym}_0, \mathbf{F}e^{-\mathbf{D}} \in \mathcal{E} \right\}, \tag{41}$$

$$f(\mathbf{F}) = \sup \left\{ p(\mathbf{F}e^{-\mathbf{D}}) - m(-\mathbf{D}) : \mathbf{D} \in \mathrm{Sym}_0, \mathbf{F}e^{-\mathbf{D}} \in \mathcal{E} \right\}. \tag{42}$$

*Then*

(i) $-\infty < f \le e < \infty$;

(ii) *for every* $\mathbf{F} \in \mathcal{E}$ *the infima and suprema in* (41) *and* (42) *are attained for* $\mathbf{D} = \mathbf{0}$ *and hence*

$$e(\mathbf{F}) = f(\mathbf{F}) = p(\mathbf{F});$$

(iii) *if* $\mathbf{F} \notin \mathcal{E}$ *then* (41) *and* (42) *hold with the condition* $\mathbf{F}e^{-\mathbf{D}} \in \mathcal{E}$ *replaced by* $\mathbf{F}e^{-\mathbf{D}} \in \partial\mathcal{E}$;

(iv) *for every process* $\pi = (\mathbf{E}(\cdot), \mathbf{P}(\cdot))$,

$$e(\mathbf{F}^{\mathrm{f}}(\mathbf{P}^{\mathrm{i}})^{-1}) - e(\mathbf{F}^{\mathrm{i}}(\mathbf{P}^{\mathrm{i}})^{-1}) \le w(\pi), \tag{43}$$

$$f(\mathbf{F}^{\mathrm{f}}(\mathbf{P}^{\mathrm{f}})^{-1}) - f(\mathbf{F}^{\mathrm{i}}(\mathbf{P}^{\mathrm{f}})^{-1}) \le w(\pi). \tag{44}$$

The proof will show that all the infima and suprema in the above theorem are taken over nonempty sets but it is not apriori clear that they are finite. The function $e$ is called the *initial extended energy* or briefly *initial energy* and $f$ the *final extended energy* or *final energy*. For a concrete material, the function $e$ is calculated in Section 1.10. Since for $\mathbf{F}$-cycles the left-hand sides of (43) and (44) vanish and thus $w(\pi) \ge 0$, we see that the existence of a function $e : \mathrm{Lin}^+ \to \mathbb{R}$ satisfying (43) or similarly the existence a function $f : \mathrm{Lin}^+ \to \mathbb{R}$ satisfying (44) implies that the material satisfies Il'yushin's condition; thus the existence of such functions is equivalent to Il'yushin's condition. Since Il'yushin's condition is strictly stronger than the cyclic second law, the dissipation inequalities (43) and (44) are strictly stronger than the dissipation inequality (9) stemming from the second law. In Remark 2 and the subsequent discussion we shall see that there are qualitative differences between the inequalities (43) and (44). It is also noted that similar potentials have been introduced by Lucchesi and Šilhavý [16], but the dissipation inequalities have been proved only for a restricted class of processes. Recently, a potential similar to $e$ has been introduced by Ortiz and Repetto [21] and Carstensen, Hackl and Mielke [4], and Mielke [18] to treat plastic materials from the variational point of view.

*Proof.* (i): If we set $\delta = \det \mathbf{F}$ and take any element $\mathbf{E}$ of $\mathcal{E}_\delta$ (see Definition 1(iv)) then by the polar decomposition theorem there exists an $\mathbf{R} \in \mathrm{Rot}$ and $\mathbf{D} \in \mathrm{Sym}_0$ such that $\mathbf{F}e^{-\mathbf{D}} = \mathbf{E}\mathbf{R}$ and hence $\mathbf{F}e^{-\mathbf{D}} \in \mathcal{E}$ for some $\mathbf{D} \in \mathrm{Sym}_0$. Thus the suprema and infima in (41) and (42) are taken over nonempty sets and hence $e(\mathbf{F}) < \infty$, $f(\mathbf{F}) > -\infty$. It remains to be proved that $f(\mathbf{F}) \le e(\mathbf{F})$. Note first that if $\tilde{\mathbf{D}}_1, \tilde{\mathbf{D}}_2, \tilde{\mathbf{D}} \in \mathrm{Sym}_0$ are such that $e^{\tilde{\mathbf{D}}_1} e^{\tilde{\mathbf{D}}_2} = \mathbf{R} e^{\tilde{\mathbf{D}}}$ for some $\mathbf{R} \in \mathrm{Rot}$, then we have the triangle inequality

$$m(\tilde{\mathbf{D}}) \le m(\tilde{\mathbf{D}}_1) + m(\tilde{\mathbf{D}}_2). \tag{45}$$

To establish (45), let $\mathbf{E}_\alpha \in \partial\mathcal{E}$, $\alpha = 1, 2$, be such that

$$\bar{\mathbf{M}}(\mathbf{E}_\alpha) \sim \tilde{\mathbf{D}}_\alpha \tag{46}$$

(see Definition 1(ii)). Let $\pi_\alpha$ be processes of duration 1 of the form $\pi_\alpha = (\mathbf{E}_\alpha, \mathbf{P}_\alpha(\cdot))$, where

$$\mathbf{P}_2(t) = e^{t\tilde{\mathbf{D}}_2}, \quad \mathbf{P}_1(t) = e^{t\tilde{\mathbf{D}}_1} e^{\tilde{\mathbf{D}}_2}, \quad t \in [0, 1].$$

Let finally $\rho$ be an elastic process connecting $\pi_2^{\mathrm{f}}$ with $\pi_1^{\mathrm{i}}$. Then $\pi = \pi_2 * \rho * \pi_1$ is a process in which

$$w^P(\pi) = \bar{\mathbf{S}}(\mathbf{E}_1) \cdot \tilde{\mathbf{D}}_1 + \bar{\mathbf{S}}(\mathbf{E}_2) \cdot \tilde{\mathbf{D}}_2$$

and we have $\mathbf{P}^{\mathrm{f}} = e^{\tilde{\mathbf{D}}_1} e^{\tilde{\mathbf{D}}_2} = \mathbf{R} e^{\tilde{\mathbf{D}}}$. Thus the lower bound for the plastic work (Theorem 1; recall $m = m^{**}$) gives

$$\bar{\mathbf{S}}(\mathbf{E}_1) \cdot \tilde{\mathbf{D}}_1 + \bar{\mathbf{S}}(\mathbf{E}_2) \cdot \tilde{\mathbf{D}}_2 \ge m(\tilde{\mathbf{D}}).$$

Using (33) we obtain (45). Next use (45) to show that if $\mathbf{F} \in \mathrm{Lin}^+$ and $\mathbf{D}_1, \mathbf{D}_2 \in \mathrm{Sym}_0$, then

$$p(\mathbf{F} e^{-\mathbf{D}_1}) - m(-\mathbf{D}_1) \le p(\mathbf{F} e^{-\mathbf{D}_2}) + m(\mathbf{D}_2).$$

Indeed, let $\mathbf{D} \in \mathrm{Sym}_0$ be such that $\mathbf{R} e^{\mathbf{D}} = e^{\mathbf{D}_2} e^{-\mathbf{D}_1}$ for some $\mathbf{R} \in \mathrm{Rot}$ so that, by (45),

$$m(-\mathbf{D}_1) + m(\mathbf{D}_2) \ge m(\mathbf{D}).$$

Condition E says

$$p(\mathbf{F} e^{-\mathbf{D}_1} e^{-\mathbf{D}}) \ge p(\mathbf{F} e^{-\mathbf{D}_1}) - m(\mathbf{D})$$

and hence

$$p(\mathbf{F} e^{-\mathbf{D}_2}) \ge p(\mathbf{F} e^{-\mathbf{D}_1}) - m(-\mathbf{D}_1) - m(\mathbf{D}_2).$$

Taking the supremum over all $\mathbf{D}_1$ such that $\mathbf{F} e^{-\mathbf{D}_1} \in \mathcal{E}$ and the infimum over all $\mathbf{D}_2$ such that $\mathbf{F} e^{-\mathbf{D}_2} \in \mathcal{E}$ completes the proof of (i). (ii): This is trivial by (36). (iii): Let

$$\bar{e}(\mathbf{F}) = \inf \left\{ p(\mathbf{F} e^{-\mathbf{D}}) + m(\mathbf{D}) : \mathbf{D} \in \mathrm{Sym}_0, \mathbf{F} e^{-\mathbf{D}} \in \partial\mathcal{E} \right\}. \tag{47}$$

This is an infimum over a smaller set than in (41) and therefore,

$$e(\mathbf{F}) \le \bar{e}(\mathbf{F}). \tag{48}$$

To prove the opposite inequality, let $\mathbf{D} \in \mathrm{Sym}_0$ be such that $\mathbf{F} e^{-\mathbf{D}} \in \mathcal{E}$. As $\mathbf{F} \notin \mathcal{E}$, there exists a $\tau \in [0, 1]$ such that $\mathbf{F} e^{-\mathbf{D}\tau} \in \partial\mathcal{E}$. Inequality (36) gives

$$p(\mathbf{F} e^{-\mathbf{D}}) = p(\mathbf{F} e^{-\mathbf{D}\tau} e^{-(1-\tau)\mathbf{D}}) \ge p(\mathbf{F} e^{-\mathbf{D}\tau}) - m(\mathbf{D}) + m(\mathbf{D}\tau),$$

where we have used the homogeneity of $m$. Thus

$$p(\mathbf{F}e^{-\mathbf{D}}) + m(\mathbf{D}) \geq p(\mathbf{F}e^{-\mathbf{D}'}) + m(\mathbf{D}'),$$

where $\mathbf{D}' := \mathbf{D}\tau \in \partial\mathcal{E}$. Thus for each $\mathbf{D}$ as in the infimum (41) there exists a $\mathbf{D}'$ as in the infimum (47) with a value of $p(\mathbf{F}e^{-\mathbf{D}'}) + m(\mathbf{D}')$ that does not exceed $p(\mathbf{F}e^{-\mathbf{D}}) + m(\mathbf{D})$. This proves the opposite inequality in (48). The assertion about $f(\mathbf{F})$ is proved similarly. (iv): If $\pi = (\mathbf{E}(\cdot), \mathbf{P}(\cdot))$ is a process and if we write $\mathbf{P}^{\mathrm{f}} = \mathbf{R}e^{\mathbf{D}}\mathbf{P}^{\mathrm{i}}$ for some $\mathbf{R} \in \mathrm{Rot}$, $\mathbf{D} \in \mathrm{Sym}_0$ then Theorem 1 (recall $m = m^{**} \geq 0$) gives

$$w(\pi) \geq p(\mathbf{E}^{\mathrm{f}}) - p(\mathbf{E}^{\mathrm{i}}) + m(\mathbf{D}). \tag{49}$$

We have

$$e(\mathbf{F}^{\mathrm{f}}(\mathbf{P}^{\mathrm{i}})^{-1}) \leq p(\mathbf{F}^{\mathrm{f}}(\mathbf{P}^{\mathrm{i}})^{-1}e^{-\mathbf{D}}) + m(\mathbf{D}) = p(\mathbf{E}^{\mathrm{f}}) + m(\mathbf{D})$$

by the definition of $e$ and

$$e(\mathbf{F}^{\mathrm{i}}(\mathbf{P}^{\mathrm{i}})^{-1}) = e(\mathbf{E}^{\mathrm{i}}) = p(\mathbf{E}^{\mathrm{i}})$$

by (ii); hence (49) leads to (43). Equation (44) is proved similarly. □

Let us estimate the extended energy functions at large values of $|\mathbf{F}|$.

**Lemma 2.** *For every* $\mathbf{U} \in \mathrm{Sym}^+$ *with* $\det \mathbf{U} = 1$ *we have*

$$\frac{1}{2\sqrt{3}}|\ln \mathbf{U}| \leq \ln |\mathbf{U}| \leq \ln \sqrt{3} + |\ln \mathbf{U}|. \tag{50}$$

*Proof.* Denote by $|\cdot|_\infty$ the maximum norm of Sym, i.e.,

$$|\mathbf{D}|_\infty := \max\{|d_1|, |d_2|, |d_3|\},$$

where $d_1 \geq d_2 \geq d_3$ are the eigenvalues of $\mathbf{D} \in \mathrm{Sym}$. Write $\mathbf{U} = e^{\mathbf{D}}$ with $\mathbf{D} \in \mathrm{Sym}_0$ since $\det \mathbf{U} = 1$. We have $d_1 + d_2 + d_3 = 0$; hence $d_3 \leq 0$ and $|\mathbf{D}|_\infty = \max\{d_1, -d_3\}$. If $|\mathbf{D}|_\infty = d_1$ then trivially $|\mathbf{D}|_\infty \leq 2d_1$. If $|\mathbf{D}|_\infty = -d_3$, then $d_1 + d_2 - |\mathbf{D}|_\infty = 0$, i.e., $|\mathbf{D}|_\infty = d_1 + d_2 \leq 2d_1$. Hence in every case $|\mathbf{D}|_\infty \leq 2d_1$ and so

$$|e^{\mathbf{D}}|^2 \geq e^{2d_1} \geq e^{|\mathbf{D}|_\infty},$$

which implies $\frac{1}{2}|\mathbf{D}|_\infty \leq \ln |e^{\mathbf{D}}|$. Combining this with $|\mathbf{D}| \leq \sqrt{3}|\mathbf{D}|_\infty$ we obtain $(50)_1$. To prove $(50)_2$, it suffices to note that

$$|e^{\mathbf{D}}|^2 \leq 3e^{2d_1} \leq 3e^{2|\mathbf{D}|_\infty},$$

i.e., $\ln |e^{\mathbf{D}}| \leq \ln \sqrt{3} + |\mathbf{D}|_\infty$. Combining this with $|\mathbf{D}|_\infty \leq |\mathbf{D}|$ we obtain the desired result. □

**Remark 2.** *Suppose that the material satisfies Il'yushin's condition and assume that for each fixed* $\delta > 0$, *the set* $\mathcal{E}_\delta$, *given by* (3), *is bounded and that* $\delta^{1/3}\mathbf{1} \in \mathcal{E}_\delta$; *moreover, assume that*

$$m(\mathbf{D}) \geq \bar{c}_0|\mathbf{D}|, \quad \mathbf{D} \in \mathrm{Sym}_0 \tag{51}$$

*for some* $\bar{c}_0 > 0$. *Then there exist positive constants* $\beta_e, \bar{\beta}_e, \beta_f, \bar{\beta}_f$ *and functions* $\gamma_e, \bar{\gamma}_e, \gamma_f, \bar{\gamma}_f : (0, \infty) \to \mathbb{R}$ *such that*

$$\bar{\gamma}_e(\delta) + \bar{\beta}_e \ln|\mathbf{F}| \le e(\mathbf{F}) \le \gamma_e(\delta) + \beta_e \ln^+|\mathbf{F}|, \tag{52}$$

$$\gamma_f(\delta) - \beta_f \ln^+|\mathbf{F}| \le f(\mathbf{F}) \le \bar{\gamma}_f(\delta) - \bar{\beta}_f \ln|\mathbf{F}| \tag{53}$$

*for every* $\mathbf{F} \in \mathrm{Lin}^+$, *where* $\delta = \det \mathbf{F}$ *and* $\ln^+$ *is the positive part of* $\ln$.

*Proof.* Let us first prove the upper bound $(52)_2$. Since $\mathcal{S}$ is bounded, there exists a $c_0 > 0$ such that

$$m(\mathbf{D}) \le c_0|\mathbf{D}| \tag{54}$$

for each $\mathbf{D} \in \mathrm{Sym}_0$. Let $\gamma_e^0 : (0, \infty) \to \mathbb{R}$ be defined by

$$\gamma_e^0(\delta) = \max\{p(\mathbf{E}) : \mathbf{E} \in \mathcal{E}_\delta\} \tag{55}$$

for each $\delta > 0$. The maximum exists and is finite since $\mathcal{E}_\delta$ is bounded $p$ is continuous. Let

$$\beta_e = 2\sqrt{3}c_0, \quad \gamma_e(\delta) := \gamma_e^0(\delta) + (2c_0/\sqrt{3})|\ln\delta| \tag{56}$$

for each $\delta > 0$. Since the functions on both sides of $(52)_2$ are objective and isotropic, it suffices to prove $(52)_2$ only if $\mathbf{F} \in \mathrm{Sym}^+$. If $\mathbf{F} \in \mathcal{E}$ then the proof is immediate. Let $\mathbf{F} \in \mathrm{Sym}^+ \setminus \mathcal{E}$. Set $\delta = \det \mathbf{F}$ and

$$\mathbf{D} := \ln(\mathbf{F}/\delta^{1/3}), \quad \text{i.e.,} \quad \mathbf{F}e^{-\mathbf{D}} = \delta^{1/3}\mathbf{1}. \tag{57}$$

Then by (41) we have $e(\mathbf{F}) \le p(\delta^{1/3}\mathbf{1}) + m(\mathbf{D})$ and using successively (57), (55), (54), (57), $(50)_1$, and (56), we obtain $(52)_2$. Let us prove $(52)_2$. We have $|\mathbf{AB}| \le m_0|\mathbf{A}||\mathbf{B}|$ for some $m_0 > 0$ and all second-order tensors $\mathbf{A}, \mathbf{B}$. Let $\mathbf{H} \in \partial\mathcal{E}$ so that $\mathbf{H}$ is invertible and $|\mathbf{H}| \ne 0$. Then $|\mathbf{F}| = |\mathbf{H}\mathbf{H}^{-1}\mathbf{F}| \le m_0|\mathbf{H}||\mathbf{H}^{-1}\mathbf{F}|$ for each $\mathbf{F} \in \mathrm{Lin}^+$ and hence

$$|\mathbf{H}^{-1}\mathbf{F}| \ge c_3|\mathbf{F}|$$

for all $\mathbf{H} \in \partial\mathcal{E}$ and $\mathbf{F} \in \mathrm{Lin}^+$, where $c_3 = \min\left\{m_0^{-1}|\mathbf{H}|^{-1} : \mathbf{H} \in \partial\mathcal{E}\right\} > 0$. The compactness assumption of $\mathcal{E}_\delta$ implies that the minimum exists and is positive. For the same reason,

$$\bar{\gamma}_e^0(\delta) := \min\{p(\mathbf{H}) : \mathbf{H} \in \partial\mathcal{E}_\delta\} \tag{58}$$

is finite. We now prove that $(52)_2$ holds with

$$\bar{\beta}_e = \bar{c}_0 > 0, \quad \bar{\gamma}_e(\delta) = \bar{\gamma}_e^0(\delta) - \bar{\beta}_e \ln\sqrt{3} + \bar{\beta}_e \ln c_3. \tag{59}$$

We can again assume that $\mathbf{F} \in \mathrm{Sym}^+$. Let $\mathbf{D} \in \mathrm{Sym}_0$ be such that $\mathbf{H} := \mathbf{F}e^{-\mathbf{D}} \in \partial\mathcal{E}$. Then using successively (58), (51), (50) and (59) we obtain

$$p(\mathbf{F}e^{-\mathbf{D}}) + m(\mathbf{D}) \ge \bar{\gamma}_e(\delta) + \bar{\beta}^{\mathrm{i}} \ln|\mathbf{F}|.$$

Equation (47) implies $(52)_2$. Inequality (53) is proved similarly. □

Remark 2 shows that there are strong differences in the behavior of $e$, $f$ for large values of $|\mathbf{F}|$ on the surfaces $\det \mathbf{F} = \delta$. We see that $f$ is not even bounded from below. For this reason, we restrict ourselves to $e$. The growth of $e$ is sublinear on surfaces of constant determinant since the set $\mathcal{S}$ of all values of the stress $\mathbf{S}$, as opposed to the Piola–Kirchhoff stress, is bounded: The assumption that

$$|\mathbf{S}| \le c < \infty \tag{60}$$

leads to the logarithmic growth due to the extra factor $\mathbf{F}^T$ in $\mathbf{S} = \mathbf{F}^T \mathrm{D}p(\mathbf{F})$. The sublinear growth is at variance with the linear growth of the energy in the infinitesimal deformation Hencky plasticity theory (see Témam [31]). There the stress relation reads $\mathbf{S} = \mathrm{D}p(\mathbf{E})$ with $\mathbf{S}$ satisfying (60), which excludes the superlinear growth.

## 1.8. Appendix: A strange conjugation

Next let us show that Condition E has little or nothing to do with the convexity properties of $p$. To this end, we introduce a transformation on the set of materials that does not change the work in appropriately changed processes but changes the signs of potentials like the elastic potential, free energy, extended energies, etc.

Consider a material $\mathcal{M} = (\Sigma, \Pi, \hat{\mathbf{F}}, \hat{\mathbf{T}})$ determined by the constitutive objects $\mathcal{E}, \bar{\mathbf{T}}, \bar{\mathbf{M}}$ and define ${}^\sharp\bar{\mathbf{T}} : \mathcal{E} \to \mathrm{Sym}$, ${}^\sharp\bar{\mathbf{M}} : \partial\mathcal{E} \to \mathrm{Sym}_0$ by

$${}^\sharp\bar{\mathbf{T}}(\mathbf{E}) = -\bar{\mathbf{T}}(\mathbf{E}), \quad \mathbf{E} \in \mathcal{E}, \quad {}^\sharp\bar{\mathbf{M}}(\mathbf{E}) = -\bar{\mathbf{M}}(\mathbf{E}), \quad \mathbf{E} \in \partial\mathcal{E}.$$

It is easily seen that $\mathcal{E}, {}^\sharp\bar{\mathbf{T}}, {}^\sharp\bar{\mathbf{M}}$ satisfy Conditions (i)–(v) in Definition 1 and thus $\mathcal{E}, {}^\sharp\bar{\mathbf{T}}, {}^\sharp\bar{\mathbf{M}}$ determine a material that we denote ${}^\sharp\mathcal{M} = ({}^\sharp\Sigma, {}^\sharp\Pi, {}^\sharp\hat{\mathbf{F}}, {}^\sharp\hat{\mathbf{T}})$. Clearly,

$${}^\sharp\Sigma = \Sigma, \quad {}^\sharp\hat{\mathbf{F}} = \hat{\mathbf{F}}, \quad {}^\sharp\hat{\mathbf{T}} = -\hat{\mathbf{T}}.$$

To determine the relationship between ${}^\sharp\Pi$ and $\Pi$, define, for every $\pi = (\mathbf{E}(\cdot), \mathbf{P}(\cdot)) \in \Pi$, a pair ${}^\sharp\pi = ({}^\sharp\mathbf{E}(\cdot), {}^\sharp\mathbf{P}(\cdot))$ by

$${}^\sharp\mathbf{E}(t) = \mathbf{E}(d_\pi - t), \quad {}^\sharp\mathbf{P}(t) = \mathbf{P}(d_\pi - t), \quad t \in [0, d_\pi].$$

It is easily seen that for every process $\pi \in \Pi$ we have ${}^\sharp\pi \in {}^\sharp\Pi$ and the operation ${}^\sharp(\cdot)$ establishes a one-to-one correspondence between $\Pi$ and ${}^\sharp\Pi$. One finds that

$$w^E({}^\sharp\pi) = w^E(\pi), \quad w^P({}^\sharp\pi) = w^P(\pi),$$

$${}^\sharp m(\mathbf{D}) = m(-\mathbf{D}), \quad \mathbf{D} \in \mathrm{Sym}_0,$$

and

$$
\begin{array}{lcl}
\mathcal{M} \text{ satisfies the cyclic second law} & \Leftrightarrow & {}^\sharp\mathcal{M} \text{ satisfies the cyclic second law,}\\
p \text{ is an elastic potential for } \mathcal{M} & \Leftrightarrow & {}^\sharp p := -p \text{ is an elastic potential for } {}^\sharp\mathcal{M},\\
\psi \text{ is a free energy for } \mathcal{M} & \Leftrightarrow & {}^\sharp\psi := -\psi \text{ is a free energy for } {}^\sharp\mathcal{M},\\
r \text{ is a residual energy for } \mathcal{M} & \Leftrightarrow & {}^\sharp r := -r \text{ is a residual energy for } {}^\sharp\mathcal{M},\\
\mathcal{M} \text{ satisfies Il'yushin's condition} & \Leftrightarrow & {}^\sharp\mathcal{M} \text{ satisfies Il'yushin's condition,}\\
\mathcal{S} \text{ is convex} & \Leftrightarrow & {}^\sharp\mathcal{S} \text{ is convex,}\\
\mathcal{M} \text{ satisfies the normality rule} & \Leftrightarrow & {}^\sharp\mathcal{M} \text{ satisfies the normality rule,}\\
e \text{ is an initial energy for } \mathcal{M} & \Leftrightarrow & {}^\sharp f := -e \text{ is a final energy for } {}^\sharp\mathcal{M},\\
f \text{ is a final energy for } \mathcal{M} & \Leftrightarrow & {}^\sharp e := -f \text{ is an initial energy for } {}^\sharp\mathcal{M}.
\end{array}
$$

## 1.9. Materials of type C

The rest of the paper is devoted to a class of materials for which the elastic potential is logarithmically convex on deformation gradients of fixed determinant. They are shown to satisfy Il'yushin's condition and the extended energy $e$ can be calculated using the double convex conjugation with respect to the logarithmic deformation. An example of a material of type C is in Section 1.10.

If $f$ is any mapping defined on $\mathrm{Lin}^+$ and $\delta > 0$, define the mapping $f_\delta$ on $\mathrm{Sym}_0$ by

$$f_\delta(\mathbf{A}) = f(\delta^{1/3}e^{\mathbf{A}}), \quad \mathbf{A} \in \mathrm{Sym}_0.$$

If $\tilde{p} : \mathrm{Lin}^+ \to \mathbb{R}$ is a continuously differentiable objective isotropic function then $\tilde{p}_\delta$ is isotropic and continuously differentiable and

$$\mathrm{D}\tilde{p}_\delta = \tilde{\mathbf{S}}_\delta, \tag{61}$$

where $\tilde{\mathbf{S}}$ is the traceless part of $\tilde{\mathbf{T}}$ defined by

$$\tilde{\mathbf{T}}(\mathbf{F}) := \mathrm{D}\tilde{p}(\mathbf{F})\mathbf{F}^T. \tag{62}$$

To establish (61), note that in view of the isotropy of both sides of (61) it suffices to verify (61) only on diagonal elements (relative to some orthonormal basis), which is trivial.

The function $\tilde{p}$ is said to be (strictly) logarithmically convex if for every $\delta > 0$, the function $\tilde{p}_\delta$ is (strictly) convex. It must be emphasized that this definition involves only symmetric tensors of constant determinant. By (61) the convexity inequality reads

$$\tilde{p}_\delta(\mathbf{B}) \geq \tilde{p}_\delta(\mathbf{A}) + \tilde{\mathbf{S}}_\delta(\mathbf{A}) \cdot (\mathbf{B} - \mathbf{A}) \tag{63}$$

for any $\mathbf{A}, \mathbf{B} \in \mathrm{Sym}_0$ with the strict inequality sign if $\tilde{p}$ is strictly logarithmically convex and $\mathbf{A} \neq \mathbf{B}$. The convexity in the logarithmic deformation has been examined by Hill [11], [12] for elastic materials and was shown to be free from undesirable consequences (in contrast to the convexity in the principal stretches); our assumption is actually weaker because of the determinant restriction. Note also that the logarithmic convexity is consistent with the polyconvexity [30]. Let us show that the logarithmic convexity implies the Baker–Ericksen inequalities.

**Lemma 3.** *Let $\tilde{p}$ be objective and isotropic, $\tilde{\mathbf{T}}$ be given by (62), and* $\mathbf{E} = \operatorname{diag}(e_1, e_2, e_3) \in \mathrm{Sym}^+$ *be diagonal so that also* $\mathbf{T} := \tilde{\mathbf{T}}(\mathbf{E}) = \operatorname{diag}(t_1, t_2, t_3)$ *is diagonal. If $\tilde{p}$ is logarithmically convex then we have the Baker–Ericksen inequalities*

$$t_i \geq t_j \quad \text{if} \quad e_i \geq e_j \tag{64}$$

*with the strict inequality sign if $\tilde{p}$ is strictly logarithmically convex and $e_i > e_j$.*

*Proof.* Write $\mathbf{E} = \delta^{1/3} e^{\mathbf{A}}$, where $\mathbf{A} = \operatorname{diag}(a_1, a_2, a_3) \in \mathrm{Sym}_0$, and fix the pair $i, j$. Apply the monotonicity $(\tilde{\mathbf{S}}_\delta(\mathbf{B}) - \tilde{\mathbf{S}}_\delta(\mathbf{A})) \cdot (\mathbf{B} - \mathbf{A}) \geq 0$ to $\mathbf{B} = \operatorname{diag}(b_1, b_2, b_3)$, where the triple $(b_1, b_2, b_3)$ is obtained from $(a_1, a_2, a_3)$ by interchanging the $i, j$-components. Then $\tilde{\mathbf{S}}_\delta(\mathbf{B})$ is a diagonal tensor that differs from $\tilde{\mathbf{S}}_\delta(\mathbf{A})$ by interchanging the $i, j$-components and one obtains

$$(t_i - t_j)(e_i - e_j) \geq 0 \tag{65}$$

with the strict inequality sign if $e_i \neq e_j$ and $\tilde{p}_\delta$ is strictly convex. Thus if $e_i > e_j$, we have $t_i \geq t_j$. Since the derivative of $\tilde{p}$ is continuous, a limit provides that $t_i \geq t_j$ also if $e_i = e_j$. □

**Lemma 4.** (i) *If $\tilde{p}$ is objective, isotropic, of class $C^1$, and strictly logarithmically convex and* $\mathbf{E}, \mathbf{M} \in \mathrm{Sym}^+$ *satisfy*

$$\tilde{p}(\mathbf{ERMR}^T) \geq \tilde{p}(\mathbf{EM})$$

*for each* $\mathbf{R} \in \mathrm{Rot}$ *then* $\mathbf{E}$ *and* $\mathbf{M}$ *commute.* (ii) *Similarly, if $f : \mathrm{Sym}_0 \to \mathbb{R}$ is strictly convex and continuously differentiable and* $\mathbf{A}, \mathbf{B} \in \mathrm{Sym}_0$ *satisfy*

$$f(\mathbf{A} + \mathbf{RBR}^T) \geq f(\mathbf{A} + \mathbf{B})$$

*for each* $\mathbf{R} \in \mathrm{Rot}$ *then* $\mathbf{A}, \mathbf{B}$ *commute.*

*Proof.* (i): If $\mathbf{W}$ is any skew tensor, then the differentiation of

$$\tilde{p}(\mathbf{E}e^{\mathbf{W}}\mathbf{M}e^{-\mathbf{W}}) = \tilde{p}(\mathbf{E}e^{\mathbf{W}}\mathbf{M}) \geq \tilde{p}(\mathbf{EM})$$

with respect to $\mathbf{W}$ at $\mathbf{W} = \mathbf{0}$ gives $\mathbf{ETE}^{-1} \cdot \mathbf{W} = 0$, where $\mathbf{T} := \tilde{\mathbf{T}}(\mathbf{F})$ and $\mathbf{F} := \mathbf{EM}$. Hence $\mathbf{ETE}^{-1}$ is symmetric which means that $\mathbf{E}^2\mathbf{T} = \mathbf{TE}^2$ and hence

$$\mathbf{ET} = \mathbf{TE}. \tag{66}$$

Since $\tilde{\mathbf{T}}(\cdot)$ is objective isotropic, we have $\mathbf{T} = \tilde{\mathbf{T}}(\mathbf{C})$, where $\mathbf{C} = (\mathbf{FF}^T)^{1/2}$. Let us show that (66) implies that

$$\mathbf{EC} = \mathbf{CE}. \tag{67}$$

In a suitable orthonormal basis we have $\mathbf{C} = \operatorname{diag}(c_1, c_2, c_3)$, $\mathbf{T} = \operatorname{diag}(t_1, t_2, t_3)$, and (67) reads (no summation)

$$(c_i - c_j)E_{ij} = 0, \quad 1 \leq i, j \leq 3, \tag{68}$$

where $E_{ij}$ are the components of $\mathbf{E}$ in our basis. If $c_i = c_j$ then (68) holds trivially. If $c_i \neq c_j$ then the strict convexity of $\tilde{p}_\delta$ implies $t_i \neq t_j$ by the strict version of the Baker–Ericksen inequalities (65) (Remark 3), and as (66) reads $(t_i - t_j)E_{ij} = 0$ we have $V_{ij} = 0$ and (68) holds again. Hence also $\mathbf{E}$ and $\mathbf{FF}^T$ commute. This leads to

$\mathbf{EM}^2 = \mathbf{M}^2\mathbf{E}$. That is, $\mathbf{E}$ commutes with $\mathbf{M}^2$ and hence also with $(\mathbf{M}^2)^{1/2} = \mathbf{M}$. (ii) is proved similarly. □

**Definition 3.** *An ideal elastic-plastic material is said to be of type C if it satisfies the following conditions:*
(i) *the material obeys the normality rule* (*see Definition* 2);
(ii) $\bar{\mathbf{T}}$ *has an elastic potential p that admits an objective, isotropic, class* $C^1$ *extension* $\tilde{p} : \mathrm{Lin}^+ \to \mathbb{R}$ *that is strictly logarithmically convex and bounded from below;*
(iii) *for every* $\delta > 0$, $\mathcal{S} = \bar{\mathbf{S}}(\mathcal{E}_\delta)$ (*see* (3) *for the definition of* $\mathcal{E}_\delta$);
(iv) $\mathbf{0}$ *is in the interior of* $\mathcal{S}$.

Note that (iv) and the boundedness of $\mathcal{S}$ imply that

$$c_1|\mathbf{D}| \le m(\mathbf{D}) \le c_2|\mathbf{D}| \tag{69}$$

for some positive constants $c_1, c_2$ and all $\mathbf{D} \in \mathrm{Sym}_0$.

**Proposition 6.** *Each material of type* C *satisfies Il'yushin's condition.*

*Proof.* Verify Condition (ii) of Theorem 2. It suffices to establish (36) for pairs $\mathbf{E}$ and $\mathbf{D}$ as in that condition and satisfying additionally $\mathbf{E} \in \mathrm{Sym}^+$. Let $\mathbf{Q}$ be a point of minimum of the function $\mathbf{R} \mapsto \tilde{p}(\mathbf{ER}e^{-\mathbf{D}}\mathbf{R}^T) = \tilde{p}(\mathbf{ER}e^{-\mathbf{D}})$, $\mathbf{R} \in \mathrm{Rot}$, so that

$$\tilde{p}(\mathbf{E}e^{-\mathbf{D}}) \ge \tilde{p}(\mathbf{EQ}e^{-\mathbf{D}}\mathbf{Q}^T). \tag{70}$$

By Lemma 4(i), $\mathbf{E}$ and $\mathbf{Q}e^{-\mathbf{D}}\mathbf{Q}^T$ commute and hence so also do $\mathbf{E}$ and $\mathbf{QDQ}^T$. Let $\delta = \det \mathbf{E}$ and $\mathbf{A} = \ln(\mathbf{E}/\delta^{1/3})$. By the commutativity, $\mathbf{E}e^{-\mathbf{QDQ}^T} = \delta^{1/3}e^{\mathbf{A}-\mathbf{QDQ}^T}$ and thus $\tilde{p}(\mathbf{E}e^{-\mathbf{QDQ}^T}) = p_\delta(\mathbf{A} - \mathbf{QDQ}^T)$; the convexity of $p_\delta$ gives

$$p_\delta(\mathbf{A} - \mathbf{QDQ}^T) \ge p_\delta(\mathbf{A}) - \tilde{\mathbf{S}}_\delta(\mathbf{A}) \cdot (\mathbf{QDQ}^T).$$

Since $\mathbf{E} \in \mathcal{E}$, this reads

$$\tilde{p}(\mathbf{EQ}e^{-\mathbf{D}}\mathbf{Q}^T) \ge p(\mathbf{E}) - \bar{\mathbf{S}}(\mathbf{E}) \cdot (\mathbf{QDQ}^T) = p(\mathbf{E}) - \bar{\mathbf{S}}(\mathbf{Q}^T\mathbf{EQ}) \cdot \mathbf{D}.$$

By Lemma 1(ii), $\bar{\mathbf{S}}(\mathbf{Q}^T\mathbf{EQ}) \cdot \mathbf{D} \le m(\mathbf{D})$ which in conjunction with (70) leads to (36). □

We now give simplified constructions of the initial energy for materials of type C.

**Theorem 4.** *Consider a material of type* C *and let* $\mathbf{F} \in \mathrm{Lin}^+$. *Then*
(i) *there exists a unique* $\mathbf{D} \in \mathrm{Sym}_0$ *such that* $\mathbf{E} := \mathbf{F}e^{-\mathbf{D}} \in \mathcal{E}$ *and*

$$e(\mathbf{F}) = p(\mathbf{E}) + m(\mathbf{D}); \tag{71}$$

(ii) *if* $\mathbf{F} \in \mathcal{E}$ *then* $\mathbf{E} = \mathbf{F}, \mathbf{D} = \mathbf{0}$;
(iii) *if* $\mathbf{F} \notin \mathcal{E}$ *then* $\mathbf{E} \in \partial\mathcal{E}$ *and*

$$\mathbf{D} \in \mathrm{N}_{\bar{\mathbf{S}}(\mathbf{E})}\mathcal{S}; \tag{72}$$

*if additionally* $\mathbf{F} \in \mathrm{Sym}^+$, *then* $\mathbf{F}, \mathbf{E}, \mathbf{D}$ *commute.*

Hence, setting

$$\mathbf{P} = e^{\mathbf{D}},$$

we have the elastic-plastic decomposition

$$\mathbf{F} = \mathbf{E}\mathbf{P} \tag{73}$$

with $\mathbf{E} \in \mathcal{E}$, $\mathbf{D} \in \mathrm{N}_{\bar{\mathbf{S}}(\mathbf{E})}\mathcal{S}$ and

$$e(\mathbf{F}) = p(\mathbf{E}) + m(\mathbf{D}).$$

For $\mathbf{F} \in \mathrm{Sym}^+$ the logarithm of (73) and the commutativity give

$$\ln \mathbf{F} = \ln \mathbf{E} + \mathbf{D},$$

which decomposes $\ln \mathbf{F}$ into its projection $\ln \mathbf{E}$ onto $\ln(\mathcal{E} \cap \mathrm{Sym}^+)$ and the complement $\mathbf{D}$ in the normal direction to $\mathcal{S}$ at the corresponding stress. If we interpret the energy $e$ as a stored energy of the associated nonlinear Hencky material, the above shows that the constuction of $e$ involves the same projections as in the small deformation theory but in the space of logarithmic deformations.

*Proof.* It is enough to give the proof in the case $\mathbf{F} \in \mathrm{Sym}^+$. Write $\mathbf{F} = \delta^{1/3} e^{\mathbf{A}}$, where $\delta = \det \mathbf{F}$, $\mathbf{A} \in \mathrm{Sym}_0$, and consider an auxilliary minimum problem

$$\bar{e} = \min \left\{ \tilde{p}_\delta(\mathbf{A} - \mathbf{D}) + m(\mathbf{D}) : \mathbf{D} \in \mathrm{Sym}_0 \right\}. \tag{74}$$

The minimum exists since $\tilde{p}$ is bounded from below and $m$ is coercive (see (69)). Let $\mathbf{D}$ be a point of minimum. The optimality conditions say that

$$\tilde{\mathbf{S}}_\delta(\mathbf{A} - \mathbf{D}) \in \partial m(\mathbf{D}). \tag{75}$$

Furthermore, we have in particular

$$\tilde{p}_\delta(\mathbf{A} - \mathbf{D}) \le \tilde{p}_\delta(\mathbf{A} - \mathbf{R}\mathbf{D}\mathbf{R}^T)$$

for each $\mathbf{R} \in \mathrm{Rot}$, and hence $\mathbf{A}, \mathbf{D}$ commute by Lemma 4(ii). We conclude from (75) and $\partial m(\mathbf{D}) \subset \mathcal{S}$ (see (34)) that $\tilde{\mathbf{S}}_\delta(\mathbf{A} - \mathbf{D}) \in \mathcal{S}$. From Definition 3(iii) we have $\bar{\mathbf{S}}(\mathcal{E}_\delta) = \tilde{\mathbf{S}}(\mathcal{E}_\delta) = \mathcal{S}$ and thus $\tilde{\mathbf{S}}_\delta(\mathbf{M}) = \tilde{\mathbf{S}}_\delta(\mathbf{A} - \mathbf{D})$ for some $\mathbf{M} \in \mathrm{Sym}_0$ such that $\delta^{1/3} e^{\mathbf{M}} \in \mathcal{E}_\delta$. The strict monotonicity of $\tilde{\mathbf{S}}_\delta$, which is a consequence of the strict convexity of $\tilde{p}_\delta$, implies that $\mathbf{M} = \mathbf{A} - \mathbf{D}$ and as $\mathbf{A}, \mathbf{D}$ commute, we obtain $\mathbf{F}e^{-\mathbf{D}} \in \mathcal{E}$. Hence

$$\bar{e} = p(\mathbf{F}e^{-\mathbf{D}}) + m(\mathbf{D})$$

with $\mathbf{F}e^{-\mathbf{D}} \in \mathcal{E}$, which implies that

$$e(\mathbf{F}) \le \bar{e}. \tag{76}$$

Let us further prove that the opposite inequality holds in (76). Note that the infimum in (41) exists as a minimum in the present case. Let $\mathbf{D}$ be the point of minimum in (41).

Let further $\mathbf{Q}$ be a point of minimum of the function $\mathbf{R} \mapsto \tilde{p}(\mathbf{F}e^{-\mathbf{RDR}^T})$, $\mathbf{R} \in \mathrm{Rot}$. Lemma 4(i) tells us that $\mathbf{F}, \mathbf{QDQ}^T$ commute and hence

$$\begin{aligned} p(\mathbf{F}e^{-\mathbf{D}}) + m(\mathbf{D}) &\geq \tilde{p}(\mathbf{F}e^{-\mathbf{QDQ}^T}) + m(\mathbf{QDQ}^T) \\ &= \tilde{p}_\delta(\mathbf{A} - \mathbf{QDQ}^T) + m(\mathbf{QDQ}^T) \\ &\geq \bar{e}, \end{aligned}$$

which proves the opposite inequality in (76). Moreover, the argument shows that $\mathbf{D}$ is a point of minimum in (41) if and only if $\mathbf{D}$ is a point of minimum in (74) and that such a point commutes with $\mathbf{F}$.

(i): To prove the uniqueness of the point of minimum in (41), it suffices to prove the uniqueness of the point of minimum in (74). Let $\mathbf{D}_\alpha$, $\alpha = 1, 2$, be two distinct points of minimum in (74) so that, in particular, $\mathbf{F}, \mathbf{A}, \mathbf{D}_\alpha$ commute,

$$\tilde{\mathbf{S}}_\delta(\mathbf{A} - \mathbf{D}_\alpha) \in \partial m(\mathbf{D}_\alpha), \tag{77}$$

and

$$\tilde{p}_\delta(\mathbf{A} - \mathbf{D}_1) + m(\mathbf{D}_1) = \tilde{p}_\delta(\mathbf{A} - \mathbf{D}_2) + m(\mathbf{D}_2). \tag{78}$$

The inclusions (77) imply

$$\tilde{\mathbf{S}}_\delta(\mathbf{A} - \mathbf{D}_\alpha) \cdot \mathbf{D}_\alpha = m(\mathbf{D}_\alpha), \tag{79}$$

and the strict convexity provides

$$\tilde{p}_\delta(\mathbf{A} - \mathbf{D}_1) > \tilde{p}_\delta(\mathbf{A} - \mathbf{D}_2) + \tilde{\mathbf{S}}_\delta(\mathbf{A} - \mathbf{D}_2) \cdot (\mathbf{D}_2 - \mathbf{D}_1),$$

which in combination with (78) and (79) leads to

$$\tilde{\mathbf{S}}_\delta(\mathbf{A} - \mathbf{D}_2) \cdot \mathbf{D}_1 > m(\mathbf{D}_1)$$

in contradiction with (32). Thus the minimizer is unique. (ii): This follows from the uniqueness and Theorem 3(ii). (iii): The inclusion $\mathbf{E} \in \partial\mathcal{E}$ follows from the uniqueness and Theorem 3(iii). The inclusion (72) follows from the above proof. □

**Remark 3.** *we consider the unique minimizer* $\mathbf{D}$ *and the* $\mathbf{E}$ *from Theorem 4 as a function of* $\mathbf{F} \in \mathrm{Lin}^+$, *written* $\mathbf{D} = \hat{\mathbf{D}}(\mathbf{F})$, $\mathbf{E} = \hat{\mathbf{E}}(\mathbf{F})$, *then the form of* (41) *implies that the functions have the following transformation properties:*

$$\hat{\mathbf{E}}(\mathbf{QF}) = \mathbf{Q}\hat{\mathbf{E}}(\mathbf{F}), \quad \hat{\mathbf{P}}(\mathbf{QF}) = \hat{\mathbf{P}}(\mathbf{F}),$$

$$\hat{\mathbf{E}}(\mathbf{FQ}^T) = \hat{\mathbf{E}}(\mathbf{F})\mathbf{Q}^T, \quad \hat{\mathbf{P}}(\mathbf{FQ}^T) = \mathbf{Q}\hat{\mathbf{P}}(\mathbf{F})\mathbf{Q}^T,$$

*for every* $\mathbf{F} \in \mathrm{Lin}^+$, $\mathbf{Q} \in \mathrm{Rot}$. *We define* $\hat{\mathbf{Z}} : \mathrm{Lin}^+ \to \mathrm{Sym}$ *by*

$$\hat{\mathbf{Z}}(\mathbf{F}) = \bar{\mathbf{T}}_0(\hat{\mathbf{E}}(\mathbf{F})),$$

*where* $\bar{\mathbf{T}}_0(\mathbf{E}) = \bar{\mathbf{T}}(\mathbf{E}) - \frac{1}{3}(\mathrm{tr}\bar{\mathbf{T}}(\mathbf{E}))\mathbf{1}$ *is the traceless part of the Kirchhoff stress. Then* $\hat{\mathbf{Z}}$ *is objective and isotropic and for every* $\delta > 0$ *we have the stress relation*

$$\hat{\mathbf{Z}}_\delta(\mathbf{A}) = \mathrm{D}e_\delta(\mathbf{A})$$

*for a.e.* $\mathbf{A} \in \mathrm{Sym}_0$. *More precisely,* $e_\delta$ *is convex and* $\hat{\mathbf{Z}}_\delta(\mathbf{A})$ *is a subgradient of* $e_\delta$ *at* $\mathbf{A}$ *for every* $\mathbf{A} \in \mathrm{Sym}_0$. *Thus the Hencky material with the stored energy e is such that the traceless part of the Kirchhoff stress always belongs to the stress range* $\mathcal{S}$.

*Proof.* Let $\mathbf{A}$ be fixed and $\mathbf{D}$ be the corresponding minimizer. If $\mathbf{B}, \bar{\mathbf{D}} \in \mathrm{Sym}_0$ then the convexity of $\tilde{p}_\delta$ says that

$$\tilde{p}_\delta(\mathbf{B} - \bar{\mathbf{D}}) \geq \tilde{p}_\delta(\mathbf{A} - \mathbf{D}) + \tilde{\mathbf{S}}_\delta(\mathbf{A} - \mathbf{D}) \cdot (\mathbf{B} - \mathbf{A} + \mathbf{D} - \bar{\mathbf{D}}) \tag{80}$$

and the convexity of $m$ that

$$m(\bar{\mathbf{D}}) \geq m(\mathbf{D}) + \tilde{\mathbf{S}}_\delta(\mathbf{A} - \mathbf{D}) \cdot (\bar{\mathbf{D}} - \mathbf{D}) \tag{81}$$

since $\tilde{\mathbf{S}}_\delta(\mathbf{A} - \mathbf{D})$ is a subgradient of $m$ at $\mathbf{D}$. A combination of (80), (81) with

$$e_\delta(\mathbf{A}) = \tilde{p}_\delta(\mathbf{A} - \mathbf{D}) + m(\mathbf{D})$$

provides

$$\tilde{p}_\delta(\mathbf{B} - \bar{\mathbf{D}}) + m(\bar{\mathbf{D}}) \geq e_\delta(\mathbf{A}) + \tilde{\mathbf{S}}_\delta(\mathbf{A} - \mathbf{D}) \cdot (\mathbf{B} - \mathbf{A}).$$

Fixing $\mathbf{B}$ and taking the infimum of the left-hand side over all $\bar{\mathbf{D}} \in \mathrm{Sym}_0$, we obtain

$$e_\delta(\mathbf{B}) \geq e_\delta(\mathbf{A}) + \tilde{\mathbf{S}}_\delta(\mathbf{A} - \mathbf{D}) \cdot (\mathbf{B} - \mathbf{A}),$$

which shows that $e_\delta$ is convex and $\tilde{\mathbf{S}}_\delta(\mathbf{A} - \mathbf{D})$ is its subgradient at $\mathbf{A}$. It follows that $e_\delta$ is locally Lipschitz continuous and the subgradient coincides with the derivative for a.e. $\mathbf{A} \in \mathrm{Sym}_0$. □

## 1.10. The duality, example

This section gives a description of the initial energy for materials of type C in terms of the convex conjugation in the logarithmic measure of deformation (Proposition 7). The dual description is the analogue, for large deformations, of the duality considerations presented in Témam [31, Chapter 1, Section 3] in the context of the small deformation theory of plasticity.

**Proposition 7.** *Consider a material of type* C. *Then for every* $\delta > 0$ *we have*

$$e_\delta = \varphi_\delta^*,$$

*where* $\varphi_\delta^*$ *is the convex conjugate of* $\varphi_\delta$ *given by*

$$\varphi_\delta(\mathbf{S}) = \begin{cases} \tilde{p}_\delta^*(\mathbf{S}) & \text{if } \mathbf{S} \in \mathcal{S}, \\ \infty & \text{otherwise,} \end{cases}$$

*where* $\tilde{p}_\delta^*$ *is the convex conjugate of* $\tilde{p}_\delta$ *on* $\mathrm{Sym}_0$.

*Proof.* The proof of Theorem 4 shows that

$$e(\mathbf{F}) = \min\left\{\tilde{p}_\delta(\mathbf{A} - \mathbf{D}) + m(\mathbf{D}) : \mathbf{D} \in \mathrm{Sym}_0\right\} \tag{82}$$

and the proof of Remark 3 that $e_\delta$ is convex on $\mathrm{Sym}_0$ for every $\delta > 0$. For a fixed $\delta$, consider the family $\{f_{\mathbf{D}} : \mathbf{D} \in \mathrm{Sym}_0\}$ of convex functions on $\mathrm{Sym}_0$ given by

$$f_{\mathbf{D}}(\mathbf{A}) = \tilde{p}_\delta(\mathbf{A} - \mathbf{D}) + m(\mathbf{D}). \tag{83}$$

Equation (82) asserts that

$$e_\delta(\mathbf{A}) = \min\{f_{\mathbf{D}}(\mathbf{A}) : \mathbf{D} \in \mathrm{Sym}_0\}, \quad \mathbf{A} \in \mathrm{Sym}_0. \tag{84}$$

The application of the general formula for the convex conjugate of a minimum of a family of convex functions [9, Chapter I, Equation (4.6)] in the present case gives

$$e_\delta^*(\mathbf{S}) = \left(\min\{f_{\mathbf{D}} : \mathbf{D} \in \mathrm{Sym}_0\}\right)^*(\mathbf{S}) = \sup\{f_{\mathbf{D}}^*(\mathbf{S}) : \mathbf{D} \in \mathrm{Sym}_0\}, \tag{85}$$

$\mathbf{S} \in \mathrm{Sym}_0$. We have

$$f_{\mathbf{D}}^*(\mathbf{S}) = \left(\tilde{p}_\delta(\cdot - \mathbf{D}) + m(\mathbf{D})\right)^*(\mathbf{S}), \tag{86}$$

where $\tilde{p}_\delta(\cdot - \mathbf{D})$ denotes the shifted function $\mathbf{A} \mapsto \tilde{p}_\delta(\mathbf{A} - \mathbf{D})$. In (86), $\mathbf{D}$ is a parameter and hence $m(\mathbf{D})$ an additive constant during the evaluation of the convex conjugate. The rules for the evaluation of the conjugate of a function shifted by an additive constant and shifted in the domain space [9, Chapter I, Equations (4.8) and (4.9)] yield in the present case

$$f_{\mathbf{D}}^*(\mathbf{S}) = \tilde{p}_\delta^*(\mathbf{S}) + \mathbf{S} \cdot \mathbf{D} - m(\mathbf{D}), \tag{87}$$

where $\tilde{p}_\delta^*$ is the conjugate of $\tilde{p}_\delta$. Inserting (87) into (85),

$$e_\delta^*(\mathbf{S}) = \tilde{p}_\delta^*(\mathbf{S}) + \sup\{\mathbf{S} \cdot \mathbf{D} - m(\mathbf{D}) : \mathbf{D} \in \mathrm{Sym}_0\} \tag{88}$$

and we note that

$$m^*(\mathbf{S}) := \sup\{\mathbf{S} \cdot \mathbf{D} - m(\mathbf{D}) : \mathbf{D} \in \mathrm{Sym}_0\}$$

is the conjugate of the convex dissipation function $m$. A combination of (35) with (88) then gives $e_\delta^*(\mathbf{S}) = \varphi_\delta(\mathbf{S})$. Since $e_\delta$ is convex, we have $e_\delta = (e_\delta)^{**} = \varphi_\delta^*$. □

## Example

Finally, an elastic-plastic material is described for which the initial energy and the Hencky-type response are calculated explicitly. Let

$$\mathcal{E} = \{\mathbf{E} := \mathbf{VR} \in \mathrm{Lin}^+ : \mathbf{V} \in \mathrm{Sym}^+, \mathbf{R} \in \mathrm{Rot}, |\mathbf{L}| \le \tau/\beta\},$$

where $\tau, \beta$ are given positive numbers and $\mathbf{L} = \ln \mathbf{V} - \frac{1}{3}\mathrm{tr}(\ln \mathbf{V})\mathbf{1}$. Define $\bar{\mathbf{T}} : \mathrm{Lin}^+ \to \mathrm{Sym}$ by

$$\bar{\mathbf{T}}(\mathbf{E}) = \alpha'(\delta)\delta\mathbf{1} + \beta\mathbf{L}$$

for each $\mathbf{E} \in \mathrm{Lin}^+$, where $\delta = \det \mathbf{E}$ and $\alpha : (0, \infty) \to \mathbb{R}$ is a continuously differentiable function. We use the same symbol $\bar{\mathbf{T}}$ for the restriction of $\bar{\mathbf{T}}$ to $\mathcal{E}$. Define $\bar{\mathbf{M}} : \partial\mathcal{E} \to \mathrm{Sym}_0$ by

$$\bar{\mathbf{M}}(\mathbf{E}) = (\beta/\tau)\mathbf{R}^T\mathbf{L}\mathbf{R}, \quad \mathbf{E} \in \partial\mathcal{E}.$$

The objects $\mathcal{E}, \bar{\mathbf{T}}, \bar{\mathbf{M}}$ determine an ideal elastic-plastic material $(\Sigma, \Pi, \hat{\mathbf{F}}, \hat{\mathbf{T}})$ (see Definition 1) and one finds that

$$\bar{\mathbf{S}}(\mathbf{E}) = \beta \mathbf{R}^T \mathbf{L} \mathbf{R}, \quad \mathbf{E} \in \mathrm{Sym}^+,$$
$$\mathcal{S} = \left\{ \mathbf{S} \in \mathrm{Sym}_0 : |\mathbf{S}| \le \tau \right\}.$$

The material is of type C. Indeed, other things being obvious, it suffices to verify Condition (ii) of Definition 3. One finds that if $p : \mathrm{Lin}^+ \to \mathbb{R}$ is defined by

$$p(\mathbf{E}) = \alpha(\delta) + \tfrac{1}{2}\beta |\mathbf{L}|^2$$

for each $\mathbf{E} \in \mathrm{Lin}^+$, where $\mathbf{E} = \mathbf{V}\mathbf{R}$ is the polar decomposition of $\mathbf{E}$, then $p$ is an elastic potential for $\bar{\mathbf{T}}$ on $\mathrm{Lin}^+$. We identify $\tilde{p}$ with $p$. One finds that

$$\tilde{p}_\delta(\mathbf{A}) = \alpha(\delta) + \tfrac{1}{2}\beta |\mathbf{A}|^2, \quad \mathbf{A} \in \mathrm{Sym}_0,$$

which is a strictly convex function. Thus the material is of type C. The dissipation function is given by

$$m(\mathbf{D}) = m^{**}(\mathbf{D}) = \tau |\mathbf{D}|, \quad \mathbf{D} \in \mathrm{Sym}_0,$$

and since it is convex, by Proposition 5, the maximal and minimal residual energies are

$$s(\mathbf{R} e^{\mathbf{D}}) = \tau |\ln \mathbf{D}|, \quad t(\mathbf{R} e^{\mathbf{D}}) = -\tau |\ln \mathbf{D}|,$$

$\mathbf{R} \in \mathrm{Rot}$, $\mathbf{D} \in \mathrm{Sym}_0$. The initial energy is given by

$$e(\mathbf{F}) = \begin{cases} \alpha(\delta) + \tfrac{1}{2}\beta |\mathbf{L}|^2 & \text{if } |\mathbf{L}| \le \tau/\beta, \\ \alpha(\delta) - \tfrac{1}{2}(\tau/\beta) + \tau |\mathbf{L}| & \text{if } |\mathbf{L}| > \tau/\beta, \end{cases} \tag{89}$$

for each $\mathbf{F} \in \mathrm{Lin}^+$, where throughout, $\mathbf{F} = \mathbf{V}\mathbf{R}$ is the polar decomposition of $\mathbf{F}$ and $\delta = \det \mathbf{F}$. The case $|\mathbf{L}| \le \tau/\beta$ in (89) follows from Theorem 3. Let us consider the case $|\mathbf{L}| > \tau/\beta$. By the objectivity and isotropy we may restrict ourselves to $\mathbf{F} = \mathbf{V} \in \mathrm{Sym}^+ \setminus \mathcal{E}$; in fact it suffices to consider an $\mathbf{F}$ that is diagonal in some orthonormal basis. By Theorem 4, it suffices to seek the infimum in (47) only on those $\mathbf{D}$ that are diagonal in some basis of eigenvectors of $\mathbf{F}$. If $\mathbf{F} e^{-\mathbf{D}} \in \partial \mathcal{E}$ then

$$p(\mathbf{F} e^{-\mathbf{D}}) = \alpha(\delta) + \tau^2/2\beta.$$

Consider an orthonormal basis of eigenvectors of $\mathbf{F}$ so that $\mathbf{F} = \mathrm{diag}(v_1, v_2, v_3)$. By (47),

$$e(\mathbf{F}) = \alpha(\delta) + \tau^2/2\beta + \tau M, \tag{90}$$

where

$$M := \inf \left\{ |\mathbf{D}| : \mathbf{D} \in \mathrm{Sym}_0, \mathbf{F} e^{-\mathbf{D}} \in \partial \mathcal{E}, \ \mathbf{D} \text{ diagonal} \right\}. \tag{91}$$

For a $\mathbf{D} = \mathrm{diag}(d_1, d_2, d_3)$ as in (91) the condition $\mathbf{F} e^{-\mathbf{D}} \in \partial \mathcal{E}$ gives $|\ln_0 \mathbf{F} - \mathbf{D}| = \tau/\beta$ and we are led to

$$M = \inf \{ |d| : d \in C \},$$

where $C = \{ d \in P, |d - c| = \tau/\beta \}$ is the circle with the center $c$, $c_i = \ln(v_i/\delta^{1/3})$, $\delta := v_1 v_2 v_3$, in the plane $P = \left\{ d \in \mathbb{R}^3 : d_1 + d_2 + d_3 = 0 \right\}$. By $|\ln_0 \mathbf{F}| > \tau/\beta$ the origin $(0, 0, 0)$ is not in the interior of $C$. Thus we are looking for the point of the circle $C$ whose distance from the origin is minimal and hence $M = |\ln_0 \mathbf{F}| - \tau/\delta$ for each $\mathbf{F} \in \mathrm{Lin}^+$. A combination with (90) gives (89). Note the linear growth of $e$ in $|\mathbf{L}|$ outside $\mathcal{E}$. A differentiation of $e$ gives

$$\hat{\mathbf{Z}}(\mathbf{F}) = \begin{cases} \beta \mathbf{L} & \text{if } |\mathbf{L}| \le \tau/\beta, \\ \tau \mathbf{L}/|\mathbf{L}| & \text{if } |\mathbf{L}| > \tau/\beta. \end{cases}$$

**Acknowledgments**

The author thanks M. Lucchesi for extensive discussions on the rate-independent theories of plasticity and D. R. Owen for useful comments on a preliminary version of the paper. This research was supported by CNUCE, Istituto di CNR, Pisa, by grant 119110 of the AV ČR, and by grant 201/00/1516 of the Czech Republic. A preliminary version appeared as CNUCE Internal Report C91-17, 1991.

## References

[1] J. M. Ball, Differentiability properties of symmetric and isotropic functions, *Duke Math. J.*, **51** (1984), 699–728.

[2] A. Bertram, Material systems: A framework for the description of material behavior, *Arch. Rational Mech. Anal.*, **80** (1982), 99–133.

[3] A. Bertram, *Axiomatische Einführung in die Kontinuumsmechanik*, Wissenschaftsverlag, Mannheim, Germany, 1989.

[4] C. Carstensen, K. Hackl, and A. Mielke, Nonconvex potentials and microstructures in finite-strain plasticity, *Proc. Roy. Soc. London Sect.* A, **458** (2002), 299–317.

[5] B. D. Coleman and D. R. Owen, A mathematical foundation for thermodynamics, *Arch. Rational Mech. Anal.*, **54** (1974), 1–104.

[6] B. D. Coleman and D. R. Owen, On thermodynamics and elastic-plastic materials, *Arch. Rational Mech. Anal.*, **59** (1975), 25–51.

[7] B. D. Coleman and D. R. Owen, On the thermodynamics of elastic-plastic materials with temperature-dependent moduli and yields stresses, *Arch. Rational Mech. Anal.*, **70** (1979), 339–354.

[8] J. Dieudonné, *Treatise on Analysis*, Vols. I, II, and IV, Academic Press, New York, 1960, 1970, 1974.

[9] I. Ekeland and R. Témam, *Convex Analysis and Variational Problems*, North-Holland, Amsterdam, 1976.

[10] R. Fosdick and E. Volkmann, Normality and convexity of the yield surface in nonlinear plasticity, *Quart. Appl. Math.*, **51** (1993), 117–127.

[11] R. Hill, Constitutive inequalities for simple materials, *J. Mech. Phys. Solids*, **16** (1968), 229–242.

[12] R. Hill, Constitutive inequalities for isotropic elastic solids under finite strain, *Proc. Roy. Soc. London Sect.* A, **314** (1970), 457–472.

[13] E. H. Lee, Elastic-plastic deformation at finite strains, *J. Appl. Mech.*, **36** (1969), 1–6.

[14] M. Lucchesi, Free-energy functions for elastic-plastic material elements, *Quart. Appl. Math.*, **51** (1993), 299–318.

[15] M. Lucchesi and P. Podio-Guidugli, Materials with elastic range: A theory with a view toward applications, Parts I and II, *Arch. Rational Mech. Anal.*, **102** (1988), 23–43 and **110** (1990), 9–42.

[16] M. Lucchesi and M. Šilhavý, Il'yushin's conditions in non-isothermal plasticity, *Arch. Rational Mech. Anal.*, **113** (1991), 121–163.

[17] M. Lucchesi and M. Šilhavý, Thermoplastic materials with combined hardening, *Internat. J. Plasticity*, **9** (1993), 291–315.

[18] A. Mielke, *Finite Elastoplasticity, Lie Groups and Geodesics on $SL(d)$*, preprint, 2000.

[19] C. B. Morrey, Jr., *Multiple Integrals in the Calculus of Variations*, Springer-Verlag, New York, 1966.

[20] W. Noll, A new mathematical theory of simple materials, *Arch. Rational Mech. Anal.*, **48** (1972), 1–50.

[21] M. Ortiz and E. A. Repetto, Nonconvex energy minimisation and dislocation in ductile single crystals, *J. Mech. Phys. Solids*, **47** (1999), 397–462.

[22] D. R. Owen, Thermodynamics of materials with elastic range, *Arch. Rational Mech. Anal.*, **31** (1968), 91–112.

[23] D. R. Owen, A mechanical theory of materials with elastic range, *Arch. Rational Mech. Anal.*, **37** (1970), 85–110.

[24] P. Perzyna and W. Kosiński, A mathematical theory of materials, *Bull. Acad. Polon. Sci. Ser. Sci. Tech.*, **21** (1973), 647–654.

[25] A. C. Pipkin and R. S. Rivlin, Mechanics of rate-independent materials, *Z. Angew. Math. Phys.*, **16** (1965), 313–326.

[26] R. T. Rockafellar, *Convex Analysis*, Princeton University Press, Princeton, NJ, 1970.

[27] M. Šilhavý, On transformation laws for plastic deformations of materials with elastic range, *Arch. Rational Mech. Anal.*, **63** (1977), 169–182.

[28] M. Šilhavý, On measures, convex cones, and foundation of thermodynamics, *Czech. J. Phys.*, **B30** (1980), 841–860 and 930–961.

[29] M. Šilhavý, *The Mechanics and Thermodynamics of Continuous Media*, Springer-Verlag, Berlin, 1997.

[30] M. Šilhavý, *On the Nonconvex Potentials for Elastic-Plastic Materials*, in preparation, 2001.

[31] R. Témam, *Mathematical Problems in Plasticity*, Gauthier–Villars, Paris, 1983.

# 2. Generalized Elastic-Plastic Decomposition in Defective Crystals

G. P. Parry*

**Abstract.** I outline ideas which allow one to prove a rigorous type of elastic-plastic decomposition between related crystal states. The context of the work is a theory of defective crystals where the microstructure is represented by fields of lattice vectors, and the construction of the "generalized elastic-plastic decomposition" is a geometrical procedure which does not involve any notion of stress. The decomposition of the change of state has the form $F_1^e F^p F_2^e$, where $F_1^e$ and $F_2^e$ are piecewise elastic deformations and $F^p$ is a rearrangement of "unit cells" of a related crystal structure. The relevant unit cells do not derive from any supposed perfect lattice–like properties (e.g., translational symmetry) of the crystal states (no such presumption is appropriate since the crystals, and corresponding unit cells, support continuously distributed defects) but are constructed by means of the piecewise elastic deformations using a self-similarity property of the crystal states. Further, each individual unit cell, with its attendant distribution of lattice vector fields and defect densities, is translated to its new position by the rearrangement $F^p$, so that the corresponding mapping of points of the body represents a "change of shape" of the crystal domain.

## 2.1. Introduction

Lee et al. [7] and Lee's [6] notion of elastic-plastic decomposition has been influential in the development of phenomenological theories of plasticity. Their decomposition is based on a thought experiment whereby a notional "unstressed state" is used to define a linear transformation $F^p$ at a particular material point. Thus, quoting from Lee et al. [7], "to achieve the unstressed state, the body must be considered to be cut up into small elements, and with the removal of the loads on the elements, the unstressed state is approached as the element size approaches zero. However, the unstressed elements do not fit together to form a continuous body." Then an elastic deformation $F$ (with components equal to the partial derivatives of a displacement field) is decomposed as $F = F^e F^p$, where $F^e$ and $F^p$ "are not in general matrices of partial derivatives, though their product . . . is." So the elastic-plastic decomposition is effectively a constitutive assumption, and common theories of plasticity proceed

*Division of Theoretical Mechanics, School of Mathematical Sciences, University of Nottingham, Nottingham NG7 2RD, UK.

by making explicit assumptions on $F^p$, for example that $F^p$ is determined in terms of "slip" on a finite set of prescribed planes compatible with some assumed crystal symmetry.

I proceed differently here, basing the analysis on a theory of crystals with microstructure proposed by Davini [1], developed in Davini et al. [2, 3], Fonseca et al. [4], Glanville et al. [5], Parry [9, 10, 11], Parry et al. [12, 13]. In the theory, a crystal state corresponds to the prescription of three smooth linearly independent "lattice" vector fields over some region $B \subseteq \mathbb{R}^3$. Convincing analogues of defect measures like the Burgers vectors and dislocation density tensor are readily available in the context of Davini's theory. However, it is also clear that there are many more objects, with properties like those of the classic Burgers vector and dislocation density tensor, whose roles one should appreciate in this theory of continuous distribution of defects. So the first few sections of the paper are devoted to constructing and examining generalizations of the "classic objects."

In fact it turns out to be productive to attempt to answer a seemingly simple question: Given two crystal states, when are they elastically related to one another? It is easy enough to discover necessary conditions which must be satisfied if two states are to be elastically related, but it seems to be much more difficult to produce conditions sufficient for this purpose. However, according to Olver [8], Cartan has provided the mathematical theory which allows one to give sufficient conditions that crystal states are *locally* elastically related (see Section 2.4 for the definition of possibly unfamiliar terms). These sufficient conditions are a subset of the above conditions necessary for states to be elastically related, and they may be expressed in terms of the dislocation density tensor and a finite number of its (directional) derivatives. Furthermore, it will be clear that the conditions which are sufficient for states to be locally elastically related are not also sufficient for those states to be elastically related. In fact, it is particularly noteworthy that the archetypical two states which are locally elastically related but not elastically related (to each other) are related by *slip* in planes determined by the lattice vector fields. In a sense, then, the slip mechanism of phenomenological plasticity theories emerges quite naturally in this microstructural theory based just on the *geometry* of relevant crystal states. For that reason, the paper is focussed on examining pairs of states which satisfy the conditions necessary for them to be elastically related, which Davini and I have called *neutrally related states*.

In Section 2.7, I provide a generalized elastic-plastic decomposition of neutrally related states. The form of the decomposition is quite explicit once such states are given; *two* piecewise elastic deformations are involved, and the plastic part of the deformation is determined geometrically as a rearrangement of certain periodic structures constructed using a self-similarity property of the relevant crystal states.

## 2.2. Preliminaries

Let the structure of the material correspond to the prescription of 3 vectors $\boldsymbol{\ell}_1$, $\boldsymbol{\ell}_2$, $\boldsymbol{\ell}_3$ at each point $\boldsymbol{x}$ of a region $B$, and assume that *smooth* vector fields are thereby defined on that region.

Define the *crystal state* $\Sigma$ by

$$\Sigma = \{\boldsymbol{\ell}_a(\cdot), B;\ a = 1, 2, 3\}. \tag{1}$$

Let $\boldsymbol{u} : B \to B^* \equiv \boldsymbol{u}(B)$ be a smooth diffeomorphism. This diffeomorphism leads to the crystal state $\Sigma^* \equiv \boldsymbol{u}\Sigma$ defined by

$$\Sigma^* = \left\{\boldsymbol{\ell}_a^*(\cdot), B^*;\ a = 1, 2, 3\right\}, \tag{2}$$

called "the crystal state elastically related to $\Sigma$ via the diffeomorphism $\boldsymbol{u}$," if the vectors $\boldsymbol{\ell}_a^*$, $a = 1, 2, 3$, are determined throughout $B^*$ by the deformation gradient $\nabla\boldsymbol{u}$ and the vectors $\boldsymbol{\ell}_a$, $a = 1, 2, 3$, via

$$\boldsymbol{\ell}_a^*(\boldsymbol{u}(\boldsymbol{x})) = \nabla\boldsymbol{u}(\boldsymbol{x})\boldsymbol{\ell}_a(\boldsymbol{x}),\ \boldsymbol{x} \in B,\ a = 1, 2, 3. \tag{3}$$

Briefly, $\Sigma^*$ is elastically related to $\Sigma$ if the vector fields $\boldsymbol{\ell}_a^*$ in $\Sigma^*$ are related to the vector fields $\boldsymbol{\ell}_a$ in $\Sigma$ by

$$\boldsymbol{\ell}_a^* = \nabla\boldsymbol{u}\,\boldsymbol{\ell}_a, \tag{4}$$

so that the vector fields "behave as infinitesimal line elements in the deformation $\boldsymbol{u}$."

Suppose the three vector fields $\boldsymbol{\ell}_1(\cdot), \boldsymbol{\ell}_2(\cdot), \boldsymbol{\ell}_3(\cdot)$ are linearly independent at each point. Then one can define dual vector fields $\boldsymbol{d}^1(\cdot), \boldsymbol{d}^2(\cdot), \boldsymbol{d}^3(\cdot)$ by

$$\boldsymbol{d}^a(\boldsymbol{x}) \cdot \boldsymbol{\ell}_b(\boldsymbol{x}) = \delta_b^a, \quad a, b = 1, 2, 3,\ \boldsymbol{x} \in B. \tag{5}$$

The dual vector fields in state $\Sigma^*$ can be shown to satisfy

$$\boldsymbol{d}^{a^*}(\boldsymbol{u}(\boldsymbol{x})) = [\nabla\boldsymbol{u}(\boldsymbol{x})]^{-\top}\boldsymbol{d}^a(\boldsymbol{x}), \quad a = 1, 2, 3, \boldsymbol{x} \in B. \tag{6}$$

Define

$$\mathcal{S}^{ab}(\boldsymbol{x}) \equiv \boldsymbol{d}^a(\boldsymbol{x}) \cdot \nabla \wedge \boldsymbol{d}^b(\boldsymbol{x}), \tag{7}$$

$$n \equiv \boldsymbol{d}^1(\boldsymbol{x}) \cdot \boldsymbol{d}^2(\boldsymbol{x}) \wedge \boldsymbol{d}^3(\boldsymbol{x}). \tag{8}$$

Note that $n$ is nonzero throughout $B$, and assume that $n > 0$. A calculation gives that

$$\left(\frac{\mathcal{S}^{ab}}{n}\right)^*(\boldsymbol{u}(\boldsymbol{x})) = \left(\frac{\mathcal{S}^{ab}}{n}\right)(\boldsymbol{x}), \quad a, b = 1, 2, 3, \boldsymbol{x} \in B. \tag{9}$$

By virtue of property (9), the quantities $\mathcal{S}^{ab/n}$ are *elastic scalar invariants*. They are quantities that are unchanged by elastic deformation, in the sense of (9), so they may be regarded as measures of *inelastic* changes of state.

## 2.3. When are two crystal states elastically related?

If one is to factorize a change of crystal states into some kind of elastic and inelastic ("plastic") components, it is a sine qua non that one must be able to recognize when two crystal states are elastically related to one another. In this section I derive certain necessary conditions which must be satisfied if two states, $\Sigma$, $\Sigma^*$ say, are to be related in that way. Thus, when states $\Sigma$, $\Sigma^*$ are given as in (1) and (2), I derive certain implications of the relation (3). The first of these follows directly from (9):

$$\operatorname*{range}_{B^*}\left(\frac{\mathcal{S}^{ab}}{n}\right)^* = \operatorname*{range}_{B}\left(\frac{\mathcal{S}^{ab}}{n}\right)^* \cdot u = \operatorname*{range}_{B}\frac{\mathcal{S}^{ab}}{n}, \tag{10}$$

since $B^* = \boldsymbol{u}(B)$ and (9) holds, respectively. So

$$\operatorname*{range}_{B}\left(\frac{\mathcal{S}^{ab}}{n}\right) \tag{11}$$

is an invariant set, unchanged by an elastic deformation of the state $\Sigma$.

Now it is plain that the essential property of the quantities $\mathcal{S}^{ab}/n$, which leads to the invariance of the set (11), is that each such quantity is an *elastic scalar invariant*, satisfying the relation (9). Thus if $\nu_\Sigma(\cdot)$ is any elastic scalar invariant of the form

$$\nu_\Sigma(\boldsymbol{x}) = \nu\left(\left\{\boldsymbol{\ell}_a(\boldsymbol{x}), \nabla\boldsymbol{\ell}_a(\boldsymbol{x}), \ldots \nabla^k\boldsymbol{\ell}_a(\boldsymbol{x})\right\}\right), \quad a = 1, 2, 3, \boldsymbol{x} \in B, \tag{12}$$

for some integer $k$ and some appropriate function $\nu$ with the property

$$\nu_{\Sigma^*}(\boldsymbol{u}(\boldsymbol{x})) = \nu_\Sigma(\boldsymbol{x}) \tag{13}$$

whenever $\Sigma^*$ is elastically related to $\Sigma$, it follows that

$$\operatorname*{range}_{B^*}\nu_{\Sigma^*} = \operatorname*{range}_{B}\nu_\Sigma. \tag{14}$$

I note now that, generally, there is an infinite number of elastic scalar invariants. For if $\nu_\Sigma$ is any one such invariant (for example, any one component of $(\mathcal{S}^{ab}/n)$), then by differentiating (13) and combining the result with (3), one obtains

$$\left[\boldsymbol{\ell}_a^* \cdot \nabla^*\nu_{\Sigma^*}\right](\boldsymbol{u}(\boldsymbol{x})) = \left[\boldsymbol{\ell}_a \cdot \nabla\nu_\Sigma\right](\boldsymbol{x}), \quad a = 1, 2, 3. \tag{15}$$

Thus, if $\nu_\Sigma$ is an elastic scalar invariant, so is each of the first-order directional derivatives $\boldsymbol{\ell}_a \cdot \nabla\nu_\Sigma$, $a = 1, 2, 3$.

Similarly, each of the $s$th-order directional derivatives

$$\left(\boldsymbol{\ell}_{a_s} \cdot \nabla\right)\left(\boldsymbol{\ell}_{a_{s-1}} \cdot \nabla\right) \ldots \left(\boldsymbol{\ell}_{a_1} \cdot \nabla\right)\nu_\Sigma, \tag{16}$$

where each of $a_1 \ldots a_{s-1}, a_s$ may take the values $1, 2, 3$, is an elastic scalar invariant, and $s$ is arbitrary. Generally, then, there is an infinite number of necessary conditions of type (14) which must be satisfied if states $\Sigma$, $\Sigma^*$ are to be elastically related. However, it will turn out that not all of these conditions are independent, for if each of a certain finite number of these conditions holds, then all of the conditions hold (see the corollary to Theorem 5). I shall refer to any necessary condition of the form (14), where $\nu_\Sigma$ is a directional derivative of $\mathcal{S}^{ab}/n$ of any finite order, as a necessary condition of type I in order to differentiate such conditions from those patterned on the invariance of the *Burgers integral*, which I now discuss and refer to subsequently as necessary conditions of type II, and also from some other conditions.

Let $\Sigma$, $\Sigma^*$, given by (1),(2), be elastically related to one another, so that (3) holds and $B^* = \boldsymbol{u}(B)$. Let $\zeta$ be an arbitrary circuit in $B$, and let $\boldsymbol{u}(\zeta)$ be the image circuit in $B^*$. Then from (6),

$$\oint_{\boldsymbol{u}(\zeta)} \boldsymbol{d}^{a*} \cdot \boldsymbol{du} = \oint_{\zeta} (\nabla \boldsymbol{u})^{-\top} \boldsymbol{d}^a \cdot (\nabla \boldsymbol{u}) \boldsymbol{dx} = \oint_{\zeta} \boldsymbol{d}^a \cdot \boldsymbol{dx}. \tag{17}$$

So $\oint_\zeta \boldsymbol{d}^a \cdot \boldsymbol{dx}$, the Burgers integral, is an elastic invariant circuit integral, unchanged by any elastic deformation of state $\Sigma$. It is clear that the essential property of the field $\boldsymbol{d}^a(\cdot)$, which leads to condition (17), is that the field satisfies (6) when state $\Sigma$ undergoes elastic deformation to state $\Sigma^*$. Evidently, one may construct an infinite number of fields which satisfy analogues of (6): Let $\nu_\Sigma$ be any nonzero elastic scalar invariant. Then

$$\left\{\nu_\Sigma^* (\boldsymbol{u}(\boldsymbol{x}))\, \boldsymbol{d}^{a*} (\boldsymbol{u}(\boldsymbol{x}))\right\} = [\nabla \boldsymbol{u}(\boldsymbol{x})]^{-\top} \left\{\nu_\Sigma(\boldsymbol{x}) \boldsymbol{d}^a(\boldsymbol{x})\right\}, \quad a = 1, 2, 3, \tag{18}$$

via (13), whenever (6) holds. So

$$\oint_\zeta \nu_\Sigma \boldsymbol{d}^a \cdot \boldsymbol{dx} \text{ is an elastic invariant circuit integral.} \tag{19}$$

I shall refer to necessary conditions of the type

$$\oint_{\boldsymbol{u}(\zeta)} \nu_{\Sigma^*} \boldsymbol{d}^{a*} \cdot \boldsymbol{du} = \oint_\zeta \nu_\Sigma \boldsymbol{d}^a \cdot \boldsymbol{dx} \tag{20}$$

as necessary conditions of type II, noting however that as condition (20) stands, it is not so useful as a condition of type I because of the explicit dependence on the deformation $\boldsymbol{u}$ (which is unknown, a priori, given states $\Sigma$, $\Sigma^*$).

Next, in an evident generalization, consider elastic invariants corresponding to integrals of functions of the lattice vector and their spatial gradients of arbitrary finite order over closed surfaces and volumes. (The restriction to closed paths and surfaces is discussed in Davini [1] and Truesdell et al. [14].) The first point to note is that it suffices to consider volume integrals of the form

$$\int_V f_\Sigma n\, dV \equiv \int_V f(\{\boldsymbol{\ell}_a, \nabla \boldsymbol{\ell}_a, \ldots, \nabla^k \boldsymbol{\ell}_a\}) n\, dV, \quad V \subseteq B, \tag{21}$$

for

- integrals over closed surfaces can be converted to volume integrals of this type (since the functions under consideration are smooth, by assumption, $k$ is arbitrary, and the divergence theorem applies),
- $n \equiv \det\{\boldsymbol{d}^a\} \neq 0$, by assumption.

Note that from (6),

$$n^*(\boldsymbol{u}(\boldsymbol{x})) = [\det(\nabla \boldsymbol{u})]^{-1} n(\boldsymbol{x}). \tag{22}$$

Hence, for example, for $V \subseteq B$, with $\Sigma$, $\Sigma^*$ elastically related

$$\int_{\boldsymbol{u}(V)} n^* dV^* = \int_V [\det(\nabla \boldsymbol{u})]^{-1} n\, [\det(\nabla \boldsymbol{u})]\, dV = \int_V n\, dV. \tag{23}$$

Similarly, using (13),

$$\int_V \nu_\Sigma n \, dV \text{ is an elastic invariant volume integral} \tag{24}$$

whenever $\nu_\Sigma$ is an elastic scalar invariant. In particular, there is an infinite number of necessary conditions of the type

$$\int_{\boldsymbol{u}(V)} \nu_{\Sigma^*} n^* dV^* = \int_V \nu_\Sigma n \, dV, \tag{25}$$

which I shall refer to as conditions of type III.

Again, conditions like (25) are not so useful, as they stand, as conditions of type I because of the explicit dependence on $\boldsymbol{u}$. However, it will be possible to convert conditions of types II and III to more useful forms later on, in Section 2.6.

## 2.4. The basic theorem on locally elastically related states

It is a natural question to ask if any of the above conditions, which are necessary in order that two crystal states be elastically related, are also sufficient for that purpose. The answer to this question turns out to be no, and relevant counterexamples will be provided later. But the conditions of type I are in themselves sufficient that the two crystal states are *locally elastically related*, where this concept is defined as follows.

**Definition 1.** *State* $\Sigma$ *is* locally elastically related *to state* $\Sigma^*$ *if, for each* $\boldsymbol{x}_0 \in B$, *there exists a diffeomorphism* $\boldsymbol{u}_{\boldsymbol{x}_0}$ *defined on a neighbourhood* $N_{\boldsymbol{x}_0}$ *of* $\boldsymbol{x}_0$ *in* $B$, *with* $\boldsymbol{u}_{\boldsymbol{x}_0}(N_{\boldsymbol{x}_0}) \subset B^*$, *such that*

$$\boldsymbol{\ell}_a^*(\boldsymbol{u}_{\boldsymbol{x}_0}(\boldsymbol{x})) = \left(\nabla \boldsymbol{u}_{\boldsymbol{x}_0}(\boldsymbol{x})\right) \boldsymbol{\ell}_a(\boldsymbol{x}), \quad a = 1, 2, 3, \ \boldsymbol{x} \in N_{\boldsymbol{x}_0}, \ \boldsymbol{x}_0 \in B. \tag{26}$$

Three more definitions are needed (cf. Olver [8, Chapter 8]).

**Definition 2.** *Let* $\boldsymbol{v}$ *be a 9-tuple whose entries are the elastic scalar invariants* $S^{ab}/n$, $a, b, = 1, 2, 3$. *The* $s$th-order classifying set *associated with the crystal state* $\Sigma$ *is the range (over* $B$*) of all* $r$*th-order directional derivatives of* $\boldsymbol{v}$ *with* $r \leq s$.

**Definition 3.** *The rank* $\rho_s$ *of the map which takes a point* $\boldsymbol{x}$ *to the set of directional derivations of* $\boldsymbol{v}$ *of order less than or equal to* $s$ *is called the* order of the $s$th-order structure map.

**Definition 4.** *Let* $\rho_\Sigma$ *be the maximum value of* $\rho_s$. *Let* $s_0$ *be the least value of* $s$ *such that* $\rho_s = \rho_\Sigma$. *Then* $s_0$ *is called* the order of the structure map. *Note that*

- $\rho_s$ *is assumed to be well defined,*
- $\rho_s$ *is strictly increasing as a function of* $s$,
- $\rho_\Sigma$ *is well defined because* $\rho_s \leq 3$ *for all* $s$ *in this context.*

One of the main results of Cartan's theory of equivalence, reinterpreted by Olver [8, p. 437] and somewhat adapted, is the following theorem.

**Theorem 5.** *If crystal states $\Sigma$, $\Sigma^*$ are such that $s_0 = s_0^*$ and the $(s_0+1)$th-order classifying sets associated with the two states are identical, then $\Sigma$ is locally elastically related to $\Sigma^*$ (and vice versa). Moreover, if $\boldsymbol{x}_0 \in B$, $\boldsymbol{x}_0^* \in B^*$ map to the same point in the $(s_0+1)$th-order classifying set, one can take $\boldsymbol{u}_{\boldsymbol{x}_0}(\boldsymbol{x}_0) = \boldsymbol{x}_0^*$, for example, in the definition of locally elastically related states.*

**Corollary 6.** *Let $S_\Sigma^4$ be the fourth-order classifying set associated with state $\Sigma$, which is the range (over $B$) of all $r$th-order directional derivatives of $(S^{ab}/n)$ with $r \leq 4$. The condition*

$$S_{\Sigma^*}^4 = S_\Sigma^4 \tag{27}$$

*is a finite subset of the necessary conditions of type* I. *Since $s_0 = s_0^* \leq 3$, it follows that* (27) *is sufficient that $\Sigma$, $\Sigma^*$ are locally elastically related to each other.*

Note further that if $\Sigma$, $\Sigma^*$ satisfy (27), so that $\Sigma$, $\Sigma^*$ are locally elastically related, the basic property (13) of scalar quantities $\nu_\Sigma$ ensures that all necessary conditions of type I are then satisfied. Indeed, if (27) holds, then given $\boldsymbol{x}_0 \in B$, there exists a neighbourhood $N_{\boldsymbol{x}_0}$ of $\boldsymbol{x}_0$ in $B$, and a mapping $\boldsymbol{u}_{\boldsymbol{x}_0} : N_{\boldsymbol{x}_0} \to \boldsymbol{u}_{\boldsymbol{x}_0}(N_{\boldsymbol{x}_0}) \subseteq B^*$ such that (26) holds. It follows from (13) that

$$\nu_{\Sigma^*}(\boldsymbol{u}_{\boldsymbol{x}_0}(\boldsymbol{x}_0)) = \nu_\Sigma(\boldsymbol{x}_0), \tag{28}$$

in particular, for an arbitrary elastic scalar invariant $\nu_\Sigma$. Therefore

$$\operatorname*{range}_{B} \nu_\Sigma \subseteq \operatorname*{range}_{B^*} \nu_{\Sigma^*}. \tag{29}$$

Since one also has the reverse inclusion, by an evident argument, it follows that (14) holds quite generally. In a sense, this argument shows that the set of up to fourth-order directional derivatives of the dislocation density tensor is a functional basis of all scalar invariants—though the set is not necessarily "minimal."

**Remark 1.** *To show that it is not necessarily true that if the conditions of the theorem are met, then $\Sigma$ and $\Sigma^*$ are elastically related to each other, consider the following example.*

*Let $\Sigma = \{\boldsymbol{e}_a, B_c;\ a = 1,2,3\}$, $\Sigma^* = \{\boldsymbol{e}_a, B_n;\ a = 1,2,3\}$, where $\boldsymbol{e}_1, \boldsymbol{e}_2, \boldsymbol{e}_3$ is the canonical basis of $\mathbb{R}^3$, $B_c$ is a cube, and $B_n$ is not a cube. Then the conditions of the theorem are met because $(S^{ab}/n) = 0$, $(S^{ab}/n)^* = 0$, and all the relevant directional derivatives are zero. One verifies the conclusions of the theorem. If $\boldsymbol{x}_0 \in B_c$, $\boldsymbol{x}_0^* \in B_n$ are arbitrary, the diffeomorphism $\boldsymbol{u}_{\boldsymbol{x}_0} \equiv \boldsymbol{x} + \boldsymbol{x}_0^* - \boldsymbol{x}_0$ is such that*

$$\boldsymbol{\ell}_a^*(\boldsymbol{u}_{\boldsymbol{x}_0}(\boldsymbol{x})) \equiv \boldsymbol{e}_a = (\nabla \boldsymbol{u}_{\boldsymbol{x}_0})\boldsymbol{\ell}_a(\boldsymbol{x}) \equiv \nabla \boldsymbol{u}_{\boldsymbol{x}_0}\boldsymbol{e}_a,\ \boldsymbol{u}_{\boldsymbol{x}_0}(\boldsymbol{x}_0) = \boldsymbol{x}_0^*. \tag{30}$$

*However, there exists* no *diffeomorphism $\boldsymbol{u}$ with the properties that*

$$\boldsymbol{e}_a = (\nabla \boldsymbol{u})\boldsymbol{e}_a \quad \textit{and} \quad \boldsymbol{u}(B_c) = B_n \tag{31}$$

*because this would imply that $(\nabla \boldsymbol{u})$ is the identity, so that $B_n$ would be a translate of the cube $B_c$, and so $B_n$ itself would be a cube, contrary to assumption.*

To conclude this section, I prove a result which will be needed later on, while acknowledging that the motivation for the theorem will not be clear at this point (so the reader could well skip this at a first reading).

**Theorem 7.** *Given a crystal state* $\Sigma$, *suppose that* $S^{ab}/n \equiv \mathcal{T}^{ab}(x^3)$, *that* $\boldsymbol{\ell}_a \cdot \boldsymbol{e}_3 \equiv \ell_{a3}(x^3)$, *and that* $\mathcal{T}^{ab}\ell_{a3} \equiv 0$. *Then* $\Sigma$ *is elastically related to a state* $\Sigma_3 = \{\tilde{\boldsymbol{\ell}}_a(\cdot), \tilde{B};\ a = 1, 2, 3\}$, *where*

$$\tilde{\boldsymbol{\ell}}_a(\boldsymbol{x}) = \boldsymbol{\ell}_a^+(x^3), \quad \boldsymbol{x} \in \tilde{B}, \tag{32}$$

*for some* $\boldsymbol{\ell}_a^+(\cdot) : x^3(\tilde{B}) \to \mathbb{R}^3$, *with* $x^3(\tilde{B}) = x^3(B)$.

That is to say, if the hypotheses of the theorem are satisfied, then $\Sigma$ is elastically related to a state $\Sigma_3$, where the lattice vectors depend only on $x^3$.

*Proof.* It is given that, for the state $\Sigma$, $S^{ab}/n \equiv \boldsymbol{d}^a \cdot \nabla \wedge \boldsymbol{d}^b/n = \mathcal{T}^{ab}(x^3)$. Hence $\nabla \wedge \boldsymbol{d}^b = n\mathcal{T}^{ab}(x^3)\boldsymbol{\ell}_a = \frac{1}{2}\mathcal{T}^{ab}(x^3)\varepsilon_{ars}\boldsymbol{d}^r \wedge \boldsymbol{d}^s$, and taking the divergence,

$$\frac{1}{2}\nabla \cdot (\mathcal{T}^{ab}\varepsilon_{ars}\boldsymbol{d}^r \wedge \boldsymbol{d}^s) = n\mathcal{T}^{ab}\varepsilon_{ars}\mathcal{T}^{sr} + \dot{\mathcal{T}}^{ab}n\ell_{a3} = 0, \tag{33}$$

when the superposed dot denotes $d/dx^3$. So the given dislocation density tensor and components $\ell_{a3}$ must satisfy the compatibility conditions

$$\dot{\mathcal{T}}^{ab}\ell_{a3} = -\mathcal{T}^{ab}\varepsilon_{ars}\mathcal{T}^{sr}. \tag{34}$$

I show that there is a crystal state $\Sigma_3 = \{\tilde{\boldsymbol{\ell}}_a(\cdot), \tilde{B}; a = 1, 2, 3\}$ with $\tilde{\boldsymbol{\ell}}_a(\boldsymbol{x}) = \boldsymbol{\ell}_a^+(x^3)$, $\ell_{a3}^+(x^3) \equiv \ell_{a3}(x^3)$ satisfying (34) and

$$\frac{\tilde{\boldsymbol{d}}^a \cdot \tilde{\nabla} \wedge \tilde{\boldsymbol{d}}^b}{\tilde{n}} = \mathcal{T}^{ab}(x^3). \tag{35}$$

To do this, it is convenient to recall from Parry et al. [12], say, that

$$\frac{\boldsymbol{d}^a \cdot \nabla \wedge \boldsymbol{d}^b}{n} = \boldsymbol{L}^a \cdot \boldsymbol{d}^b, \quad a, b = 1, 2, 3, \tag{36}$$

where, for example, $\boldsymbol{L}^1$ is defined as $(\boldsymbol{\ell}_2 \cdot \nabla)\boldsymbol{\ell}_3 - (\boldsymbol{\ell}_3 \cdot \nabla)\boldsymbol{\ell}_2$, the Lie bracket of the vector fluids $\boldsymbol{\ell}_2(\cdot)$ and $\boldsymbol{\ell}(\cdot)$. Therefore, with $\mathcal{T}^{ab}$ and $\ell_{a3}^+$ the prescribed functions of $x^3$, it is enough to show that

$$\boldsymbol{L}^{a+} = \mathcal{T}^{ab}(x^3)\boldsymbol{\ell}_b^+ \tag{37}$$

has a solution for $\tilde{\ell}_{a\alpha} = \ell_{a\alpha}^+(x^3)$, $a = 1, 2, 3$, $\alpha = 1, 2$. These three vector equations read (dropping the $^+$)

$$\left.\begin{aligned} \ell_{23}\dot{\ell}_{3i} - \ell_{33}\dot{\ell}_{2i} &= \mathcal{T}^{1a}\ell_{ai}, \\ \ell_{33}\dot{\ell}_{1i} - \ell_{13}\dot{\ell}_{3i} &= \mathcal{T}^{2a}\ell_{ai}, \\ \ell_{13}\dot{\ell}_{2i} - \ell_{23}\dot{\ell}_{1i} &= \mathcal{T}^{3a}\ell_{ai} \end{aligned}\right\}. \tag{38}$$

The equations with $i = 3$ are identically satisfied, since $\ell_{a3}^+ \equiv \ell_{a3}$ and the corresponding equations hold in $\Sigma$. At least one of $\ell_{13}, \ell_{23}, \ell_{33}$ is nonzero; suppose for the sake of argument that $\ell_{13} \neq 0$. Then equations $(38)_2$ and $(38)_3$ imply that $(38)_1$ holds, by virtue of the hypothesis $\mathcal{T}^{ia}\ell_{i3} = 0$. So (38) is equivalent to four linear equations for the six unknown quantities $\ell_{i\alpha}(x^3)$, $i = 1, 2, 3$, $\alpha = 1, 2$, given $\ell_{i3}$, $i = 1, 2, 3$ and $\mathcal{T}^{ab}$, $a, b, = 1, 2, 3$. Choose $\ell_{11} = \ell_{12} = 0$. Then $(38)_2$ and $(38)_3$ become

$$\left.\begin{aligned} -\ell_{13}\dot{\ell}_{3i} &= \mathcal{T}^{2a}\ell_{ai}, \quad i = 1, 2, \\ \ell_{13}\dot{\ell}_{2i} &= \mathcal{T}^{3a}\ell_{ai}, \quad i = 1, 2, \end{aligned}\right\} \tag{39}$$

and these equations have a unique solution, given appropriate initial conditions.

So one can construct a state $\Sigma_3$ of the required form, satisfying (34) and with the same dislocation density tensor as $\Sigma$. Recalling the definitions given earlier in this section, $\Sigma$, $\Sigma_3$ are rank zero states with the same first-order classifying sets. (The only thing that is not immediate is that $(\boldsymbol{\ell}_a \cdot \nabla)\mathcal{T}^{rs}$ depends only on $x^3$ in state $\Sigma$, but this quantity equals $\ell_{a3}\dot{\mathcal{T}}^{rs}$, which is a given function of $x^3$.) Hence $\Sigma$ and $\Sigma_3$ are locally elastically related by Theorem 5. Also by the theorem, the (local) elastic deformations leave invariant the planes $x^3 = $ constant. I assume that $\tilde{B} = \boldsymbol{\psi}(B)$ for some elastic deformation $\boldsymbol{\psi}$ (even though this restricts the topology, and size, of $B$). □

## 2.5. Self-similarity of states with constant dislocation density

It will turn out, later, that generalized elastic-plastic decompositions can be divided into two classes, according as to whether or not the dislocation density tensor is constant in the two states under consideration. (This statement will become clearer in the section dealing with neutral deformations.) So, as a necessary preliminary, I discuss geometrical properties of a state $\Sigma$ which has all components of $\mathcal{S}^{ab}/n$ constant throughout $B$. In this case, all directional derivations of order greater than or equal to one are zero. Theorem 5 may be applied to states $\Sigma$, $\Sigma^* \equiv \Sigma$, since it is clear that $s_0 = s_0^*$ and that the associated classifying sets are identical (as $\Sigma$, $\Sigma^*$ are the same state). Therefore, let $\boldsymbol{x}_0 \in B, \boldsymbol{x}_0^* \in B (\equiv B^*)$ be arbitrary points of $B$. Then $\boldsymbol{x}_0$, $\boldsymbol{x}_0^*$ map to the same point in the $(s_0 + 1)$th-order classifying set, since $(\mathcal{S}^{ab}/n)^* \equiv (\mathcal{S}^{ab}/n)$ are constant in $B$. The theorem gives the existence of a diffeomorphism $\boldsymbol{u}_{\boldsymbol{x}_0}$ defined in a neighbourhood $N_{\boldsymbol{x}_0}$ of $\boldsymbol{x}_0$ such that

$$\boldsymbol{\ell}_a\left(\boldsymbol{u}_{\boldsymbol{x}_0}(\boldsymbol{x})\right) = \left[\nabla \boldsymbol{u}_{\boldsymbol{x}_0}(\boldsymbol{x})\right]\boldsymbol{\ell}_a(\boldsymbol{x}), \quad \boldsymbol{x} \in N_{\boldsymbol{x}_0}, \quad \text{and} \quad \boldsymbol{u}_{\boldsymbol{x}_0}(\boldsymbol{x}_0) = \boldsymbol{x}_0^*. \tag{40}$$

This is the self-similarity to which I have referred earlier: Lattice vectors are determined *throughout* the state $\Sigma$ by knowledge of the lattice vectors in an arbitrary small neighbourhood of an (arbitrary) fixed point. In fact, there is a Lie group structure underlying this self-similarity, and the local elastic deformation $\boldsymbol{u}_{\boldsymbol{x}_0}$ in (40) may be related to the composition function in the group, but I do not consider this aspect in detail here (see Parry [11]).

As an example, consider the state $\Sigma = \{\boldsymbol{\ell}_a(\cdot), B;\ a = 1, 2, 3\}$ with $\boldsymbol{d}^1 = (1, -x^3, 0)$, $\boldsymbol{d}^2 = \boldsymbol{e}^2$, $\boldsymbol{d}^3 = \boldsymbol{e}^3$, and $B = [0, 1]^3$. Then

$$\mathcal{S}^{ab}/n = \delta_1^a \delta_1^b \tag{41}$$

is constant in $B$. One calculates, via (40), that

$$\boldsymbol{u}_{\boldsymbol{x}_0} \equiv \boldsymbol{x} + (x^2 - x_0^2)(x_0^{3*} - x_0^3)\boldsymbol{e}_1 + \boldsymbol{x}_0^* - \boldsymbol{x}_0, \tag{42}$$

has the properties

$$d^a\left(u_{x_0}(x)\right) = \left[\nabla u_{x_0}\right]^{-\mathsf{T}} d^a(x), \; x \in N_{x_0} \text{ and } u_{x_0}(x_0) = x_0^* \tag{43}$$

whenever $N_{x_0} \subset B$ is such that $u_{x_0}\left(N_{x_0}\right) \subset B$.

## 2.6. Neutral deformations

Now I continue the examination of the necessary conditions of types II and III. Consider, to begin with, the conditions of type II, corresponding to circuit integrals for which

$$\oint_{u(\zeta)} \nu_{\Sigma^*} d^{a*} \cdot du = \oint_{\zeta} \nu_{\Sigma} d^a \cdot dx. \tag{44}$$

The question of interest is whether the three types of necessary condition, taken together, are sufficient that states $\Sigma$, $\Sigma^*$ are elastically related. Suppose then, in particular, that (44) holds for some choice of the mapping $u : B \to B^*$ but that it is not known whether $\Sigma$, $\Sigma^*$ are elastically related. Now introduce state $\Sigma'$ by elastic deformation of state $\Sigma^*$ via the mapping $u^{-1} : B^* \to B$, so that $\Sigma' = \{\ell^{a'}(\cdot), B;\ a = 1, 2, 3\}$, where

$$\ell_a'\left(u^{-1}(y)\right) = \left[\left(\nabla u^{-1}\right)(y)\right] \ell_a^*(y), \quad y \in B^* \equiv u(B). \tag{45}$$

By the basic invariance property of the integral $\oint_\zeta \nu_\Sigma d^a \cdot dx$, it follows that

$$\oint_{u(\zeta)} \nu_{\Sigma^*} d^{a*} \cdot du = \oint_{\zeta} \nu_{\Sigma'} d^{a'} \cdot dx. \tag{46}$$

So (44) holds for some mapping $u : B \to B^*$ if and only if there exists a state $\Sigma^*$ such that

$$\oint_{\zeta} \nu_{\Sigma'} d^{a'} \cdot dx = \oint_{\zeta} \nu_{\Sigma} d^a \cdot dx, \tag{47}$$

for each circuit $\zeta \subset B$. It follows that, if $\Sigma$ is *not* elastically related to $\Sigma^*$ but (44) holds for some mapping $u : B \to B^*$, then there is a state $\Sigma' \neq \Sigma$ such that $\Sigma'$ is elastically related to $\Sigma^*$ and (47) holds.

Now, since $\zeta$ is arbitrary, (47) implies

$$\nabla \wedge \left(\nu_{\Sigma'} d^{a'}\right) = \nabla \wedge \left(\nu_{\Sigma} d^a\right). \tag{48}$$

Likewise, the conditions of type III lead to the relations

$$\nu_{\Sigma'} n' = \nu_{\Sigma} n. \tag{49}$$

Thus equations (48) and (49) must have solutions $\Sigma' \neq \Sigma$ with $\Sigma'$ elastically related to $\Sigma^*$ if $\Sigma$ is *not* elastically related to $\Sigma^*$ but (44) and (45) hold for some mapping $u : B \to B^*$.

Clearly, the constant elastic scalar invariant $\nu_\Sigma \equiv 1$ is an admissible choice of $\nu_\Sigma$ in both (48) and (49). So, in particular, (49) implies that for arbitrary elastic scalar invariants $\nu_\Sigma$,

$$n = n', \quad \nu_{\Sigma'} = \nu_\Sigma. \tag{50}$$

Since $(50)_2$ implies that

$$\operatorname*{range}_B \nu_{\Sigma'} = \operatorname*{range}_B \nu_\Sigma, \tag{51}$$

it follows that all conditions of type I hold (for states $\Sigma$, $\Sigma'$) when (49) holds. Moreover, since $\nu_\Sigma$ is arbitrary in (51), that equation implies that the classifying sets of arbitrary order $s$ corresponding to $\Sigma$, $\Sigma'$ are identical. Therefore (50) (or (49)) implies that $\Sigma$, $\Sigma'$ are locally elastically related to one another.

Also, taking $\nu_\Sigma \equiv \nu_{\Sigma'} \equiv 1$ in (48) one deduces

$$\nabla \wedge \boldsymbol{d}^{a'} = \nabla \wedge \boldsymbol{d}^a, \quad \nabla\nu_{\Sigma'} \wedge \boldsymbol{d}^{a'} + \nu_{\Sigma'}\nabla \wedge \boldsymbol{d}^{a'} = \nabla\nu_\Sigma \wedge \boldsymbol{d}^a + \nu_\Sigma\nabla \wedge \boldsymbol{d}^a, \tag{52}$$

and since $(50)_2$ implies $\nabla\nu_{\Sigma'} = \nabla\nu_\Sigma$, $(52)_2$ holds only if

$$\nabla\nu_\Sigma \wedge (\boldsymbol{d}^{a'} - \boldsymbol{d}^a) = \boldsymbol{0}. \tag{53}$$

It follows that the necessary conditions of types I, II, and III hold for states $\Sigma$, $\Sigma'$ if and only if

$$\nabla \wedge \boldsymbol{d}^{a'} = \nabla \wedge \boldsymbol{d}^a, \quad \nabla\nu_\Sigma \wedge (\boldsymbol{d}^{a'} - \boldsymbol{d}^a) = \boldsymbol{0}, \quad \nu_{\Sigma'} = \nu_\Sigma, \quad n' = n, \tag{54}$$

for arbitrary scalars $\nu_\Sigma$.

The rest of this section is devoted to showing that (54) holds if and only if the choice of $\nu_\Sigma$ is restricted to the set of directional derivatives of $\mathcal{S}^{ab}/n$ of order less than or equal to one. So set

$$\mathcal{F} = \{(\mathcal{S}^{bc}/n),\ (\boldsymbol{\ell}_a \cdot \nabla)(\mathcal{S}^{bc}/n);\ a, b, c = 1, 2, 3\}. \tag{55}$$

Reiterating, the equations

$$\nabla \wedge \boldsymbol{d}^{a'} = \nabla \wedge \boldsymbol{d}^a, \quad \nabla\nu_\Sigma \wedge (\boldsymbol{d}^{a'} - \boldsymbol{d}^a) = 0, \quad \nu_{\Sigma'} = \nu_\Sigma, \quad n' = n, \quad \nu_\Sigma \in \mathcal{F}, \tag{56}$$

will be the focus of the rest of the paper. To emphasize this fact, states $\Sigma'$ such that (56) holds, given state $\Sigma$, will be called states *neutrally related* to $\Sigma$. Mappings of $\Sigma$ to states $\Sigma'$ neutrally related to $\Sigma$ will be called *neutral deformations*.

Next, I show that (56) implies (54). First notice that $\nabla \wedge \boldsymbol{d}^{a'} = \nabla \wedge \boldsymbol{d}^a$ implies that $\boldsymbol{d}^{a'} = \boldsymbol{d}^a + \nabla\tau^a$ for some potentials $\tau^a$, $a = 1, 2, 3$, not all of them constant if $\Sigma' \neq \Sigma$. Then $(56)_2$ gives $\nabla\nu_\Sigma \wedge \nabla\tau^a = \boldsymbol{0}$, $\nu_\Sigma \in \mathcal{F}$, $a = 1, 2, 3$, so in particular $\nu_\Sigma = \nu_\Sigma(\tau^a)$ for a nonconstant potential $\tau^a$, $\nu_\Sigma \in \mathcal{F}$. Now either all the $\mathcal{S}^{ab}/n$ are constant, or there exists one which is nonconstant; call it $\theta$. In the latter case, $\theta = \theta(\tau^a)$ and I shall assume that this relationship is invertible. It follows that

$\nu_\Sigma = \nu_\Sigma(\theta)$, $\nu_\Sigma \in \mathcal{F}$. Furthermore, in particular, $(\boldsymbol{\ell}_b \cdot \nabla)(\mathcal{S}^{cd}/n)$ is a function of $\theta$; call it $\bar{\nu}(\theta)$, since first-order directional derivatives are in $\mathcal{F}$. Hence

$$(\boldsymbol{\ell}_a \cdot \nabla)(\boldsymbol{\ell}_b \cdot \nabla)(\mathcal{S}^{cd}/n) = (\boldsymbol{\ell}_a \cdot \nabla)\bar{\nu}(\theta) = (\boldsymbol{\ell}_a \cdot \nabla\theta)\frac{d\bar{\nu}}{d\theta} \tag{57}$$

is a function of $\theta$, since $\boldsymbol{\ell}_a \cdot \nabla\theta \in \mathcal{F}$. Likewise, by induction, directional derivatives of $\mathcal{S}^{ab}/n$ of arbitrary finite order are functions of the single variable $\theta$.

Recall now the definitions introduced in Section 2.4, and divide the discussion into two cases according as to whether or not all the components of $\mathcal{S}^{ab}/n$ are constant:

(i): *Each component of $\mathcal{S}^{ab}/n$ constant*. In this case, the order of the $s$th-order structure map $\rho_s$ is zero for all choices of $s$, and each $s$th-order classifying set consists of a single point. Using theorem 5, one sees that $(56)_3$ implies that the first-order classifying sets corresponding to states $\Sigma$, $\Sigma'$ are identical, so (56) implies that states $\Sigma$, $\Sigma'$ are locally elastically related to each other.

(ii): *There is a nonconstant component of $\mathcal{S}^{ab}/n$, call it $\theta$*. It has been shown that all directional derivatives of $\mathcal{S}^{ab}/n$ depend on the single variable $\theta$ in this case. Let $\boldsymbol{p}^s(\theta)$ denote a "vector" whose components are the elements of the set of directional derivatives of $\mathcal{S}^{ab}/n$ of order less than or equal to $s$. Then $\nabla(\boldsymbol{p}^s(\theta)) = \frac{d\boldsymbol{p}^s}{d\theta} \otimes \nabla\theta$ has rank 1, so $\rho_s = 1$ for all $s$ and the order of the structure map is zero once again. So (56) implies that $\Sigma$, $\Sigma'$ are locally elastically related to each other in this case, too.

These remarks allow a proof of the fact that (56) implies (the integral version of) (54).

**Theorem 8.** *If* (56) *holds, then*

$$\oint_\zeta \nu_{\Sigma'}\boldsymbol{d}^{a'} \cdot d\boldsymbol{x} = \oint_\zeta \nu_\Sigma \boldsymbol{d}^a \cdot d\boldsymbol{x}, \quad \int_V \nu_{\Sigma'} n' dV = \int_V \nu_\Sigma n \, dV$$

*for arbitrary scalars $\nu_\Sigma$, arbitrary circuits $\zeta$, arbitrary volumes $V$.*

*Proof.*

- Since (56) holds, $\Sigma$, $\Sigma'$ are locally elastically related, and by virtue of Theorem 5, the relevant local elastic deformation $\boldsymbol{u}_{\boldsymbol{x}_0}$ has the property $\boldsymbol{u}_{\boldsymbol{x}_0}(\boldsymbol{x}_0) = \boldsymbol{x}_0$. Let $\nu_\Sigma$ be an arbitrary scalar. Then putting $\boldsymbol{u} \equiv \boldsymbol{u}_{\boldsymbol{x}_0}$, $\boldsymbol{x} = \boldsymbol{x}_0$, $\Sigma = \Sigma'$ in (13), one obtains

$$\nu_{\Sigma'}(\boldsymbol{x}_0) = \nu_\Sigma(\boldsymbol{x}_0), \quad \boldsymbol{x}_0 \in B. \tag{58}$$

- Suppose each $\mathcal{S}^{ab}/n$ is constant. Let $\boldsymbol{x}_0, \boldsymbol{x}_0^* \in B$ be arbitrary. Then (40) holds and $\boldsymbol{u}_{\boldsymbol{x}_0}(\boldsymbol{x}_0) = \boldsymbol{x}_0^*$. Let $\nu_\Sigma$ be an arbitrary scalar and set $\Sigma_N = \{\boldsymbol{\ell}_a(\cdot), N_{\boldsymbol{x}_0};\ a = 1,2,3\}$, $\Sigma_N^* = \{\boldsymbol{\ell}_a(\cdot), \boldsymbol{u}_{\boldsymbol{x}_0}(N_{\boldsymbol{x}_0});\ a = 1,2,3\}$, so that $\Sigma_{N^*}$ is elastically related to $\Sigma_N$. Then since $\boldsymbol{x}_0^* \in \boldsymbol{u}_{\boldsymbol{x}_0}(N_{\boldsymbol{x}_0})$ and $\boldsymbol{x}_0 \in N_{\boldsymbol{x}_0}$, by (13),

$$\nu_\Sigma(\boldsymbol{x}_0^*) = \nu_{\Sigma_{N^*}}(\boldsymbol{x}_0^*) = \nu_{\Sigma_N}(\boldsymbol{x}_0) = \nu_\Sigma(\boldsymbol{x}_0). \tag{59}$$

So all scalars are constant in this case.

- Similarly, if $\theta$ is a nonconstant component of $\mathcal{S}^{ab}/n$, one shows that all scalars depend on the single variable $\theta$ (since (40) holds for points $\boldsymbol{x}_0$, $\boldsymbol{x}_0^*$ such that $\theta(\boldsymbol{x}_0) = \theta(\boldsymbol{x}_0^*)$ and $\boldsymbol{u}_{\boldsymbol{x}_0}(\boldsymbol{x}_0) = \boldsymbol{x}_0^*$).

  Thus, $\int_V \nu_{\Sigma'} n' \, dV = \int_V \nu_\Sigma n \, dV$ for arbitrary $\nu_\Sigma$ because $(56)_4$ and (58) hold. Also, with $\zeta = \partial \mathcal{S}$, for arbitrary $\nu_\Sigma$,

$$\oint_\zeta (\nu_{\Sigma'} \boldsymbol{d}^{a'} - \nu_\Sigma \boldsymbol{d}^a) \cdot d\boldsymbol{x} = \oint_\zeta \nu_\Sigma (\boldsymbol{d}^{a'} - \boldsymbol{d}^a) \cdot d\boldsymbol{x}, \quad \text{by (58)},$$
$$= \oint_{\mathcal{S}} \nabla \nu_\Sigma \wedge (\boldsymbol{d}^{a'} - \boldsymbol{d}^a) \cdot d\boldsymbol{S}, \quad \text{by } (56)_1.$$

  But this last surface integral is zero in the case that each $\mathcal{S}^{ab}/n$ is constant because each scalar $\nu_\Sigma$ is also constant in that case. And when $\theta$ is a nonconstant component of $\mathcal{S}^{ab}/n$, by the above remarks,

$$\nabla \nu_\Sigma \wedge (\boldsymbol{d}^{a'} - \boldsymbol{d}^a) = \frac{d\nu_\Sigma}{d\theta} \nabla \theta \wedge (\boldsymbol{d}^{a'} - \boldsymbol{d}^a) = \boldsymbol{0}.$$

  by $(56)_2$ with the choice $\nu_\Sigma = \theta \in \mathcal{F}$. □

## 2.7. Generalized elastic-plastic decomposition of neutrally related crystal states

First, notice that if (56) holds, it is not necessarily true that states $\Sigma$, $\Sigma'$ are elastically related. For if $\boldsymbol{d}^a \equiv \boldsymbol{e}^a$, $\boldsymbol{d}^{a'} = \nabla \psi^a$ for some potentials $\psi^a$, $a = 1, 2, 3$, with $\det(\nabla \psi^a) = 1$, then (56) holds (with $\nabla \wedge \boldsymbol{d}^{a'} = \nabla \wedge \boldsymbol{d}^a \equiv \boldsymbol{0}$, each $\nu_\Sigma \in \mathcal{F}$ zero, and $n' = n \equiv 1$). An elastic deformation of $\Sigma'$ via the mapping $\boldsymbol{\psi} \equiv (\psi^a) : B \to \boldsymbol{\psi}(B)$ gives the state $\Sigma^* = \{\boldsymbol{e}_a, \boldsymbol{\psi}(B);\ a = 1, 2, 3\}$ with $\operatorname{vol} \boldsymbol{\psi}(B) = \operatorname{vol} B$. But by an argument similar to that given in the remarks after Corollary 6, there is no elastic deformation which maps $\Sigma^*$ to $\Sigma$ unless $\boldsymbol{\psi}(B)$ is a translation of $B$. Therefore it is worthwhile to try to understand the structure of solutions of (56) for states $\Sigma'$ given the state $\Sigma$, and a generalized notion of elastic-plastic decomposition will result from this endeavour. This generalized elastic-plastic decomposition will be confined to pairs of neutrally related states, though, and so it is limited in that respect.

It would be an interesting project to try to construct a canonical procedure whereby arbitrary states, defined over the same domain but not satisfying the conditions (56), are deformed one into the other; this might require an appreciation of which of the quantities that appear in (56) are independent of each other. For example, suppose that two states with different constant values of $\mathcal{S}^{ab}/n$, defined over the same domain, are given. Note that a state with given constant $\mathcal{S}^{ab}/n$ is defined to within elastic deformation, so a canonical state with the prescribed constant dislocation density tensor may be chosen. Likewise, one may construct a canonical state with given constant $(\mathcal{S}^{ab}/n)' \neq (\mathcal{S}^{ab}/n)$. The desired procedure would deform the first canonical state to the second. Even more specifically, take $(\mathcal{S}^{ab}/n) = \boldsymbol{a} \otimes \boldsymbol{b}$, $(\mathcal{S}^{ab}/n)' = \boldsymbol{a}' \otimes \boldsymbol{b}'$, $\boldsymbol{a}, \boldsymbol{b}, \boldsymbol{a}', \boldsymbol{b}' \in \mathbb{R}^3$. Then one may take as the first canonical state $\Sigma = \{\boldsymbol{\ell}_a(\cdot),\ B;\ a = 1, 2, 3\}$ with $\boldsymbol{d}^a = \boldsymbol{e}^a + b^a \boldsymbol{a} \wedge \boldsymbol{x} \phi(t)$, where $t \equiv \boldsymbol{a} \wedge \boldsymbol{x} \cdot \boldsymbol{b}$ and

$\phi(x) \equiv (e^t - 1 - t)/t^2$, and as the second canonical state $\Sigma'$, where primed quantities replace the analogous objects in state $\Sigma$.

Now let $\Sigma_p = \{\boldsymbol{\ell}_a^p(\cdot), B;\ a = 1, 2, 3\}$, $p \in [0, 1]$, where $\boldsymbol{d}^{ap} = \boldsymbol{e}^a + b^{ap}\boldsymbol{a}^p \wedge \boldsymbol{x}\phi(t^p)$, $t^p = \boldsymbol{a}^p \wedge \boldsymbol{x} \cdot \boldsymbol{b}^p$, and $\boldsymbol{a}^p \equiv p\boldsymbol{a} + (1-p)\boldsymbol{a}'$, $\boldsymbol{b}^p = p\boldsymbol{b} + (1-p)\boldsymbol{b}'$. One has $\Sigma_0 = \Sigma$, $\Sigma_1 = \Sigma'$, and an inelastic deformation between the two canonical states is prescribed. (In fact, one should be a little more careful than this.)

However, the general problem seems to be rather difficult, and so attention is confined to solutions of (56) in this paper. Even so, the structure of solutions of (56) will turn out to be wide enough to provide an analogue of the slip mechanism of phenomenological plasticity theories, and it will have the merit that the decomposition of relevant changes of state is a rigorous geometrical procedure.

As before, it is efficient to divide the discussion into the two "usual" cases:

(i): *All components of $S^{ab}/n$ constant*. The aim is to characterize the solutions of (56) in (more or less) the following way.

- Consider states $\Sigma$, $\Sigma'$, where the corresponding lattice vector fields solve (56).
- Apply appropriate elastic deformations to $\Sigma$, $\Sigma'$ so as to produce canonical states $\Sigma_c$, $\Sigma_c'$.
- Cut states $\Sigma_c$, $\Sigma_c'$ into pieces determined by the self-similarity property of Section 2.5, and apply (local) elastic deformations to each of these pieces so as to produce a collection of *copies* of one distinguished piece of each of $\Sigma_c$, $\Sigma_c'$.
- It will turn out that the one distinguished piece of $\Sigma_c$ can be chosen so as to be a copy of the one distinguished piece of $\Sigma_c'$.
- Rearrange the pieces that derive from $\Sigma_c$ to correspond exactly to the pieces that derive from $\Sigma_c'$. (It will turn out that one has the correct number of pieces for this purpose.)

This shows that the solutions of (56), given state $\Sigma$, may be characterized in terms of a sequence of piecewise elastic deformations and rearrangement, and this is the decomposition that was advertised in the introduction, in this case. The argument will be further split into three subcases, depending on the rank of the matrix $(S^{ab}/n)$. Note that rank $(S^{ab}/n) < 3$ by virtue of the fact that $(56)_{1,3}$, with $\nu_\Sigma = (S^{ab}/n)$, implies that $\nabla \wedge \boldsymbol{d}^a \cdot (\boldsymbol{d}^{b'} - \boldsymbol{d}^b) = 0$ has nontrivial solutions for $(\boldsymbol{d}^{b'} - \boldsymbol{d}^b)$, given $\Sigma$, provided $\Sigma' \neq \Sigma$. So I consider the cases rank $(S^{ab}/n) = 0, 1, 2$:

(a): *Rank* $(S^{ab}/n) = 0$. In this case (56) is equivalent to

$$\nabla \wedge \boldsymbol{d}^{a'} = \nabla \wedge \boldsymbol{d}^a = \boldsymbol{0}, \quad n = n'. \tag{60}$$

From $(60)_1$, $\boldsymbol{d}^{a'} \equiv \nabla\psi^a$, $\boldsymbol{d}^a \equiv \nabla\zeta^a$ for some potentials $\boldsymbol{\psi} = (\psi^a)$, $\boldsymbol{\zeta} = (\zeta^a)$ with $\det \nabla\boldsymbol{\psi} = \det \nabla\boldsymbol{\zeta}$ by $(60)_2$. Construct canonical states $\Sigma_c$, $\Sigma_c'$ by applying elastic deformations $\boldsymbol{\psi}$, $\boldsymbol{\zeta}$ to $\Sigma$, $\Sigma'$, respectively (so $\boldsymbol{d}_c^{a'} = (\nabla\boldsymbol{\psi})^{-\mathsf{T}}\nabla\psi^a = \boldsymbol{e}^a$, for example). Then

$$\Sigma_c = \{\boldsymbol{e}_a,\ \boldsymbol{\psi}(B);\ a = 1, 2, 3\}, \quad \Sigma_c' = \{\boldsymbol{e}_a,\ \boldsymbol{\zeta}(B);\ a = 1, 2, 3\}, \tag{61}$$

where vol $\psi(B) = \text{vol}\,\zeta(B)$. Note that $\psi$, $\zeta$ are determined only to within an arbitrary constant. These two constants may be chosen so that $\psi(B)$ and $\zeta(B)$ have an interior point in common; call it $\mathbf{0}$. Choose a (small enough) cube surrounding the point $\mathbf{0}$, and note that the corresponding "piece" of $\Sigma_c$ that is so distinguished is identical to the corresponding "piece" of $\Sigma_c'$. Cut states $\Sigma_c$, $\Sigma_c'$ into pieces determined by translation of the small enough cube (in the process constructing a lattice whose unit cell is this cube). Note that the self-similarity of state $\Sigma_c$, say, determined by solutions of (40) with $\boldsymbol{\ell}_a \equiv \boldsymbol{e}_a$, is translation in this case; $\boldsymbol{u}_{\boldsymbol{x}_0} = \boldsymbol{x} + \boldsymbol{c}(\boldsymbol{x}_0)$, for some $\boldsymbol{c}(\cdot)$. I assume that the two sets $\psi(B), \zeta(B)$, with vol $\psi(B) = \text{vol}\,\zeta(B)$, may be approximated sufficiently well by collections of copies of the small enough unit cell, the same number of unit cells for $\psi(B)$ as for $\zeta(B)$. (Of course, the approximation relates to boundary effects.) Then, a rearrangement of the finite number of cells that make up the approximation to $\psi(B)$ produces those cells that make up the approximation to $\zeta(B)$. In the limit as the size of the small enough unit cell approaches zero, $\Sigma_c'$ is reconstructed more and more accurately from $\Sigma_c$.

(b): *Rank* $(S^{ab}/n) = 1$. Bearing in mind that $(S^{ab}/n) =$ constant, the equations of interest are

$$\nabla \wedge \boldsymbol{d}^{a'} = \nabla \wedge \boldsymbol{d}^{a}, \quad (S^{ab}/n)' = (S^{ab}/n), \qquad n' = n. \tag{62}$$

It has been remarked already that knowledge of the neutral deformations for states elastically related to $\Sigma$ determines the neutral deformations of $\Sigma$, so that one is entitled to choose a canonical state which renders the discussion of solutions of (62) as transparent as possible. There is, in fact, another property of (56) which allows one to simplify the relevant calculations. Given $\Sigma, \Sigma'$, let $\Sigma_\gamma, \Sigma_\gamma'$ be defined as $\{\gamma_a^b \boldsymbol{\ell}_b(\cdot),\ B;\ a = 1,2,3\}$, $\{\gamma_a^b \boldsymbol{\ell}_b'(\cdot),\ B;\ a = 1,2,3\}$, respectively, where $(\gamma_a^b) \in M^{3\times 3}$ is such that $\det(\gamma_a^b) \neq 0$. Then it is easy to show that (56) holds if and only if, when $\boldsymbol{\ell}_a$ and $\boldsymbol{\ell}_a'$ are replaced by $\gamma_a^b \boldsymbol{\ell}_b$ and $\gamma_a^b \boldsymbol{\ell}_b'$, respectively, the corresponding equations also hold. It follows that one can choose $(\gamma_a^b)$ so that $(\gamma^{-1})_r^a (S^{rs}/n)(\gamma^{-1})_s^b \det \gamma$ takes a particularly convenient form (where $\gamma^{-1}$ is the matrix inverse of $\gamma \equiv (\gamma_a^b)$). In the case in hand, where $(S^{ab}/n) = \boldsymbol{a} \otimes \boldsymbol{b}$, $\boldsymbol{a}, \boldsymbol{b} \in \mathbb{R}^3$, it is easy to show that

- if $\boldsymbol{a}$ is parallel to $\boldsymbol{b}$, one can assume that $(S^{ab}/n) = \boldsymbol{e}_1 \otimes \boldsymbol{e}_1$,
- if $\boldsymbol{a}$ is not parallel to $\boldsymbol{b}$, one can assume that $(S^{ab}/n) = \boldsymbol{e}_1 \otimes \boldsymbol{e}_2$.

(Also, one checks a posteriori that this reduction does not affect the nature of the generalized elastic-plastic decomposition.) Here, I give the relevant construction just for the first of these two cases, since the second case is not much different to the first.

Note that the state $\Sigma_c = \{\boldsymbol{\ell}_a(\cdot),\ B;\ a = 1,2,3\}$, with $\boldsymbol{d}^1 = (1, -x^3, 0)$, $\boldsymbol{d}^2 = \boldsymbol{e}^2$, $\boldsymbol{d}^3 = \boldsymbol{e}^3$ has $(S^{ab}/n) \equiv \boldsymbol{e}_1 \otimes \boldsymbol{e}_1$. The rank of the structure map for states $\Sigma$, $\Sigma_c$ is zero, and the first-order classifying set for $\Sigma_c$ is identical to that of $\Sigma$. Hence one may deal with $\Sigma_c$ instead of $\Sigma$. Similarly if $\Sigma_c'$ is neutrally related to $\Sigma_c$, an elastic deformation $\boldsymbol{\mu}$ of $\Sigma_c'$ gives a state $\Sigma_c^* \equiv \{\boldsymbol{\ell}_a(\cdot),\ B^*;\ a = 1,2,3\}$, where $\boldsymbol{\ell}_a(\cdot)$ are the same fields as above (extended in the obvious way), and vol $B^* = \text{vol}\,B$ because $n^*(\boldsymbol{\mu}(\boldsymbol{x})) = n_c'(\boldsymbol{x}) = n_c(\boldsymbol{x}) = 1$, $\boldsymbol{x} \in B$, implies that $\det(\nabla \boldsymbol{\mu}) \equiv 1$.

The task is to reconstruct $\Sigma_c^*$ from $\Sigma_c$ by a division of the states into collections of "unit cells" which are congruent to one another after piecewise elastic deformation, followed by a rearrangement of those unit cells. Recalling the example of Section 2.5, apply the following piecewise elastic deformation to $\Sigma_c$,

$$\boldsymbol{u}_n^k(\boldsymbol{x}) = \boldsymbol{x} + x^2(z_0 - z_k)\boldsymbol{e}_1, \quad z_k \leq x^3 < z_{k+1}, \tag{63}$$

where the points $\ldots, z_{-k}, \ldots, z_{-1}, z_0 = 0, z_1, \ldots, z_k, \ldots$ are equally spaced on $\mathbb{R}$, with separation $2^{-n}$, say. This deformation has the property

$$(\nabla \boldsymbol{u}_n^k)\boldsymbol{\ell}_a(\boldsymbol{x}) = \boldsymbol{\ell}_a(u_n^{k3} - z_k + z_0), \tag{64}$$

and since $u_n^{k3} = x^3 \in [z_k, z_{k+1})$, it follows that the deformation constructs a *periodic* set of lattice vector fields, with period $2^{-n}\boldsymbol{e}_3$. As $n \to \infty$, the image of $B$ under the mapping $\boldsymbol{u}_n^k$ approaches $\boldsymbol{u}_\infty(B)$, where $\boldsymbol{u}_\infty = \boldsymbol{x} - x^2 x^3 \boldsymbol{e}_1$, and vol $\boldsymbol{u}_\infty(B) =$ vol $B$.

Now apply the same piecewise elastic deformation to $\Sigma_c^*$ to produce (the obvious extension of) the same *periodic* set of lattice vector fields defined over the domain $\boldsymbol{u}_\infty(\boldsymbol{\mu}(B))$, and note that vol $\boldsymbol{u}_\infty(\boldsymbol{\mu}(B)) = \mathrm{vol}(\boldsymbol{\mu}(B)) = \mathrm{vol}(B)$. Finally, a rearrangement of the (congruent) unit cells that approximately make up the state defined as $\boldsymbol{u}_\infty(B)$ then gives the unit cells that approximately make up the state defined on $\boldsymbol{u}_\infty(\boldsymbol{\mu}(B))$, and the generalized elastic-plastic deformation is accomplished as $n \to \infty$.

(c): *Rank* $(S^{ab}/n) = 2$. The relevant equations are (62), again. From $(62)_1$, $\boldsymbol{d}^{a'} = \boldsymbol{d}^a + \nabla\tau^a$, $a = 1, 2, 3$, for some potentials $\tau^a$, and then $(62)_2$ gives $(\nabla \wedge \boldsymbol{d}^a) \cdot \nabla\tau^b = 0$, $a, b = 1, 2, 3$. Since rank $(S^{ab}/n) = 2$ implies that rank $(\nabla \wedge \boldsymbol{d}^a) = 2$, it follows that $\nabla\tau^a = \lambda^a \boldsymbol{b}^1 \wedge \boldsymbol{b}^2$ for some functions $\lambda^1, \lambda^2, \lambda^3$, where $\boldsymbol{b}^a$ denotes $\nabla \wedge \boldsymbol{d}^a$ and $\boldsymbol{b}^1, \boldsymbol{b}^2$ (say) is a choice of two linearly independent vectors of the set $\boldsymbol{b}^1, \boldsymbol{b}^2, \boldsymbol{b}^3$. (Hence the fields $\boldsymbol{b}^1, \boldsymbol{b}^2, \boldsymbol{b}^3$ are integrable in the sense each that $\boldsymbol{b}^a$ is tangent to a surface $\tau^b =$ constant, $b = 1, 2, 3$.) Clearly, $\nabla\tau^1 \wedge \nabla\tau^2 = \boldsymbol{0}$, etc., so that each of the potentials $\tau^1, \tau^2, \tau^3$ may be expressed in terms of any one of the potentials $\tau^a$ which is such that $\nabla\tau^a \neq \boldsymbol{0}$ locally. Choose an elastic deformation that deforms the surface $\tau^a =$ constant (corresponding to a potential with $\nabla\tau^a \neq \boldsymbol{0}$) to the plane $x^3 =$ constant. It follows that one can assume that

$$\boldsymbol{d}^{a'} = \boldsymbol{d}^a + f^a(x^3)\boldsymbol{e}^3. \tag{65}$$

Then $\nabla \wedge \boldsymbol{d}^b \cdot (\boldsymbol{d}^{a'} - \boldsymbol{d}^a) = \nabla \wedge \boldsymbol{d}^b \cdot \boldsymbol{e}^3 f^a(x^3) = 0$, so

$$(\nabla \wedge \boldsymbol{d}^b \cdot \boldsymbol{d}^a)(\boldsymbol{\ell}_a \cdot \boldsymbol{e}^3) = S^{ab}\ell_{a3} = 0, \tag{66}$$

since at least one of the functions $f^a$ is to be nonzero. Also, since rank $(S^{ab}) = 2$, (66) implies that $\ell_{a3} = p(\boldsymbol{x})\nu_a$, where $\boldsymbol{\nu} \equiv (\nu^a) \in \mathbb{R}^3$ is a unit left eigenvector of $(S^{ab})$, and $p(\boldsymbol{x})$ is arbitrary. One can arrange, through a further elastic deformation which preserves the planes $x^3 =$ constant, to show that $\ell_{33} = n^{-1}$, where $n$ is a function of $x^3$ only (see Davini et al. [3]). Hence each $\ell_{a3}$ depends only on $x^3$ and the hypotheses of Theorem 7 as satisfied. So one can assume that lattice vectors depend only on $x^3$ in the canonical state under consideration.

That is, it is sufficient to consider canonical states of the form

$$\Sigma_c = \left\{ \tilde{\boldsymbol{\ell}}_a(\cdot),\ B;\ a = 1, 2, 3 \right\}, \tag{67}$$

where $\tilde{\boldsymbol{\ell}}_a(\boldsymbol{x}) = \boldsymbol{\ell}_a^+(x^3)$, for some $\boldsymbol{\ell}_a^+ \colon \mathbb{R} \to \mathbb{R}^3$, $a = 1, 2, 3$. Let $\Sigma_c'$ be neutrally related to $\Sigma_c$ so (cf. (65))

$$\boldsymbol{d}^{a'} = \tilde{\boldsymbol{d}}^a + f^a(x^3)\boldsymbol{e}^3, \tag{68}$$

and note that $(62)_3$ implies

$$f^a \tilde{\boldsymbol{\ell}}_a \cdot \boldsymbol{e}^3 = 0. \tag{69}$$

In this case, there is a slightly simpler procedure than that which was outlined at the beginning of this section which produces the required decomposition. Define

$$\boldsymbol{x}^* \equiv \boldsymbol{x} - \int^{x^3} f^a(s)\tilde{\boldsymbol{\ell}}_a(s)ds, \tag{70}$$

and calculate that (using(69))

$$\nabla \boldsymbol{x}^* = \mathbb{I} - f^a \tilde{\boldsymbol{\ell}}_a \otimes \boldsymbol{e}^3,\ (\nabla \boldsymbol{x}^*)^{-\top} = \mathbb{I} + \boldsymbol{e}^3 \otimes f^a \tilde{\boldsymbol{\ell}}_a,\ \det(\nabla \boldsymbol{x}^*) = 1. \tag{71}$$

To construct a state $\Sigma_c^*$, apply the elastic deformation $\boldsymbol{x}^*$ to state $\Sigma_c$ and calculate that

$$\boldsymbol{d}^{a*}(\boldsymbol{x}^*) = (\mathbb{I} + \boldsymbol{e}^3 \otimes f^b \tilde{\boldsymbol{\ell}}_a)\tilde{\boldsymbol{d}}^a = \tilde{\boldsymbol{d}}^a + f^a \boldsymbol{e}^3 = \boldsymbol{d}^{a'}(\boldsymbol{x}). \tag{72}$$

Note that $x^{*3} = x^3$, by virtue of (69) and (70), so $\boldsymbol{d}^{a*}$ depends only on $x^{*3}$ by (67), (68), and (72). So the deformation $\boldsymbol{x}^*$ leaves planes $x^3$ = constant invariant and preserves volumes. The lattice vector fields in state $\Sigma_c^*$ are identical to (the obvious extensions) of the lattice vector fields in $\Sigma_c'$:

$$\Sigma_c^* = \left\{ \boldsymbol{\ell}_a'(\cdot), \boldsymbol{x}^*(B);\ a = 1, 2, 3 \right\}, \operatorname{vol} \boldsymbol{x}^*(B) = \operatorname{vol}(B). \tag{73}$$

Cut states $\Sigma_c^*$, $\Sigma_c'$ into pieces determined by a cubic lattice with one set of lattice planes perpendicular to $\boldsymbol{e}^3$. It is clear that $\Sigma_c'$ may be approximately reconstructed from $\Sigma_c^*$ by rearrangement (i.e., translation in planes $x^3$ = constant) of cells in appropriate layers of the lattice. Again, in the limit as the size of the unit cells approaches zero, $\Sigma_c'$ is reconstructed more and more accurately from state $\Sigma_c^*$.

(ii): *There is a nonconstant component of $S^{ab}/n$; call it $\theta$.* First, an elastic deformation allows one to assume that $\theta(\boldsymbol{x}) = x^3$. Then, using (56), one finds that

$$\boldsymbol{d}^{a'} = \boldsymbol{d}^a + f^a(x^3)\boldsymbol{e}^3,\ f^a \ell_{a3} = 0,\ \nabla \wedge \boldsymbol{d}^a \cdot \boldsymbol{e}^3 = S^{ab}\ell_{a3} = 0. \tag{74}$$

From (74), one proceeds exactly as in case (i)c above to construct the generalized elastic-plastic decomposition.

**Acknowledgments**

I am very grateful for the support of the Royal Society of London and the Leverhulme Trust.

## References


[1] C. Davini, A proposal for a continuum theory of defective crystals, *Arch. Rational Mech. Anal.*, **96** (1986), 295–317.

[2] C. Davini and G. P. Parry, On defect-preserving deformations in crystals, *Internat. J. Plasticity*, **5** (1989), 337–369.

[3] C. Davini and G. P. Parry, A complete list of invariants for defective crystals, *Proc. Roy. Soc. London Sect.* A, **432** (1991), 341–365.

[4] I. Fonseca and G. P. Parry, Equilibrium configurations of defective crystals, *Arch. Rational Mech. Anal.*, **120** (1992), 245–283.

[5] M. Glanville and G. P. Parry, Elastic invariants in complex materials, in P. M. Mariano, V. Sepe, and M. Lacagnina, eds., *AIMETA*'01: XV *Congressa AIMETA di Meccanica Teorica e Applicata*, Tipografia dell' Universita', Catania, Italy, 2001.

[6] E. H. Lee, Elastic-plastic deformation at finite strains, *J. Appl. Mech.*, **36** (1969), 1–6.

[7] E. H. Lee and D. T. Liu, Finite-strain elastic-plastic theory with application to plane-wave analysis, *J. Appl. Phys.*, **38** (1967), 19–27.

[8] P. J. Olver, *Equivalence, Invariants, and Symmetry*, Cambridge University Press, Cambridge, UK, 1996.

[9] G. P. Parry, Defects and rearrangements in crystals, in L. M. Brock, ed., *Defects and Anelasticity in the Characterization of Crystalline Solids*, ASME, New York, 1992, 117–132.

[10] G. P. Parry, Elastic-plastic decompositions of defect-preserving changes of state in solid crystals, in M. D. P. Monteiro Marques and J. F. Rodrigues, eds., *Trends in Applications of Mathematics to Mechanics*, Longman, Harlow, UK, 1994, 154–163.

[11] G. P. Parry, The moving frame, and defects in crystals, *Internat. J. Solids Structures*, **38** (2001), 1071–1087.

[12] G. P. Parry and M. Šilhavý, Elastic invariants in the theory of defective crystals, *Proc. Roy. Soc. London Sect.* A, **455** (1999), 4333–4346.

[13] G. P. Parry and M. Šilhavý, Invariant line integrals in the theory of defective crystals, *Rend. Mat. Acc. Lincei.* 9, **11** (2000), 111–140.

[14] C. Truesdell and R. A. Toupin, *The Classical Field Theories*, Handbuch der Physik Band III/1, Springer-Verlag, Berlin, Heidelberg, New York, 1960.

# 3. Cocycles, Compatibility, and Poisson Brackets for Complex Fluids

Hernán Cendra*, Jerrold Marsden†, and Tudor S. Ratiu††

*Dedicated to Alan Weinstein on the occasion of his* 60*th birthday.*

**Abstract.** Motivated by Poisson structures for complex fluids, such as the Poisson structure for spin glasses given in Holm and Kupershmidt [1988], we investigate a general construction of Poisson brackets with cocycles. Connections with the construction of compatible brackets found in the theory of integrable systems are also briefly discussed.

## 3.1. Introduction

PURPOSE OF THIS PAPER. The goal of this paper is to explore a general construction of Poisson brackets with cocycles and to link it to questions of compatibility similar to those encountered in the theory of integrable systems. The study is motivated by the specific brackets for spin glasses studied in Holm and Kupershmidt [1988].

In the body of the paper, we first describe several examples of Poisson brackets as reduced versions of, in principle, simpler brackets. Then we will show how this technique can be applied to study the compatibility of Poisson brackets. In the last section, we give the example of spin glasses and show that the Jacobi identity for the bracket given in Holm and Kupershmidt [1988] is a result of a cocycle identity.

BACKGROUND. It is a remarkable fact that many important examples of Poisson brackets are reduced versions of canonical Poisson brackets. Perhaps the most basic case of this may be described as follows. Start with a configuration manifold $Q$ with a Lie group $G$ acting on $Q$ in such a way that $Q \to Q/G$ is a principal bundle (for instance, assume that the action is free and proper). Consider the cotangent space $T^*Q$

*Departamento de Matemática, Universidad Nacional del Sur, Av. Alem 1254, 8000 Bahia Blanca, Argentina and CONICET, Argentina, uscendra@criba.edu.ar. The research of this author was partially supported by the EPFL.

†Control and Dynamical Systems, California Institute of Technology, Pasadena, CA, USA, marsden@cds.caltech.edu. The research of this author partially supported by NSF/KDI grant ATM-9873133.

††Département de Mathématiques, École Polytechnique Fédérale de Lausanne, CH-1015 Lausanne, Switzerland, tudor.ratiu@epfl.ch. The research of this author was partially supported by the European Commission and the Swiss Federal Government through funding for the Research Training Network *Mechanics and Symmetry in Europe* (MASIE) as well as the Swiss National Science Foundation.

with its canonical symplectic and hence Poisson structure. As with general quotients of Poisson manifolds (see, for instance, Marsden and Ratiu [1999] for the relevant basic background material), the quotient space $(T^*Q)/G$ inherits a Poisson structure from $T^*Q$ in a natural way.

This construction of $(T^*Q)/G$ along with its quotient Poisson structure is a very important and rather general method for finding Poisson structures in mathematical physics, including fluid and plasma systems, and for obtaining brackets that satisfy the Jacobi identity by the nature of their construction (whereas verifying the Jacobi identity "by hand" could be quite arduous); see, for instance, Marsden and Weinstein [1982], Marsden and Weinstein [1983], Marsden, Ratiu, and Weinstein [1984a,b], Holm, Marsden, Ratiu, and Weinstein [1985], and Cendra, Holm, Hoyle, and Marsden [1998].

The detailed mathematical structure of the reduced Poisson brackets on quotients $(T^*Q)/G$ has itself a long and interesting history going back to the work of Sternberg [1977], Weinstein [1978], and Montgomery [1984], Montgomery [1986] on particles in Yang–Mills fields and the bundle picture in mechanics. Recent works in this area, such as Cendra, Marsden, and Ratiu [2001a] and Marsden and Perlmutter [2000] take the view that one should choose a connection on the *shape space bundle* $Q \to Q/G$ and then realize $(T^*Q)/G$ as the Whitney sum bundle $T^*(Q/G) \oplus \tilde{\mathfrak{g}}^*$, where $\mathfrak{g}$ is the Lie algebra of $G$, $\mathfrak{g}^*$ is its dual space, and where $\tilde{\mathfrak{g}}^*$ is the coadjoint bundle associated to the given $G$-action on $Q$ and the coadjoint action of $G$ on $\mathfrak{g}^*$. The reduced Hamilton equations on this space are called the Hamilton–Poincaré equations and are introduced and studied in Cendra, Marsden, Pekarsky, and Ratiu [2003].

However, the construction of brackets on quotients of cotangent bundles is not the only case in which the idea of reduction is important to unify the study of otherwise mysterious Poisson brackets, some of which are well known from an algebraic point of view, namely that for spin glasses in Holm and Kupershmidt [1988]. A key feature of such brackets is the presence of *cocycle terms*.

GOALS OF THIS PAPER. In this paper we shall present a fairly general construction of Poisson brackets with cocycles and show a simple reduction context for their study. We then show that the brackets of Holm and Kupershmidt [1988] may be viewed as examples of this construction. We will also make some suggestive links with compatible Poisson structures in the same sense as are found in the theory of integrable systems.

FUTURE DIRECTIONS. One of the key ideas that will be further developed in the future, especially in the context of complex fluids and similar examples, is to realize these cocycle terms as curvature (magnetic) terms associated with the reduction process itself. This point of view was explored in some detail in Marsden, Misiolek, Perlmutter, and Ratiu [1998] and Marsden, Misiolek, Ortega, Perlmutter, and Ratiu [2002] in the context of reduction by stages. The corresponding theory of Lagrangian reduction by stages was developed in Cendra, Marsden, and Ratiu [2001a] and it was used in Holm [2002] to study the variational approach to complex fluids. A similar reduction by stages method was used in Patrick [1999], who studied the Landau–Lifschitz equations, which occurs in magnetic materials.

It is our hope that our results will help provide a better understanding of not only spin glasses but also other important examples of complex fluids. The reader should realize that even the example of spin glasses has some obviously interesting additional and fundamental properties that require a deeper understanding, such as the following: When the spin glass bracket is "reduced" from a connection description to a "curvature" description, the cocycle disappears, as Holm and Kupershmidt [1988] point out. This seemingly miraculous property suggests that the overall reduction procedure is simply a Lie–Poisson reduction from the cotangent bundle of a group to the dual of its Lie algebra and that this one-step reduction may be realized as a two-stage reduction (the first by a normal subgroup). The cocycle is introduced at the first step of the reduction and then disappears in the second step, as it must since the final bracket is Lie–Poisson.

Another idea that requires further development is the connection of the global Poisson point of view presented here with the local, variational multisymplectic field theoretic point of view (see, for instance, Gotay, Isenberg, and Marsden [1997], Marsden, Patrick, and Shkoller [1998], and Castrillón López, Ratiu, and Shkoller [2000]). In particular, it would be very interesting to explore the issue of cocycles and reduction by stages in the multisymplectic context and to link this theory to that of complex fluids. This is a relatively ambitious project that we do not address directly in this paper.

Parts of this paper are pedagogical in that we sometimes give two proofs of a result, sometimes by a direct coordinate computation; we feel that this may be beneficial for readers who want to see the same result from another viewpoint or who prefer an emphasis on coordinate calculations.

PRELIMINARIES AND NOTATION. Let $(M, \{,\})$ be a Poisson manifold (see, for example, Marsden and Ratiu [1999] for an exposition and further references). It is well known that the Poisson structure $\{,\}$, regarded as a skew-symmetric bilinear mapping on $\mathcal{F}(M) \times \mathcal{F}(M)$, where $\mathcal{F}(M)$ denotes the space of smooth real-valued functions on $M$, may alternatively be given in terms of a contravariant skew-symmetric 2-tensor field $C \in \mathcal{T}_0^2(M)$ (the space of all 2-tensors with indices up) on $M$. The relation between $\{,\}$ and $C$ is given by the identity

$$\{f, g\} = C(df, dg) \tag{1}$$

for any $f, g \in \mathcal{F}(M)$. Let us choose local coordinates $x^i$ on $M$, where $i = 1, \ldots, m$, and $m$ is the dimension of $M$. Then the Poisson bracket may be written as follows

$$\{f, g\} = C^{ij} \frac{\partial f}{\partial x^i} \frac{\partial g}{\partial x^j},$$

where $C^{ij}$ are the components of $C$ in the chosen coordinate system and where the summation convention is in force. The Jacobi identity for the Poisson structure is, as is well known, equivalent to the coordinate condition

$$C^{ir}\frac{\partial C^{jk}}{\partial x^r} + C^{jr}\frac{\partial C^{ki}}{\partial x^r} + C^{kr}\frac{\partial C^{ij}}{\partial x^r} = 0 \tag{2}$$

for all $i, j, k = 1, \ldots, m$.

## 3.2. A bracket associated to a 2-form on a symplectic manifold

**Definition 3.2.1.** *Let $(P, \omega)$ be a symplectic manifold, and let $\Sigma$ be a given 2-form on $P$. For any given $f, g \in \mathcal{F}(P)$ the* phase space $\Sigma$-bracket *of $f$ and $g$ is defined by*

$$\{f, g\}_\Sigma = \Sigma(X_f, X_g), \tag{3}$$

*where $X_f$ and $X_g$ are the Hamiltonian vector fields associated to $f$ and $g$, respectively.*

At this point we do not claim anything about the phase space $\Sigma$-bracket being a Poisson structure. We shall address this issue in due course.

Recall that the *Lagrange bracket* of two vector fields $X$ and $Y$ on a symplectic manifold $(P, \omega)$ is defined by the expression $[\![X, Y]\!] := \omega(X, Y)$ (see Abraham and Marsden [1978], p. 196). Thus the phase space $\Sigma$-bracket of $f$ and $g$ may be viewed as a generalization of the Lagrange bracket of $X_f$ and $X_g$ to an arbitrary 2-form on $P$. Note also that $\{f, g\}_\omega = \{f, g\}$ is the usual Poisson bracket associated to a symplectic manifold.

Let $P = T^*Q$ and denote by $\omega$ the canonical symplectic form on $P$. Let $\Sigma_0$ be a 2-form on $Q$. If $\pi_Q : T^*Q \to Q$ is the canonical projection, $\Sigma = \pi_Q^*\Sigma_0$ is a 2-form on $T^*Q$ and one can then form the corresponding phase space $\Sigma$-bracket using (3).

It is clear that the phase space $\Sigma$-bracket is skew-symmetric and is a derivation on each factor. The natural question is therefore to find the conditions on $\Sigma$ for $\{\, ,\}_\Sigma$ to satisfy the Jacobi identity. We shall give below intrinsic as well as coordinate proofs of our assertions and shall study this problem in an increasing degree of generality.

THE PHASE SPACE $\Sigma$-BRACKET ON THE COTANGENT BUNDLE OF A BANACH SPACE. We study the simplest case of a general construction by first considering the case of vector spaces. Let $V$ be a Banach space and $\Sigma_0 : V \times V \to \mathbb{R}$ be a bilinear skew-symmetric map. The lift of this bilinear form to a bilinear form on $V \times V^*$ by means of the projection map $\pi_V : V \times V^* \to V$ is given by

$$\Sigma\left((v_1, \mu_1), (v_2, \mu_2)\right) = \Sigma_0\left(v_1, v_2\right).$$

Consider the cotangent bundle $T^*V \equiv V \times V^*$. Note that the canonical projection $\pi_V : V \times V^* \to V$ is the projection on the first factor. To compute the phase space $\Sigma$-bracket, let $f \in \mathcal{F}(V \times V^*)$, the space of real-valued functions on $V \times V^*$.

Recall that the functional derivative of $f \in \mathcal{F}(V^*)$ at a point $\mu \in V^*$ is the element $\delta f/\delta\mu \in V$ defined by the identity

$$\mathbf{D}f(\mu) \cdot \nu = \left\langle \nu, \frac{\delta f}{\delta \mu} \right\rangle \quad \text{for all} \quad \mu, \nu \in V^*, \tag{4}$$

where $\mathbf{D}f(\mu)$ denotes the Fréchet derivative of $f : V^* \to \mathbb{R}$.

Since the Hamiltonian structure is canonical, $X_f(v, \mu) = (\delta f/\delta\mu, -\delta f/\delta v)$, and so we get

$$\{f, g\}_\Sigma (v, \mu) = \Sigma\left(X_f, X_g\right)(v, \mu) = \Sigma_0\left(\frac{\delta f}{\delta\mu}, \frac{\delta g}{\delta\mu}\right)$$

for all $v \in V$ and $\mu \in V^*$. This phase space $\Sigma$-bracket is actually a Poisson bracket, as the next theorem shows.

**Theorem 3.2.2.** *Let $V$ be a Banach space, $\Sigma_0 : V \times V \to \mathbb{R}$ be a bilinear skew-symmetric map, and $\Sigma$ be the induced bilinear skew-symmetric form on $V \times V^*$ defined above. For given $f, g : V^* \to \mathbb{R}$, the phase space $\Sigma$-bracket on $V \times V^*$*

$$\{f, g\}_\Sigma(v, \mu) = \Sigma_0\left(\frac{\delta f}{\delta\mu}, \frac{\delta g}{\delta\mu}\right), \quad \mu \in V^*, \tag{5}$$

*satisfies the Jacobi identity. In particular, for $f$ and $g$ independent of the first factor, we have a Poisson bracket on $V^*$.*

A more general theorem will be proved in Theorem 3.2.6 below.

THE MOMENTUM $\Sigma$-BRACKET ON THE DUAL OF A BANACH SPACE. Interestingly, one can also get a Poisson structure only on $V^*$ by a similar construction.

**Definition 3.2.3.** *Let $V$ be a Banach space and $\Sigma : V \times V \to \mathbb{R}$ be a bilinear skew-symmetric form. For given $f, g : V^* \to \mathbb{R}$, the* momentum $\Sigma$-bracket *on functions on $V^*$ is defined by*

$$\{f, g\}_\Sigma(\mu) = \Sigma\left(\frac{\delta f}{\delta\mu}, \frac{\delta g}{\delta\mu}\right), \quad \mu \in V^*. \tag{6}$$

We use the term "momentum $\Sigma$-bracket" since we think of elements of $V^*$ as momentum variables and the functions under consideration depend on $\mu \in V^*$ and not on $(v, \mu) \in T^*V$, as was the case in the preceding theorem.

Remarkably, the momentum $\Sigma$-bracket is also a Poisson bracket.

**Theorem 3.2.4.** *The momentum $\Sigma$-bracket given in Definition* 3.2.3 *satisfies the Jacobi identity and therefore defines a Poisson bracket on $V^*$.*

We now give three proofs of Theorem 3.2.4, starting with a local one in finite dimensions.

*Calculus-based proof.* Assume that $V$ is finite dimensional. An immediate corollary of (2) is that any skew-symmetric *constant* tensor field $C \in \mathcal{T}_0^2(V)$ on a vector space defines a Poisson bracket. In particular, the bracket given by (6) is a Poisson bracket. □

The following second proof also works for infinite dimensional Banach spaces.

*Direct proof*. Of course, as usual, the only thing we need to worry about is the Jacobi identity. Denote by $\Sigma^\flat : V \to V^*$ the map given by $\langle \Sigma^\flat(\xi), \eta\rangle = \Sigma(\xi, \eta)$, for any $\xi, \eta \in V$, where $\langle\, , \rangle : V^* \times V \to \mathbb{R}$ denotes the pairing between the vector space $V$ and its dual $V^*$. The following formula, obtained by differentiating (4) relative to $\mu$, will be used in the computation below:

$$\left\langle \rho, \mathbf{D}\left(\mu \mapsto \frac{\delta f}{\delta \mu}\right)(\mu)\cdot \nu \right\rangle = \left\langle \nu, \mathbf{D}\left(\mu \mapsto \frac{\delta f}{\delta \mu}\right)(\mu)\cdot \rho \right\rangle = \mathbf{D}^2 f(\mu)(\nu, \rho), \tag{7}$$

for all $\mu, \nu, \rho \in V^*$. Thus, one has

$$\frac{\delta}{\delta \mu}\{f, g\}_\Sigma(\mu) = -\mathbf{D}^2 f(\mu)\left(\Sigma^\flat\left(\frac{\delta g}{\delta \mu}\right), \cdot\right) + \mathbf{D}^2 g(\mu)\left(\Sigma^\flat\left(\frac{\delta f}{\delta \mu}\right), \cdot\right) \tag{8}$$

so that for any $f, g, h : V^* \to \mathbb{R}$, it follows that

$$\begin{aligned}&\{\{f, g\}_\Sigma, h\}_\Sigma(\mu)\\ &\quad = \mathbf{D}^2 f(\mu)\left(\Sigma^\flat\left(\frac{\delta g}{\delta \mu}\right), \Sigma^\flat\left(\frac{\delta h}{\delta \mu}\right)\right) - \mathbf{D}^2 g(\mu)\left(\Sigma^\flat\left(\frac{\delta f}{\delta \mu}\right), \Sigma^\flat\left(\frac{\delta h}{\delta \mu}\right)\right).\end{aligned}$$

It readily follows that if one adds the other two terms obtained by circular permutations and uses the symmetry of the second derivative, the sum will vanish. This proves that $\{\, , \}_\Sigma$ is a Poisson bracket on $V^*$. □

The third proof is by reduction. It may seem the most complicated at first, but in fact, it is the one that gives the most insight into generalizations of the result.

*Proof by reduction.* Start with the cotangent bundle $T^*V = V \times V^*$. Let $V$ act on $V \times V^*$ as follows (cotangent lift of translations together with a momentum shift):

$$\eta \cdot (\xi, \mu) = \left(\xi + \eta, \mu + \frac{1}{2}\Sigma^\flat(\eta)\right).$$

One checks that this action is Poisson with respect to the canonical cotangent bracket

$$\{f, g\}_{\text{can}}(\xi, \mu) = \left\langle \frac{\delta f}{\delta \xi}, \frac{\delta g}{\delta \mu}\right\rangle - \left\langle \frac{\delta g}{\delta \xi}, \frac{\delta f}{\delta \mu}\right\rangle.$$

Now we simply Poisson reduce. The quotient space $(V \times V^*)/V$ is isomorphic to $V^*$ by the quotient map

$$\pi : (\xi, \mu) \mapsto \mu - \frac{1}{2}\Sigma^\flat(\xi).$$

Since for $f : V^* \to \mathbb{R}$ we have

$$\frac{\delta(\pi^* f)}{\delta \xi} = \frac{1}{2}\Sigma^\flat\left(\frac{\delta f}{\delta(\mu - \frac{1}{2}\Sigma^\flat(\xi))}\right) \quad \text{and} \quad \frac{\delta(\pi^* f)}{\delta \mu} = \frac{\delta f}{\delta(\mu - \frac{1}{2}\Sigma^\flat(\xi))},$$

it follows that

$$\begin{aligned}\{\pi^* f, \pi^* g\}_{\text{can}}(\xi, \mu) &= \left\langle \frac{\delta(\pi^* f)}{\delta\xi}, \frac{\delta(\pi^* g)}{\delta\mu} \right\rangle - \left\langle \frac{\delta(\pi^* g)}{\delta\xi}, \frac{\delta(\pi^* f)}{\delta\mu} \right\rangle \\ &= \Sigma\left( \frac{\delta f}{\delta(\mu - \frac{1}{2}\Sigma^\flat(\xi))}, \frac{\delta g}{\delta(\mu - \frac{1}{2}\Sigma^\flat(\xi))} \right) \\ &= \{f, g\}_\Sigma \left(\mu - \frac{1}{2}\Sigma^\flat(\xi)\right) \\ &= \left(\pi^*\{f, g\}_\Sigma\right)(\xi, \mu).\end{aligned}$$

Therefore, $\{f, g\}_\Sigma$ is the reduced bracket and hence is a Poisson bracket on $V^*$. □

THE $\Sigma$-BRACKET ON $T^*Q$. Let $Q$ be any given manifold and let $\Sigma_0$ be a 2-form on $Q$. Let $\pi_Q : T^*Q \to Q$ be the natural projection. Then $\Sigma = \pi_Q^* \Sigma_0$ is a 2-form on $T^*Q$ and thus it has a phase space $\Sigma$-bracket associated to it, as explained before.

For any given 2-form $\Sigma_0$ on $Q$, one has a well-defined map

$$\pi^{\Sigma_0} : T^*TQ \to T^*Q$$

given in local coordinates by

$$\pi^{\Sigma_0}(q^i, \delta q^i, p_i, \delta p_i) := \left( q^i, \delta p_i - \frac{1}{2}\Sigma_0(q)_{ji}\delta q^j \right).$$

The map $\pi^{\Sigma_0}$ is defined at an element $w \in T^*_{v_q}TQ$ and paired with an element $u_q \in T_qQ$, by

$$\left\langle \pi^{\Sigma_0}(w), u_q \right\rangle = \left\langle \left( \tau_{T^*Q} \circ \kappa_Q^{-1} \right)(w), u_q \right\rangle - \frac{1}{2}\Sigma_0(v_q, u_q), \tag{9}$$

where $v_q \in T_qQ$. Here, $\tau_{T^*Q} : TT^*Q \to T^*Q$ denotes the tangent bundle projection and $\kappa_Q : TT^*Q \to T^*TQ$ is the canonical isomorphism. These maps are related to some constructions given by Tulczyjew [1977], but for expository clarity, we shall recall the details here.

INTRINSIC CHARACTERIZATION OF THE MAP $\kappa_Q$. First of all, choose a local trivialization of $Q$, in which $Q$ is represented as an open set $U$ in a Banach space $E$. With such a choice, $TT^*Q$ is represented by $(U \times E^*) \times (E \times E^*)$, while $T^*TQ$ is represented by $(U \times E) \times (E^* \times E^*)$. In this representation, the map $\kappa_Q$ will turn out to be given by

$$(q, p, \delta q, \delta p) \mapsto (q, \delta q, \delta p, p).$$

We will show that the map $\kappa_Q$ is the unique map that intertwines two sets of maps. These maps are given as follows:

1. *First set of maps*. Consider the following two maps:

$$\begin{aligned}T\pi_Q &: TT^*Q \to TQ, \\ \pi_{TQ} &: T^*TQ \to TQ,\end{aligned}$$

which are the obvious maps (recall that we write $\pi_R : T^*R \to R$ for the cotangent projection). The first commutation condition that will be used to define $\kappa_Q$ is that

$$\pi_{TQ} \circ \kappa_Q = T\pi_Q.$$

Thus, the diagram in Figure 1 should commute.

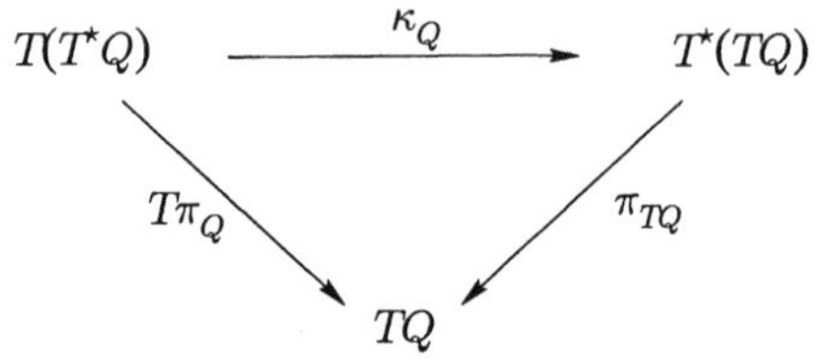


FIGURE I. Commutative diagram for the cotangent analogue of the canonical involution.

2. *Second set of maps.* The second set of maps is the following:

$$\tau_{T^*Q} : TT^*Q \to T^*Q,$$
$$\pi^0 : T^*TQ \to T^*Q.$$

Recall that the tangent bundle projection of $TR$ to $R$ is denoted $\tau_R : TR \to R$.

We need to explain the map $\pi^0$; let $\alpha_{v_q} \in T^*_{v_q}TQ$ and let $u_q \in T_qQ$. Then

$$\left\langle \pi^0(\alpha_{v_q}), u_q \right\rangle = \left\langle \alpha_{v_q}, \mathrm{ver}(u_q, v_q) \right\rangle,$$

where

$$\mathrm{ver}(u_q, v_q) = \left.\frac{d}{dt}\right|_{t=0} (v_q + tu_q) \in T_{v_q}TQ$$

denotes the vertical lift of $u_q$ along $v_q$.

The second commutation condition is that

$$\pi^0 \circ \kappa_Q = \tau_{T^*Q}.$$

In other words, the diagram in Figure 2 commutes.

In a natural local trivialization, these four maps are readily checked to be given by

$$T\pi_Q(q, p, \delta q, \delta p) = (q, \delta q),$$
$$\pi_{TQ}(q, \delta q, p, \delta p) = (q, \delta q),$$
$$\tau_{T^*Q}(q, p, \delta q, \delta p) = (q, p),$$
$$\pi^0(q, \delta q, p, \delta p) = (q, \delta p),$$

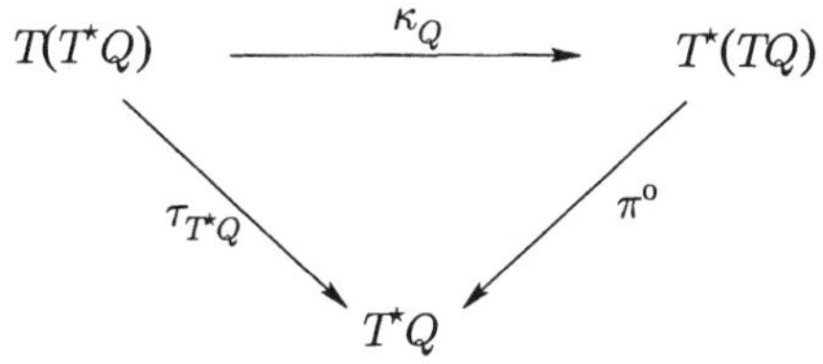


FIGURE 2. The second commutative diagram for the cotangent analogue of the canonical involution.

from which it is easy to see that the commutation conditions are satisfied with the coordinate formula for $\kappa_Q$. It is clear that this uniquely characterizes the map $\kappa_Q$. We summarize what we have proved in the following.

**Proposition 3.2.5.** *For any manifold $Q$, there is a unique diffeomorphism*

$$\kappa_Q : TT^*Q \to T^*TQ$$

*such that the above two diagrams commute.*

In summary, $\pi^{\Sigma_0}$ is defined by equation (9), where $\kappa_Q$ is the map defined in the preceding proposition.

COTANGENT BUNDLE $\Sigma$-BRACKET. Now we are ready to give the generalization of Theorem 3.2.2.

**Theorem 3.2.6.** *Let $Q$ be a manifold and let $\Sigma_0$ be a 2-form on $Q$. Endow $T^*TQ$ with the canonical cotangent bundle symplectic structure. There is a unique Poisson structure on $T^*Q$ obtained by declaring $\pi^{\Sigma_0} : T^*TQ \to T^*Q$ to be a Poisson map; in fact, this bracket on $T^*Q$ is given by the* cotangent $\Sigma_0$-bracket

$$\{f, g\}_{\Sigma_0}(q, p) = \Sigma_0(q)\left(\frac{\delta f}{\delta p}, \frac{\delta g}{\delta p}\right),$$

*where $f, g \in \mathcal{F}(T^*Q)$ and where the* fiber functional derivative *$\delta f/\delta p \in T_qQ$ is defined by*

$$\left\langle (q, p'), \frac{\delta f}{\delta p}\right\rangle = \left.\frac{d}{dt}\right|_{t=0} f(q, p + tp')$$

*for all $(q, p') \in T_q^*Q$.*

We give two proofs. The first is a brute force coordinate proof and the second is by reduction.

*Coordinate proof.* Let the 2-form $\Sigma_0$ be written in components as

$$\Sigma_0 = \Sigma_{ij} dq^i \otimes dq^j,$$

where the sum is understood to be over all $i, j$ (no restriction on indices, such as $i < j$, which would introduce factors of 2) and where $\Sigma_{ij}$ is a function of $q$ alone. Then we have

$$\{f, g\}_{\Sigma_0} = \Sigma_{ij} \frac{\partial f}{\partial p_i} \frac{\partial g}{\partial p_j}.$$

The only axiom that is not obvious is the Jacobi identity. To check it, we compute directly in coordinates as follows:

$$\begin{aligned}\{f, \{g, h\}_{\Sigma_0}\}_{\Sigma_0} &= \Sigma_{kl} \frac{\partial f}{\partial p_k} \frac{\partial}{\partial p_l} \left( \Sigma_{ij} \frac{\partial g}{\partial p_i} \frac{\partial h}{\partial p_j} \right) \\ &= \Sigma_{kl} \Sigma_{ij} \frac{\partial f}{\partial p_k} \frac{\partial^2 g}{\partial p_l \partial p_i} \frac{\partial h}{\partial p_j} + \Sigma_{kl} \Sigma_{ij} \frac{\partial f}{\partial p_k} \frac{\partial g}{\partial p_i} \frac{\partial^2 h}{\partial p_l \partial p_j}.\end{aligned}$$

Therefore, we get

$$\begin{aligned}&\{f, \{g, h\}_{\Sigma_0}\}_{\Sigma_0} + \{g, \{h, f\}_{\Sigma_0}\}_{\Sigma_0} + \{h, \{f, g\}_{\Sigma_0}\}_{\Sigma_0} \\ &\quad = \Sigma_{kl} \Sigma_{ij} \frac{\partial f}{\partial p_k} \frac{\partial^2 g}{\partial p_l \partial p_i} \frac{\partial h}{\partial p_j} + \Sigma_{kl} \Sigma_{ij} \frac{\partial f}{\partial p_k} \frac{\partial g}{\partial p_i} \frac{\partial^2 h}{\partial p_l \partial p_j} \\ &\qquad + \Sigma_{kl} \Sigma_{ij} \frac{\partial g}{\partial p_k} \frac{\partial^2 h}{\partial p_l \partial p_i} \frac{\partial f}{\partial p_j} + \Sigma_{kl} \Sigma_{ij} \frac{\partial g}{\partial p_k} \frac{\partial h}{\partial p_i} \frac{\partial^2 f}{\partial p_l \partial p_j} \\ &\qquad + \Sigma_{kl} \Sigma_{ij} \frac{\partial h}{\partial p_k} \frac{\partial^2 f}{\partial p_l \partial p_i} \frac{\partial g}{\partial p_j} + \Sigma_{kl} \Sigma_{ij} \frac{\partial h}{\partial p_k} \frac{\partial f}{\partial p_i} \frac{\partial^2 g}{\partial p_l \partial p_j}.\end{aligned}$$

The sum of the first and the last term is

$$\begin{aligned}&\Sigma_{kl} \Sigma_{ij} \frac{\partial f}{\partial p_k} \frac{\partial^2 g}{\partial p_l \partial p_i} \frac{\partial h}{\partial p_j} + \Sigma_{kl} \Sigma_{ij} \frac{\partial f}{\partial p_i} \frac{\partial^2 g}{\partial p_l \partial p_j} \frac{\partial h}{\partial p_k} \\ &\quad = \Sigma_{kl} \Sigma_{ij} \frac{\partial f}{\partial p_k} \frac{\partial^2 g}{\partial p_l \partial p_i} \frac{\partial h}{\partial p_j} + \Sigma_{jl} \Sigma_{ki} \frac{\partial f}{\partial p_k} \frac{\partial^2 g}{\partial p_l \partial p_i} \frac{\partial h}{\partial p_j} \\ &\quad = \left( \Sigma_{kl} \Sigma_{ij} + \Sigma_{ji} \Sigma_{kl} \right) \frac{\partial f}{\partial p_k} \frac{\partial^2 g}{\partial p_l \partial p_i} \frac{\partial h}{\partial p_j},\end{aligned}$$

where in the first equality we renamed the summation indices and in the second we used symmetry of the mixed partial derivatives. This expression vanishes since $\Sigma_{ij} = -\Sigma_{ji}$. There are two similar pairs whose sum is also 0 by the same argument. Thus, Jacobi's identity holds.

Now we shall prove that $\pi^{\Sigma_0} : (T^*TQ, \{ , \}) \to (T^*Q, \{ , \}_{\Sigma_0})$ is a Poisson map. Let $f, g : T^*Q \to \mathbb{R}$ and denote by $(q^i, \delta q^i, p_i, \delta p_i)$ the canonical coordinates on $T^*TQ$. By definition of pull-back and the map $\pi^{\Sigma_0}$, we have

$$\left(\pi^{\Sigma_0}\right)^* f(q^i, \delta q^i, p_i, \delta p_i) = f\left(q^i, \delta p_i - \frac{1}{2} \Sigma_{ji} \delta q^j\right).$$

Therefore, by the chain rule, we have the following identities:

$$\frac{\partial\left(\pi^{\Sigma_0}\right)^* f}{\partial q^i} = \frac{\partial f}{\partial q^i} - \frac{1}{2}\frac{\partial f}{\partial p_j}\frac{\partial \Sigma_{kj}}{\partial q^i}\delta q^k,$$

$$\frac{\partial\left(\pi^{\Sigma_0}\right)^* f}{\partial (\delta q)^i} = -\frac{1}{2}\Sigma_{ik}\frac{\partial f}{\partial p_k},$$

$$\frac{\partial\left(\pi^{\Sigma_0}\right)^* f}{\partial p_i} = 0,$$

$$\frac{\partial f}{\partial (\delta p)_i} = \frac{\partial f}{\partial p_i},$$

where the right-hand sides are evaluated at the point $(q^i, \delta p_i - \frac{1}{2}\Sigma_{ji}\delta q^j)$. Therefore, the canonical Poisson bracket of $(\pi^{\Sigma_0})^* f$ and $(\pi^{\Sigma_0})^* g$ is given by

$$\begin{aligned}
\left\{\left(\pi^{\Sigma_0}\right)^* f, \left(\pi^{\Sigma_0}\right)^* g\right\} &= \frac{\partial\left(\pi^{\Sigma_0}\right)^* f}{\partial q^i}\frac{\partial\left(\pi^{\Sigma_0}\right)^* g}{\partial p_i} + \frac{\partial\left(\pi^{\Sigma_0}\right)^* f}{\partial (\delta q)^i}\frac{\partial\left(\pi^{\Sigma_0}\right)^* g}{\partial (\delta p)_i} \\
&\quad - \frac{\partial\left(\pi^{\Sigma_0}\right)^* g}{\partial q^i}\frac{\partial\left(\pi^{\Sigma_0}\right)^* f}{\partial p_i} - \frac{\partial\left(\pi^{\Sigma_0}\right)^* g}{\partial (\delta q)^i}\frac{\partial\left(\pi^{\Sigma_0}\right)^* f}{\partial (\delta p)_i} \\
&= -\frac{1}{2}\Sigma_{ik}\frac{\partial f}{\partial p_k}\frac{\partial g}{\partial p_i} + \frac{1}{2}\Sigma_{ik}\frac{\partial g}{\partial p_k}\frac{\partial f}{\partial p_i} \\
&= \Sigma_{ik}\frac{\partial g}{\partial p_k}\frac{\partial f}{\partial p_i} = \left(\pi^{\Sigma_0}\right)^* \{f, g\}_{\Sigma_0}.
\end{aligned}$$

Uniqueness follows since $(\pi^{\Sigma_0})^*$ is a surjective submersion. □

*Proof by reduction.* The strategy of the proof is to reduce the canonical Poisson structure on $T^*TQ$ to $T^*Q$ by means of the map $\pi^{\Sigma_0}$. Let us first recall some general facts about how one does such a procedure.

To motivate the construction, we note that if $\pi : P \to R$ is a submersive Poisson map from the Poisson manifold $P$ to the Poisson manifold $R$, and if $f, g \in \mathcal{F}(P)$ are such that for $z \in P$, $df(z), dg(z)$ vanish on $\ker T\pi(z)$, then $d\{f, g\}(z)$ vanishes on $\ker T\pi(z)$ as well. In fact, the condition on a 1-form $\alpha \in T_z^*P$ needed for it to push down to a well-defined 1-form at $\pi(z) \in R$ is precisely that $\alpha$ vanish on the kernel of $T_z\pi$.

Now suppose that we want $R$ to inherit a Poisson structure from $P$. The condition needed for the Poisson structure to push down from $P$ to $R$ under a submersion $f : P \to R$ is exactly that if $f, g \in \mathcal{F}(P)$ are such that for $z \in P$, $df(z), dg(z)$ vanish on $\ker T\pi(z)$, then $d\{f, g\}(z)$ vanishes on $\ker T\pi(z)$. If this condition holds, then it is a straightforward procedure to compute the reduced bracket on $R$, as in the last part of the preceding proof. This technique has been developed in a more general context by Marsden and Ratiu [1986] and has been used in interesting ways in the study of integrable systems (see, for instance, Pedroni [1995]).

In our case, consider the foliation of $T^*TQ$ whose leaves are the level sets of the map $\pi^{\Sigma_0}$. To establish the theorem it is enough to show that for all functions

$f, g : T^*TQ \to \mathbb{R}$, the condition $df | \ker T\pi^{\Sigma_0} = 0$ and $dg | \ker T\pi^{\Sigma_0} = 0$ implies $d\{f, g\} | \ker T\pi^{\Sigma_0} = 0$. Differentiating along a curve $(q^i(t), \delta q^i(t))$, we find that

$$\left.\frac{d}{dt}\right|_{t=0} \Sigma_{ij}(q)\delta q^i = \frac{\partial \Sigma_{ij}(q)}{\partial q^k}\dot{q}^k \delta q^i + \Sigma_{ij}(q)\delta \dot{q}^i.$$

It follows that the tangent of the map $\pi^{\Sigma_0}$ is given by

$$\begin{aligned} &T\pi^{\Sigma_0}(q^i, \delta q^i, p_i, \delta p_i, \dot{q}, \delta\dot{q}^i, \dot{p}_i, \delta\dot{p}_i) \\ &\quad = \left(q^j, \delta p_j - \frac{1}{2}\Sigma_{ij}(q)\delta q^i, \dot{q}^j, \delta\dot{p}_j - \frac{1}{2}\left(\frac{\partial \Sigma_{ij}(q)}{\partial q^k}\dot{q}^k\delta q^i + \Sigma_{ij}(q)\delta\dot{q}^i\right)\right). \end{aligned}$$

Thus, a typical tangent vector $(q^i, \delta q^i, p_i, \delta p_i, \dot{q}, \delta\dot{q}^i, \dot{p}_i, \delta\dot{p}_i)$ to the foliation can be written in the form

$$\left(q^i, \delta q^i, p_i, \delta p_i, 0, \delta\dot{q}^i, \dot{p}_i, \frac{1}{2}\Sigma_{ij}(q)\delta\dot{q}^i\right).$$

From this it follows that, for a given $f(q^i, \delta q^i, p_i, \delta p_i)$, the condition

$$df(q^i, \delta q^i, p_i, \delta p_i) \cdot \left(0, \delta\dot{q}^i, \dot{p}_i, \frac{1}{2}\Sigma_{ij}(q)\delta\dot{q}^i\right) = 0$$

is equivalent to

$$\frac{\partial f}{\partial p_i} = 0, \tag{10}$$

$$\frac{\partial f}{\partial \delta q^i} + \frac{1}{2}\frac{\partial f}{\partial \delta p_j}\Sigma_{ij}(q) = 0. \tag{11}$$

Assume that $f$ and $g$ satisfy (10) and (11). We must show that the canonical Poisson bracket on $T^*TQ$ given by

$$\{f, g\} = \frac{\partial f}{\partial q^i}\frac{\partial g}{\partial p_i} + \frac{\partial f}{\partial \delta q^i}\frac{\partial g}{\partial \delta p_i} - \left(\frac{\partial g}{\partial q^i}\frac{\partial f}{\partial p_i} + \frac{\partial g}{\partial \delta q^i}\frac{\partial f}{\partial \delta p_i}\right)$$

also satisfies (10) and (11). One can easily check that if both $f$ and $g$ satisfy (10) and (11), then

$$\begin{aligned} \{f, g\} &= \frac{1}{2}\Sigma_{ij}(q)\left(-\frac{\partial f}{\partial \delta p_j}\frac{\partial g}{\partial \delta p_i}\right) + \frac{1}{2}\Sigma_{ij}(q)\frac{\partial g}{\partial \delta p_j}\frac{\partial f}{\partial \delta p_i} \\ &= \Sigma_{ij}(q)\frac{\partial f}{\partial \delta p_i}\frac{\partial g}{\partial \delta p_j}. \end{aligned}$$

Using this expression it follows directly that (10) holds for $\{f, g\}$, that is,

$$\frac{\partial \{f, g\}}{\partial p_i} = 0.$$

It remains to show that (11) holds for $\{f, g\}$, that is,

$$\frac{\partial}{\partial \delta q^i}\Sigma_{lk}(q)\frac{\partial f}{\partial \delta p_l}\frac{\partial g}{\partial \delta p_k} + \frac{1}{2}\Sigma_{ij}(q)\frac{\partial}{\partial \delta p_j}\Sigma_{lk}(q)\frac{\partial f}{\partial \delta p_l}\frac{\partial g}{\partial \delta p_k} = 0.$$

Let us write, for short,

$$\frac{\partial}{\partial \delta q^i} \Sigma_{lk}(q) \frac{\partial f}{\partial \delta p_l} \frac{\partial g}{\partial \delta p_k} = \alpha_i$$

and

$$\frac{1}{2} \Sigma_{ij}(q) \frac{\partial}{\partial \delta p_j} \Sigma_{lk}(q) \frac{\partial f}{\partial \delta p_l} \frac{\partial g}{\partial \delta p_k} = \beta_i.$$

The $k$th components of $\alpha$ and $\beta$ are

$$\begin{aligned}\alpha_k &= \frac{\partial}{\partial \delta q^k} \Sigma_{ij}(q) \left( \frac{\partial f}{\partial \delta p_i} \frac{\partial g}{\partial \delta p_j} \right) \\ &= \Sigma_{ij} \frac{\partial^2 f}{\partial \delta q^k \partial \delta p_i} \frac{\partial g}{\partial \delta p_j} + \Sigma_{ij} \frac{\partial f}{\partial \delta p_i} \frac{\partial^2 g}{\partial \delta q^k \partial \delta p_j}\end{aligned}$$

and

$$\begin{aligned}\beta_k &= \frac{1}{2} \Sigma_{kl}(q) \left( \frac{\partial}{\partial \delta p_l} \left( \Sigma_{ij} \frac{\partial f}{\partial \delta p_i} \frac{\partial g}{\partial \delta p_j} \right) \right) \\ &= \frac{1}{2} \Sigma_{kl}(q) \left( \Sigma_{ij} \frac{\partial^2 f}{\partial \delta p_l \partial \delta p_i} \frac{\partial g}{\partial \delta p_j} \right) + \frac{1}{2} \Sigma_{kl}(q) \left( \Sigma_{ij} \frac{\partial f}{\partial \delta p_i} \frac{\partial^2 g}{\partial \delta p_l \partial \delta p_j} \frac{\partial f}{\partial \delta q^k} \right).\end{aligned}$$

We have, by (11),

$$\frac{\partial^2 f}{\partial \delta p_i \partial \delta q^k} = \frac{1}{2} \Sigma_{lk} \frac{\partial^2 f}{\partial \delta p_i \partial \delta p_l}$$

and, similarly,

$$\frac{\partial^2 g}{\partial \delta p_i \partial \delta q^k} = \frac{1}{2} \Sigma_{lk} \frac{\partial^2 g}{\partial \delta p_i \partial \delta p_l}.$$

Substituting these expressions for $\partial^2 f / \partial \delta p_i \partial \delta q^k$ and $\partial^2 g / \partial \delta p_i \partial \delta q^k$ into the expression for $\alpha_k$ gives that $\alpha_k + \beta_k = 0$. □

## 3.3. Compatible brackets

Recall (see, for instance, Marsden and Weinstein [1983] or Marsden and Ratiu [1999]) that if $\mathfrak{g}$ is a Lie algebra, its dual $\mathfrak{g}^*$ is a Poisson manifold relative to the Lie–Poisson bracket

$$\{f, g\}(\mu) = \left\langle \mu, \left[ \frac{\delta f}{\delta \mu}, \frac{\delta g}{\delta \mu} \right] \right\rangle, \tag{12}$$

where $f, g : \mathfrak{g}^* \to \mathbb{R}$, $\mu \in \mathfrak{g}^*$, and $\langle\, , \rangle : \mathfrak{g}^* \times \mathfrak{g} \to \mathbb{R}$ denotes the canonical pairing between the Lie algebra and its dual. We now prove that the Lie–Poisson bracket and the momentum $\Sigma$-bracket are compatible exactly when $\Sigma$ is a cocycle.

**Theorem 3.3.1.** *Let $\mathfrak{g}$ be a Lie algebra and $\Sigma : \mathfrak{g} \times \mathfrak{g} \to \mathbb{R}$ be a bilinear skew-symmetric map. For $f, g : \mathfrak{g}^* \to \mathbb{R}$, the Lie–Poisson bracket and the $\Sigma$-bracket are* compatible, *that is,*

$$\{f, g\}^\Sigma := \{f, g\} + \{f, g\}_\Sigma$$

*is also a Poisson bracket on $\mathfrak{g}^*$, if and only if $\Sigma$ is a* 2-cocycle, *that is, $\Sigma$ satisfies the identity*

$$\Sigma([\xi, \eta], \zeta) + \Sigma([\eta, \zeta], \xi) + \Sigma([\zeta, \xi], \eta) = 0 \quad \textit{for all} \quad \xi, \eta, \zeta \in \mathfrak{g}. \tag{13}$$

**Remarks**

1. As an important special case of a cocycle $\Sigma$, one can take an arbitrary coboundary, which is also known as the *modified* or *constant Lie–Poisson bracket*; that is, for each given $\nu \in \mathfrak{g}^*$ define

$$\Sigma_\nu(\xi, \eta) = \langle \nu, [\xi, \eta] \rangle, \qquad \xi, \eta \in \mathfrak{g}$$

and consider the associated Poisson bracket $\{ , \}^{\Sigma_\nu}$. This Poisson bracket appears in connection with several integrable systems and gives rise, via the associated recursion operator, to the commuting integrals of motion (see, for instance, Ratiu [1980] for details).

2. On p. 213 (Chapter IV, Section 5.1 of the 1987 translation) of Libermann and Marle [1987], there is a note on the sum of the canonical bracket on the dual of a Lie algebra plus a cocycle bracket which gives essentially the content of the preceding theorem. However, we give very direct proofs and, in addition, we give a more general version, not only because we do it in infinite dimensions but also because of the more general Theorem 3.3.3, from which Theorem 3.3.1 follows as a special case.

There are three interesting ways to prove the preceding theorem.

*Coordinate proof*. Let $\{\xi_1, \ldots, \xi_n\}$ be a basis of the Lie algebra $\mathfrak{g}$ and denote by $C_{ij}^k$ the structure constants; that is, $C_{ij}^k$ are defined by

$$[\xi_i, \xi_j] = C_{ij}^k \xi_k.$$

The Jacobi identity for the Lie bracket is equivalent to

$$C_{ir}^s C_{jk}^r + C_{jr}^s C_{ki}^r + C_{kr}^s C_{ij}^r = 0 \tag{14}$$

for all $i, j, k, s = 1, \ldots, n$. The Lie–Poisson bracket has the coordinate expression

$$\{f, g\}(\mu) = \left\langle \mu, \left[ \frac{\delta f}{\delta \mu}, \frac{\delta g}{\delta \mu} \right] \right\rangle = \mu_k C_{ij}^k \frac{\partial f}{\partial \mu_i} \frac{\partial g}{\partial \mu_j},$$

where $\mu = \mu_i \xi^i$ and $\{\xi^1, \ldots, \xi^n\}$ is the dual basis of $\{\xi_1, \ldots, \xi_n\}$ in $\mathfrak{g}^*$.

The momentum $\Sigma$-bracket has the coordinate expression

$$\{f, g\}_\Sigma(\mu) = \Sigma_{ij} \frac{\partial f}{\partial \mu_i} \frac{\partial g}{\partial \mu_j},$$

where, as before, $\Sigma = \Sigma_{ij}\xi^i \otimes \xi^j$ (with a sum over all $i, j$). Therefore, the coordinate expression of the sum bracket is

$$\{f, g\}^{\Sigma}(\mu) = \left(\mu_k C^k_{ij} + \Sigma_{ij}\right) \frac{\partial f}{\partial \mu_i} \frac{\partial g}{\partial \mu_j}.$$

Recall that (2) shows how to express the Jacobi identity on a Poisson manifold $(P, \{\,,\})$ in terms of the components of the Poisson tensor. For our case, the quantity we must show vanishes for the validity of the Jacobi identity, is

$$\begin{aligned}
&\left(\mu_s C^s_{ir} + \Sigma_{ir}\right) \frac{\partial}{\partial \mu_r}\left(\mu_s C^s_{jk} + \Sigma_{jk}\right) + \left(\mu_s C^s_{jr} + \Sigma_{jr}\right) \frac{\partial}{\partial \mu_r}\left(\mu_s C^s_{ki} + \Sigma_{ki}\right) \\
&\quad + \left(\mu_s C^s_{kr} + \Sigma_{kr}\right) \frac{\partial}{\partial \mu_r}\left(\mu_s C^s_{ij} + \Sigma_{ij}\right) \\
&= \left(\mu_s C^s_{ir} + \Sigma_{ir}\right) C^r_{jk} + \left(\mu_s C^s_{jr} + \Sigma_{jr}\right) C^r_{ki} + \left(\mu_s C^s_{kr} + \Sigma_{kr}\right) C^r_{ij} \\
&= \mu_s \left(C^s_{ir} C^r_{jk} + C^s_{jr} C^r_{ki} + C^s_{kr} C^r_{ij}\right) + \Sigma_{ir} C^r_{jk} + \Sigma_{jr} C^r_{ki} + \Sigma_{kr} C^r_{ij}.
\end{aligned}$$

The term $(C^s_{ir} C^r_{jk} + C^s_{jr} C^r_{ki} + C^s_{kr} C^r_{ij})$ vanishes by (14). The rest vanishes if and only if the cocycle identity (13) holds, since

$$\Sigma(\eta, [\zeta, \theta]) = \Sigma(\eta^i \xi_i, C^r_{jk} \zeta^j \theta^k \xi_r) = \Sigma_{ir} C^r_{jk} \eta^i \zeta^j \theta^k,$$

for $\eta, \zeta, \theta \in \mathfrak{g}$. □

*Direct proof*. To show that the two Poisson brackets are compatible we must establish Jacobi's identity, so we proceed to a direct computation of the term

$$\{\{f, g\}^{\Sigma}, h\}^{\Sigma} = \{\{f, g\}, h\} + \{\{f, g\}_{\Sigma}, h\} + \{\{f, g\}, h\}_{\Sigma} + \{\{f, g\}_{\Sigma}, h\}_{\Sigma}.$$

The sum of the first term and the other two obtained by circular permutations of $(f, g, h)$ vanishes because the Lie–Poisson bracket satisfies the Jacobi identity. The same is true for the last term because the $\Sigma$-bracket satisfies the Jacobi identity, as we have already proved. To compute the two cross terms we shall need the following formula which is readily checked:

$$\frac{\delta}{\delta \mu}\{f, g\}(\mu) = \left[\frac{\delta f}{\delta \mu}, \frac{\delta g}{\delta \mu}\right] - \mathbf{D}^2 f(\mu)\left(\operatorname{ad}^*_{\delta g/\delta \mu} \mu, \cdot\right) + \mathbf{D}^2 g(\mu)\left(\operatorname{ad}^*_{\delta f/\delta \mu} \mu, \cdot\right). \tag{15}$$

Therefore, using (8) and (15) yields

$$
\begin{aligned}
&\{\{f, g\}_\Sigma, h\}(\mu) + \{\{f, g\}, h\}_\Sigma(\mu) \\
&= \left\langle \mu, \left[ \frac{\delta}{\delta\mu}\{f, g\}_\Sigma, \frac{\delta h}{\delta\mu} \right] \right\rangle + \Sigma\left( \frac{\delta}{\delta\mu}\{f, g\}, \frac{\delta h}{\delta\mu} \right) \\
&= -\left\langle \mathrm{ad}^*_{\delta h/\delta\mu}\, \mu, \frac{\delta}{\delta\mu}\{f, g\}_\Sigma \right\rangle - \left\langle \Sigma^\flat\left(\frac{\delta h}{\delta\mu}\right), \frac{\delta}{\delta\mu}\{f, g\} \right\rangle \\
&= \mathbf{D}^2 f(\mu)\left( \Sigma^\flat\left(\frac{\delta g}{\delta\mu}\right), \mathrm{ad}^*_{\delta h/\delta\mu}\, \mu \right) - \mathbf{D}^2 g(\mu)\left( \Sigma^\flat\left(\frac{\delta f}{\delta\mu}\right), \mathrm{ad}^*_{\delta h/\delta\mu}\, \mu \right) \\
&\quad + \Sigma\left( \left[ \frac{\delta f}{\delta\mu}, \frac{\delta g}{\delta\mu} \right], \frac{\delta h}{\delta\mu} \right) \\
&\quad + \mathbf{D}^2 f(\mu)\left( \mathrm{ad}^*_{\delta g/\delta\mu}\, \mu, \Sigma^\flat\left(\frac{\delta h}{\delta\mu}\right) \right) - \mathbf{D}^2 g(\mu)\left( \mathrm{ad}^*_{\delta f/\delta\mu}\, \mu, \Sigma^\flat\left(\frac{\delta h}{\delta\mu}\right) \right).
\end{aligned}
$$

The sum of the four terms involving second derivatives and the terms obtained by circular permutations of $(f, g, h)$ vanishes in view of the symmetry of the second derivatives.

The sum of the third term and those obtained by circular permutations of $(f, g, h)$ vanishes if and only if the cocycle identity (13) holds. □

*Reduction proof*. This will be discussed after we prove the more general Theorem 3.3.3.

Affine Poisson structures. The following calculations are similar to some of the previous coordinate calculations, but reveal that, instead of dealing with the dual of a Lie algebra, which is a Poisson manifold on a vector space in which the Poisson tensor depends *linearly* on $\mu \in \mathfrak{g}^*$, one can as well start with a slightly more general situation, in which the bracket depends *affinely* on $\mu$.

Let $(M, \{\,,\})$ be a Poisson manifold and assume for the moment that $M$ is a *vector space*. The Poisson structure $\{\,,\}$ defines a tensor field $C \in \mathcal{T}^2_0(M)$ such that, for any given $f, g \in \mathcal{F}(M)$, (1) is satisfied. Let us choose linear coordinates $x^i$ on $M$, where $i = 1, \ldots, m$ and $m$ is the dimension of $M$. Then the Poisson structure is written as

$$
\{f, g\} = C^{ij} \frac{\partial f}{\partial x^i} \frac{\partial g}{\partial x^j},
$$

where $C^{ij}$ are the components of $C$ in the chosen coordinate system. An immediate corollary of (2), which we have already utilized, is that for any skew-symmetric constant tensor $C \in \mathcal{T}^2_0(M)$ the structure on $M$ defined by (1) defines a Poisson bracket.

Now let us consider a skew-symmetric tensor field $C \in \mathcal{T}^2_0(M)$ depending linearly on $x$, that is,

$$
C = C_s x^s. \tag{16}
$$

Then the Jacobi identity (2) becomes

$$
x^s (C^{ir}_s C^{jk}_r + C^{jr}_s C^{ki}_r + C^{kr}_s C^{ij}_r) = 0. \tag{17}
$$

By differentiation with respect to $x^s$ we obtain the equivalent condition

$$C_s^{ir}C_r^{jk} + C_s^{jr}C_r^{ki} + C_s^{kr}C_r^{ij} = 0, \tag{18}$$

which, of course, coincides with (14). Let $a \in M$ be fixed and consider the constant Poisson structure $C(a) = C_s a^s$. Then consider the structure $C + C(a)$, whose components are $C^{ij} + C^{ij}(a) = (x^s + a^s)C_s^{ij}$. The Jacobi identity (2) becomes

$$(x^s + a^s)(C_s^{ir}C_r^{jk} + C_s^{jr}C_r^{ki} + C_s^{kr}C_r^{ij}) = 0, \tag{19}$$

which is satisfied because of (18). We have therefore proven that $C + C(a)$ is a Poisson structure that depends affinely on $x \in M$.

**Example.** Let $M = \mathfrak{g}^*$, where $\mathfrak{g}$ is a Lie algebra. Then, as we have remarked already above, the Lie–Poisson bracket is given by

$$\{f, g\} = \left\langle \mu, \left[\frac{\delta f}{\delta \mu}, \frac{\delta g}{\delta \mu}\right]\right\rangle = \mu_k C_{ij}^k \frac{\partial f}{\partial \mu_i}\frac{\partial g}{\partial \mu_j}, \tag{20}$$

where $C_{ij}^k$ are the structure constants of the Lie algebra $\mathfrak{g}$, which depends linearly on $\mu$. From the previous results we can conclude that, for any $\mu_0$, the expression

$$\{f, g\} = \left\langle \mu, \left[\frac{\delta f}{\delta \mu}, \frac{\delta g}{\delta \mu}\right]\right\rangle + \left\langle \mu_0, \left[\frac{\delta f}{\delta \mu}, \frac{\delta g}{\delta \mu}\right]\right\rangle \tag{21}$$

defines a Poisson bracket.

We must remark that any linear Poisson structure on a vector space, in the sense of (16), comes from a Lie algebra structure on the dual of the vector space.

This discussion proves the following theorem.

**Theorem 3.3.2.** *If $C$ is a Poisson structure on the vector space $M$ that depends linearly on $x$ and if $a \in M$ is fixed then $C + C(a)$ is a Poisson structure.*

COMPATIBILITY OF BRACKETS ON $T^*Q$. Now we shall prove the following theorem, from which Theorem 3.3.1 can be deduced by reduction, as we will show in a moment.

**Theorem 3.3.3.** *Let $Q$ be a manifold and $\Sigma_0$ a 2-form on $Q$. The bracket*

$$\{\,,\}^{\Sigma_0} = \{\,,\}_0 + \{\,,\}_{\Sigma_0}$$

*on $T^*Q$, where $\{\,,\}_0$ is the canonical Poisson bracket and where $\{\,,\}_{\Sigma_0}$ is the bracket defined in Theorem 3.2.6, satisfies the Jacobi identity if and only if $\Sigma_0$ is closed.*

*Proof.* By working in a canonical (Darboux) chart, we can assume without loss of generality that $Q$ is an open subset of $\mathbb{R}^n$. A point of $T^*Q$ is denoted in the usual way as $(q, p) = (q^1, \ldots, q^n, p_1, \ldots, p_n)$. However, in order to use the notation introduced in the previous paragraph, we shall denote $(q^1, \ldots, q^n, p_1, \ldots, p_n) \equiv (x^1, \ldots, x^n, \ldots, x^{2n})$.

The canonical Poisson structure on $T^*Q$ is given by the constant tensor field defined by $C_0^{ij} = -\delta^{i,j-n}$ if $j > n$, $C_0^{ij} = \delta^{i-n,j}$ if $i > n$, and $C_0^{ij} = 0$ otherwise.

Now let $\Sigma_0$ be a 2-form on $Q$. Then we have the bracket $\{F, K\}_{\Sigma_0}$ defined in Theorem 3.2.6, whose associated tensor field, using the notation $x^i$ instead of

$q^i$ or $p_i$ for the coordinates of a point of $T^*Q$, is given by $C_\Sigma^{ij} = 2\Sigma_{i-n,j-n}$ if $i > n$ and $j > n$; $C_\Sigma^{ij} = 0$, otherwise. Define the tensor field $C^{ij} = C_0^{ij} + C_\Sigma^{ij}$. It is antisymmetric because $C_0^{ij}$ and $C_\Sigma^{ij}$ are antisymmetric. Now let us see which condition $\Sigma_0$ must satisfy in order for $C^{ij}$ to be the associated tensor field to a Poisson bracket. This can be achieved using the coordinate expression of the Jacobi identity (2). Taking into account that $C_\Sigma^{ij}$ satisfies the Jacobi identity and the fact that the tensor field $C_0^{ij}$ is constant, it can be easily shown that the Jacobi identity reduces to

$$C_0^{ir}\frac{\partial C_\Sigma^{jk}}{\partial x^r} + C_0^{jr}\frac{\partial C_\Sigma^{ki}}{\partial x^r} + C_0^{kr}\frac{\partial C_\Sigma^{ij}}{\partial x^r} = 0, \tag{22}$$

from which, using the special form of $C_0^{ij}$, one can easily deduce the equivalent condition

$$\frac{\partial \Sigma_{jk}}{\partial q^i} + \frac{\partial \Sigma_{ki}}{\partial q^j} + \frac{\partial \Sigma_{ij}}{\partial q^k} = 0. \tag{23}$$

This last identity simply states that $\mathbf{d}\Sigma_0 = 0$; that is, $\Sigma_0$ is a closed 2-form. □

*Proof of Theorem* 3.3.1 *by reduction.* If $Q \equiv G$ is a Lie group and $\Sigma_0$ is a left-invariant closed 2-form on $G$, then $\{\,,\}^{\Sigma_0}$ is a Poisson bracket on $T^*G$ if and only if $\Sigma_0$ is closed, which is equivalent to the statement that the restriction $\Sigma$ of $\Sigma_0$ to the Lie algebra $\mathfrak{g}$ of $G$ is a 2-cocycle. (This is easily seen by writing the standard formula for the 3-form $\mathbf{d}\Sigma_0$ acting on three left invariant vector fields as a sum of six terms. Three of these vanish by left invariance and the remaining terms are the terms that are involved in the cocycle identity.) From this and the previous theorem we obtain easily a proof of Theorem 3.3.1, by reduction. □

## 3.4. Spin glasses

Holm and Kupershmidt [1988] show that the evolution equations for spin glasses is governed by an affine bracket on the dual of a Lie algebra. Formula (2.26a) of that article gives

$$\partial_t \begin{pmatrix} P_i \\ \rho \\ G_\alpha \\ A_i^\alpha \end{pmatrix} = - \begin{pmatrix} P_k\partial_i + \partial_k P_i & \rho\partial_i & G_\beta\partial_i & \partial_k A_i^\beta - A_{k,i}^\beta \\ \partial_k\rho & 0 & 0 & 0 \\ \partial_k G_\alpha & 0 & t_{\alpha\beta}^\gamma G_\gamma & \delta_\alpha^\beta\partial_k + t_{\alpha\gamma}^\beta A_k^\gamma \\ A_k^\alpha\partial_i + A_{i,k}^\alpha & 0 & \delta_\beta^\alpha\partial_i + t_{\beta\gamma}^\alpha A_i^\gamma & 0 \end{pmatrix} \begin{pmatrix} \delta H/\delta P_k \\ \delta H/\delta\rho \\ \delta H/\delta G_\beta \\ \delta H/\delta A_k^\beta \end{pmatrix}.$$

The Hamiltonian matrix in this equation is given by the Lie–Poisson structure on the dual of the Lie algebra $\mathfrak{X} \circledS \left(\Lambda^0 \oplus \left((\Lambda^0 \otimes \mathfrak{g}) \circledS (\Lambda^{n-1} \otimes \mathfrak{g}^*)\right)\right)$ augmented by a 2-cocycle. We shall show below how this bracket is obtained as a corollary of Theorem 3.3.1 and in the process introduce the dynamic variables and explain the above notation.

Consider two Lie algebras $\mathfrak{a}$, $\mathfrak{b}$ and a vector space $U$. Assume that $U$ is a representation space for $\mathfrak{a}$ and that $\mathfrak{a}$ acts on $\mathfrak{b}$, as a Lie algebra; i.e., there is a Lie algebra homomorphism $\mathfrak{a} \to \mathrm{Der}(\mathfrak{b})$, where $\mathrm{Der}(\mathfrak{b})$ denotes the derivations of the

Lie algebra $\mathrm{Der}(\mathfrak{b})$. Consider $U \oplus \mathfrak{b}$ as a direct sum of Lie algebras, $U$ with the trivial Lie algebra structure. Then the diagonal action of $\mathfrak{a}$ on $U \oplus \mathfrak{b}$ is a Lie algebra action of $\mathfrak{a}$ on $U \oplus \mathfrak{b}$ and one can form the semidirect product Lie algebra $\mathfrak{a}\,Ⓢ(U \oplus \mathfrak{b})$ whose bracket is given by

$$\begin{aligned}&[(\xi_1, u_1, \eta_1), (\xi_2, u_2, \eta_2)]\\ &\qquad = ([\xi_1, \xi_2], \xi_1 \cdot u_2 - \xi_2 \cdot u_1, \xi_1 \cdot \eta_2 - \xi_2 \cdot \eta_1 + [\eta_1, \eta_2]),\end{aligned} \tag{24}$$

for $\xi_i \in \mathfrak{a}$, $\eta_i \in \mathfrak{b}$, and $u_i \in U$.

Now let us assume that $\mathfrak{b} = \mathfrak{c}\,Ⓢ\, W$, for $\mathfrak{c}$ a Lie algebra and $W$ a representation space of $\mathfrak{c}$. In addition, assume that $\mathfrak{a}$ acts on $\mathfrak{c}$ as a Lie algebra, that $W$ is also an $\mathfrak{a}$-module, and that these three actions are compatible in the sense that the following identity holds:

$$\xi \cdot (\sigma \cdot w) = (\xi \cdot \sigma) \cdot w + \sigma \cdot (\xi \cdot w)$$

for $\xi \in \mathfrak{a}$, $\sigma \in \mathfrak{c}$, and $w \in W$. Then the diagonal action of $\mathfrak{a}$ on $\mathfrak{c}\,Ⓢ\, W$, given by $\xi \cdot (\sigma, w) = (\xi \cdot \sigma, \xi \cdot w)$, $\xi \in \mathfrak{a}$, $\sigma \in \mathfrak{c}$, and $w \in W$, is a Lie algebra action of $\mathfrak{a}$ on $\mathfrak{c}\,Ⓢ\, W$. Then, taking $\mathfrak{b} = \mathfrak{c}\,Ⓢ\, W$, in (24) we get

$$\begin{aligned}&[(\xi_1, u_1, \sigma_1, w_1), (\xi_1, u_1, \sigma_1, w_1)]\\ &\quad = ([\xi_1, \xi_2], \xi_1 \cdot u_2 - \xi_2 \cdot u_1, \xi_1 \cdot \sigma_2 - \xi_2 \cdot \sigma_1 + [\sigma_1, \sigma_2],\\ &\qquad \xi_1 \cdot w_2 - \xi_2 \cdot w_1 + \sigma_1 \cdot w_2 - \sigma_2 \cdot w_1).\end{aligned} \tag{25}$$

As a particular case, we shall take

- $\mathfrak{a} = \mathfrak{X}(\mathbb{R}^n)$, the Lie algebra of all vector fields on $\mathbb{R}^n$,
- $W = \Omega \otimes V$, where $\Omega$ is the space of all tensor fields on $\mathbb{R}^n$ of a given type and $V$ is a given vector space,
- $\mathfrak{c} = \Lambda^0 \otimes \mathfrak{g}$, where $\Lambda^0$ denotes the space of smooth functions in $\mathbb{R}^n$ and $\mathfrak{g}$ is a given finite dimensional Lie algebra, with bracket

  $$[(f_1 \otimes \xi_1), (f_2 \otimes \xi_2)] = f_1 f_2 \otimes [\xi_1, \xi_2],$$

  for $f_i \in \Lambda^0$ and $\xi_i \in \mathfrak{g}$, $i = 1, 2$.

In addition, we assume that the $\mathfrak{a}$-representation on $W$ is given by

$$X \cdot (\omega \otimes v) = \pounds_X \omega \otimes v,$$

for $X \in \mathfrak{X}(\mathbb{R}^n)$, $\omega \in \Omega$, $v \in V$ and where $\pounds$ denotes the Lie derivative. The Lie algebra action of $\mathfrak{a}$ on $\mathfrak{c}$ is given by

$$X \cdot (f \otimes \xi) = \pounds_X f \otimes \xi,$$

for $X \in \mathfrak{X}(\mathbb{R}^n)$, $f \in \Lambda^0$, $\xi \in \mathfrak{g}$. The $\mathfrak{c}$-representation on $W$ is given by

$$(f \otimes \xi) \cdot (\omega \otimes v) = f\omega \otimes \xi \cdot v$$

for $\xi \in \mathfrak{g}$, $f \in \Lambda^0$, $v \in V$, $\omega \in \Omega$. Using the derivation property of the Lie derivative, a direct verification shows that the compatibility identity (25), which in this case becomes

$$X \cdot ((f \otimes \xi) \cdot (\omega \otimes v)) = (X \cdot (f \otimes \xi)) \cdot (\omega \otimes v) + (f \otimes \xi) \cdot (X \cdot (\omega \otimes v)),$$

is satisfied. Therefore, we can form the semidirect product Lie algebra

$$\mathfrak{X}(\mathbb{R}^n) \circledS \left[\Lambda^0 \oplus \left((\Lambda^0 \otimes \mathfrak{g}) \circledS (\Omega \otimes V)\right)\right]$$

whose bracket is thus given by

$$\begin{aligned}
&[(X_1, g_1, f_1 \otimes \xi_1, \omega_1 \otimes v_1), (X_2, g_2, f_2 \otimes \xi_2, \omega_2 \otimes v_2)] \\
&\quad = \big([X_1, X_2],\ \pounds_{X_1} g_2 - \pounds_{X_2} g_1,\ \pounds_{X_1} f_2 \otimes \xi_2 - \pounds_{X_2} f_1 \otimes \xi_1 + f_1 f_2 \otimes [\xi_1, \xi_2], \\
&\qquad \pounds_{X_1} \omega_2 \otimes v_2 - \pounds_{X_2} \omega_1 \otimes v_1 + f_1 \omega_2 \otimes \xi_1 \cdot v_2 - f_2 \omega_1 \otimes \xi_2 \cdot v_1\big). \qquad (26)
\end{aligned}$$

This Lie algebra bracket coincides with the one given by formula (3.3) in Holm and Kupershmidt [1988].

If $V = \mathfrak{g}^*$, $\Omega = \Lambda^{n-1}(\mathbb{R}^n)$, the space of $(n-1)$-forms on $\mathbb{R}^n$, and the representation of $\mathfrak{g}$ on $\mathfrak{g}^*$ is the coadjoint representation $\xi \cdot \mu = -\operatorname{ad}^*_\xi \mu$, for $\xi \in \mathfrak{g}$ and $\mu \in \mathfrak{g}^*$, then the dual of the Lie algebra above is the phase space of spin glasses. Let us denote the variables in

$$\left(\mathfrak{X}(\mathbb{R}^n) \circledS \left[\Lambda^0 \oplus \left((\Lambda^0 \oplus \mathfrak{g}) \circledS (\Lambda^{n-1} \otimes \mathfrak{g}^*)\right)\right]\right)^* = \Lambda^1 \times \Lambda^n \times (\Lambda^n \otimes \mathfrak{g}^*) \times (\Lambda^1 \otimes \mathfrak{g})$$

by $(P, \rho, G, A)$; here, $\Lambda^k$ denotes the space of $k$-forms on $\mathbb{R}^n$. In coordinates, $P_i$ is dual to $X_i$, $\rho$ is dual to $1 \in \Lambda^0$, $G_\alpha$ is dual to $1 \otimes e^\alpha$, where $e^\alpha$ is a basis of $\mathfrak{g}^*$, $A_i^\alpha$ is dual to $(\partial_i \mathbin{\lrcorner} d^n x) \otimes e^\alpha \in \Lambda^{n-1} \otimes \mathfrak{g}^*$.

Consider the skew-symmetric bilinear map given by

$$\begin{aligned}
&\Sigma\left((X_1, g_1, f_1 \otimes \xi_1, \omega_1 \otimes \mu_1), (X_2, g_2, f_1 \otimes \xi_2, \omega_2 \otimes \mu_2)\right) \\
&\quad = \int (df_1 \wedge \omega_2)\langle \mu_2, \xi_1 \rangle - (df_2 \wedge \omega_1)\langle \mu_1, \xi_2 \rangle. \qquad (27)
\end{aligned}$$

A direct verification, using the Stokes theorem and the usual exterior differential calculus, shows that $\Sigma$ satisfies the cocycle identity, which in this case comes down to proving that the sum of all cyclic permutations of

$$\begin{aligned}
&\Sigma\big(([X_1, X_2],\ X_1[g_2] - X_2[g_1],\ X_1[f_2] \otimes \xi_2 - X_2[f_1] \otimes \xi_1 + f_1 f_2[\xi_1, \xi_2], \\
&\qquad \pounds_{X_1} \omega_2 \otimes \mu_2 - \pounds_{X_2} \omega_1 \otimes \mu_1 + f_1 \omega_2 \otimes \xi_1 \cdot \mu_2 - f_2 \omega_1 \otimes \xi_2 \cdot \mu_1), \\
&\qquad (X_3, g_3, f_3 \otimes \xi_3, \omega_3 \otimes \mu_3)\big)
\end{aligned}$$

is zero. An alternative way to express (27) is the following:

$$\begin{aligned}
&\Sigma\left((X_1, g_1, f_1 \otimes \xi_1, \omega_1 \otimes \mu_1), (X_2, g_2, f_2 \otimes \xi_2, \omega_2 \otimes \mu_2)\right) \\
&\quad = \int \left((\partial_k \omega_1^k) f_2 \langle \mu_1, \xi_2 \rangle + (\partial_i f_1) \omega_2^i \langle \mu_2, \xi_1 \rangle\right) d^n x,
\end{aligned}$$

which gives

$$\Sigma^\flat(X, g, f \otimes \xi, \omega \otimes \mu) = \begin{pmatrix} 0 & 0 & 0 & 0 \\ 0 & 0 & 0 & 0 \\ 0 & 0 & 0 & \delta_\alpha^\beta \partial_k \\ 0 & 0 & \delta_\beta^\alpha \partial_i & 0 \end{pmatrix} \begin{pmatrix} X^i \\ g \\ f \xi^\beta \\ \omega^k \mu_\beta \end{pmatrix} = \begin{pmatrix} 0 \\ 0 \\ \partial_k \omega^k \mu_\alpha \\ (\partial_i f) \xi^\alpha \end{pmatrix}.$$

Applying Theorem 3.3.1 it follows that the bracket given by the operator matrix

$$-\begin{pmatrix} P_k\partial_i + \partial_k P_i & \rho\partial_i & G_\beta\partial_i & \partial_k A_i^\beta - A_{k,i}^\beta \\ \partial_k\rho & 0 & 0 & 0 \\ \partial_k G_\alpha & 0 & t^\gamma_{\alpha\beta}G_\gamma & \delta^\beta_\alpha\partial_k + t^\beta_{\alpha\gamma}A^\gamma_k \\ A^\alpha_k\partial_i + A^\alpha_{i,k} & 0 & \delta^\alpha_\beta\partial_i + t^\alpha_{\beta\gamma}A^\gamma_i & 0 \end{pmatrix},$$

which coincides with the Hamiltonian matrix in formula (2.26a) of Holm and Kupershmidt [1988], defines a Poisson bracket.

CONCLUSIONS. In this paper we have given a glimpse at a possible connection of constructions of Poisson brackets with cocycles and brackets that one finds in the theory of complex fluids. We explicitly considered one example, namely that of spin glasses, following Holm and Kupershmidt [1988], but the approach hopefully is generalizable to other systems, such as those found in Holm [2002]. One of the longer term objectives of this endeavor would be to gain a deeper insight into the origins of cocycles. One approach to this is an algebraic one developed by Kuperschmidt [1985], but another, as mentioned earlier, is to view cocycles as magnetic terms that arise by reduction as the curvature of a connection. For example, the Bott 2-cocycle can be viewed as arising this way and there is a rather general approach to this theory on both the Hamiltonian and Lagrangian sides, as shown in Marsden, Misiolek, Ortega, Perlmutter, and Ratiu [2002] and Cendra, Marsden, and Ratiu [2001a], respectively. We hope to pursue this line of thinking in the future and to make links with multisymplectic geometry and reduction.

**Acknowledgments**

We are grateful to Larry Bates, Darryl Holm, and Hans-Christian Öttinger for helpful comments.

## References

R. Abraham and J. E. Marsden, *Foundations of Mechanics*, Benjamin Cummings, San Francisco, 1978; 2nd ed. (reprinted and updated), Perseus Publishing, Cambridge, MA, 1985.

R. Abraham, J. E. Marsden, and T. S. Ratiu, *Manifolds, Tensor Analysis and Applications*, 2nd ed., Applied Mathematical Sciences 75, Springer-Verlag, New York, 1988.

V. I. Arnold, *Mathematical Methods of Classical Mechanics*, 1st and 2nd eds., Graduate Texts in Mathematics 60, Springer-Verlag, 1978 and 1989.

M. Castrillón López, T. S. Ratiu, and S. Shkoller, Reduction in principal fiber bundles: Covariant Euler–Poincaré equations, *Proc. Amer. Math. Soc.*, **128** (2000), 2155–2164.

H. Cendra, D. D. Holm, M. J. W. Hoyle, and J. E. Marsden, The Maxwell–Vlasov equations in Euler–Poincaré form, *J. Math. Phys.*, **39** (1998), 3138–3157.

H. Cendra, J. E. Marsden, S. Pekarsky, and T. S. Ratiu, Variational principles for Lie–Poisson and Hamilton–Poincaré equations, *Moscow Math. J.*, special issue for the 65th birthday of V. Arnold, to appear (2003).

H. Cendra, J. E. Marsden, and T. S. Ratiu, *Lagrangian Reduction by Stages*, Memoirs of the American Mathematical Society 152, AMS, Providence, RI, 2001a.

H. Cendra, J. E. Marsden, and T. S. Ratiu, Geometric mechanics, Lagrangian reduction, and nonholonomic systems. in B. Enquist and W. Schmid, eds., *Mathematics Unlimited:* 2001 *and Beyond*, Springer-Verlag, New York, 2001b, 221–273.

M. Gotay, J. Isenberg, and J. E. Marsden, *Momentum Maps and the Hamiltonian Structure of Classical Relativistic Field Theories* I, 1997; available online from `http://www.cds.caltech.edu/~marsden/`.

D. D. Holm, Euler–Poincaré dynamics of perfect complex fluids, in P. Newton, P. Holmes, and A. Weinstein, *Geometry, Mechanics, and Dynamics*, Springer-Verlag, New York, 2002, 113–168.

D. D. Holm and B. A. Kupershmidt, The analogy between spin glasses and Yang–Mills fluids, *J. Math. Phys.*, **29** (1988), 21–30.

D. D. Holm, J. E. Marsden, T. S. Ratiu, and A. Weinstein, Nonlinear stability of fluid and plasma equilibria, *Phys. Rep.*, **123** (1985), 1–196.

B. A. Kupershmidt, Discrete Lax equations and differential-difference calculus, *Astérisque*, **212** (1985).

P. Libermann and C. M. Marle, *Symplectic Geometry and Analytical Mechanics*, Kluwer Academic Publishers, Dordrecht, The Netherlands, 1987.

J. Marsden, G. Misiolek, J. P. Ortega, M. Perlmutter, and T. Ratiu, *Symplectic Reduction by Stages*, preprint, 2002.

J. Marsden, G. Misiolek, M. Perlmutter, and T. Ratiu, Symplectic reduction for semidirect products and central extensions, *Differential Geom. Appl.*, **9** (1998), 173–212.

J. E. Marsden, G. W. Patrick, and S. Shkoller, Multisymplectic geometry, variational integrators and nonlinear PDEs, *Comm. Math. Phys.*, **199** (1998), 351–395.

J. E. Marsden and M. Perlmutter, The orbit bundle picture of cotangent bundle reduction, *C. R. Math. Acad. Sci. Soc. Roy. Canada*, **22** (2000), 33–54.

J. E. Marsden and T. Ratiu, Reduction of Poisson manifolds, *Lett. Math. Phys.*, **11** (1986), 161–170.

J. E. Marsden and T. S. Ratiu, *Introduction to Mechanics and Symmetry*, 1st and 2nd eds., Texts in Applied Mathematics 17, Springer-Verlag, New York, 1994 and 1999.

J. E. Marsden, T. S. Ratiu, and A. Weinstein, Semi-direct products and reduction in mechanics, *Trans. Amer. Math. Soc.*, **281** (1984a), 147–177.

J. E. Marsden, T. S. Ratiu, and A. Weinstein, Reduction and Hamiltonian structures on duals of semidirect product Lie algebras, *Contemp. Math.*, **28** (1984b), 55–100.

J. E. Marsden and A. Weinstein, The Hamiltonian structure of the Maxwell–Vlasov equations, *Phys.* D, **4** (1982), 394–406.

J. E. Marsden and A. Weinstein, Coadjoint orbits, vortices and Clebsch variables for incompressible fluids, *Phys.* D, **7** (1983), 305–323.

R. Montgomery, Canonical formulations of a particle in a Yang–Mills field, *Lett. Math. Phys.*, **8** (1984), 59–67.

R. Montgomery, *The Bundle Picture in Mechanics*, Ph.D. thesis, University of California at Berkeley, Berkeley, CA, 1986.

H. C. Öttinger, Modeling complex fluids with a tensor and a scalar as structural variables, *Rev. Mexicana Fis.*, **48** (2002), supl. 1, 220–229.

G. W. Patrick, The Landau–Lifshitz equation by semidirect product reduction, *Lett. Math. Phys.*, **50** (1999), 177–188.

M. Pedroni, Equivalence of the Drinfeld–Sokolov reduction to a bi-Hamiltonian reduction, *Lett. Math. Phys.*, **35** (1995), 291–302.

T. S. Ratiu, Involution theorems, in G. Kaiser and J. Marsden, eds., *Geometric Methods in Mathematical Physics*, Springer Lecture Notes 775, Springer-Verlag, Berlin, New York, Heidelberg, 1980, 219–257.

S. Sternberg, Minimal coupling and the symplectic mechanics of a classical particle in the presence of a Yang–Mills field, *Proc. Nat. Acad. Sci.*, **74** (1977), 5253–5254.

W. M. Tulczyjew, The Legendre transformation, *Ann. Inst. H. Poincaré*, **27** (1977), 101–114.

A. Weinstein, A universal phase space for particles in Yang–Mills fields, *Lett. Math. Phys.*, **2** (1978), 417–420.

# 4. Defect-Induced Transitions as Mechanisms of Plasticity and Failure in Multifield Continua

O. B. Naimark*

**Abstract.** The multifield statistical approach for the description of collective properties of mesodefect ensembles is developed to allow the determination of the specific nonlinear form of the evolution equation for the macroscopic tensor parameter of defect density. Characteristic self-similar solutions of this equation are found that describe the transition from damage to damage localization, strain localization, change of symmetry properties due to the generation of collective modes of mesodefects. This approach is applied to the study of stochastic crack dynamics, scaling effects in failure, the resonance excitation of failure (failure waves), the structure of plastic waves and instability in shocked condensed matter.

## 4.1. Introduction

During the last few decades the interrelation between the structure and properties of solids has been one of the key problems in physics and mechanics. An extensive study of transformations of structures stimulates a bridging of the gap between general approaches of physics and mechanics of solids. This tendency provides a deeper insight into general laws of plasticity and failure, the latter being treated as structural transitions induced by defects.

Real solids are complex in structure bound to a hierarchy of different scale levels. Solids under loading show changes on all structural levels. These changes have been qualified as plastic deformation and damage processes, realized by nucleation, evolution, and interaction of defects on appropriate structural levels as well as by the interaction of defects between levels. Until recently, no unified multifield theory of solids has been developed to describe the variety, complexity, and interplay of processes commonly observed on all levels of structure. Thus, to construct an adequate model, prime attention should be given to the choice of physical level of microstructure and, consequently, to the type of defects. Furthermore, each structural level is influenced by the effects of processes occurring at the smaller scales.

*Institute of Continuous Media Mechanics, Urals Division, Russian Academy of Sciences, Perm 614013, Russia.

Experimental study of material responses for a large range of loading rates reveals some unresolved puzzles in failure and plasticity and shows the link of solid behavior with the evolution of the ensemble of typical mesoscopic defects (dislocation substructures, microcracks, microshears). This feature is particularly pronounced for dynamic and shock wave loading, when the internal times of the ensemble evolution at different structural levels are approaching the characteristic loading times. As a consequence, the widely used assumption in phenomenology of plasticity and failure concerning the subjective role of structural variables to stress–strain variables (adiabatic limit) can not be generally applied. Moreover, in this case another fundamental problem arises concerning the defect ensemble behavior when the multifield nature of the defect interaction and the specific role of defects are taken into account. These problems are related to the role of the defect ensemble that could be considered in the local view, in the continuum limit, as the localized change of the symmetry of the displacement field and in the global view as the change of the system symmetry under the generation of new collective modes. These modes could influence the system behavior and determine the relaxation ability (plasticity) and failure. The first change of symmetry follows physically from the dislocation nature of defects (formally from the gauge field theory) and must be reflected in the corresponding structure of internal variables (the order parameters). The second symmetry change is the consequence of the group properties of the evolution equations for the order parameters with specific nonlinearity, which leads to the spatial-temporal localization in defect distribution and the generation of the structures responsible for new internal times scales, and finally scenarios of failure and plasticity. In this view the multifield theory of mesoscopic defects was developed and was applied for the explanation of the effects under dynamic crack propagation, failure dynamics, and the flow instability in shocked condensed matter.

## 4.2. Mesodefect properties

### 4.2.1. Dislocation substructures

It is well known that the dislocation density increases as a consequence of plastic deformation; correspondingly, changes of dislocation substructures are observed under the active loading, fatigue, creep, dynamic, and shock wave loading. The structure evolves in a regular manner and despite the variety of the deformed materials only a few types of dislocation substructures are observed: As was shown in [1] the transitions in dislocation substructures belong to a general scenario for polycrystalline materials and monocrystals. The succession in these transitions depends on dislocation interactions and temperature. Sometimes, the transition from one type of dislocation substructure to another leads to sharp changes in mechanical properties of metals and alloys. The main mechanisms of dislocation friction (viscoplastic material responses) and deformation hardening are related to the reconstruction of dislocation substructures.

Each type of dislocation substructure exists in a specific range of dislocation density and it is important that these ranges are stable for large classes of materials. The reason of such universality is related to the inherent property of dislocation ensembles as essentially nonequilibrium systems with self-similarity features, i.e., of characteristic nonlinear responses. This universality in the behavior of dislocation systems appears in the experiments as a low sensitivity of the evolution of dislocation structures to the external stress, but a high sensitivity to structural stresses induced by dislocation interaction.

The increase of the dislocation density is accompanied by a decrease of the distance between dislocations and by stresses due to dislocation interaction. The collective properties in dislocation ensembles play the leading role in these transitions and the substructure formation. The driving agency for the reconstruction of dislocation ensembles is the tendency to reach a relative minimum of total energy through the creation of dislocation substructures [2]. The energy of the dislocation substructure includes two parts: the dislocation self-energy and the energy of the dislocation interaction. The reconstruction of the dislocation substructure leads to the change of both parts. As a consequence, the energy of a newly formed dislocation substructure is less than the energy of the preceding substructure.

The main part in the energy of the dislocation substructure belongs to the dislocation self-energy [2]

$$\Delta U = \frac{\rho G b^2}{2\pi} \ln \frac{L}{r_0}, \tag{1}$$

where $\rho$ is the dislocation density in the dislocation substructure, $b$ is the Burgers vector, $G$ is the shear elastic modulus, $r_0$ is the radius of the dislocation nucleus, and $L$ is the screening radius of the elastic field created by the dislocations. The last scale plays the important role in the evolution of dislocation substructures: The increase of the dislocation density leads to the decrease of the $L$-scale in the order of the substructure sequence. The typical dislocation substructures that are experimentally observed are chaotic: tangles, walls of the cells, subboundaries of strip substructures. The value of $L$ is close to the grain size of chaotic substructures and is of the order of the width of cell walls or of high density dislocation areas in strip substructures.

Under the transition to misoriented substructures dislocation charges arise, which lead to a decrease of the energy: charge signs change their spatial distribution and, simultaneously, substructure parameters become finer.

Such evolution reflects the self-organizing tendency of reconstruction in the dislocation ensemble. The existence of a critical dislocation density corresponding to the creation of dislocation substructures was discussed in [1]. Typical values of critical dislocation density are presented in Table 1 for copper alloys.

As usual, the appearance of the local fluctuation of dislocation density proceeds to the transformation of one substructure into another. The growth of these fluctuations leads to the change of the distribution function for the dislocation density caused by the second mode generation. These regularities of the evolution of defect substructures allow one to consider the latter as independent subsystems during deformation. In this

TABLE 1. Types of transition in dislocation substructures [1].

| **Type of transition** | $\langle\rho_c\rangle 10^{-10}\mathrm{cm}^{-2}$ |
|---|---|
| chaotic substructure → tangles | 0.2 |
| tangles → oriented cells | 0.2–0.5 |
| oriented cells → disoriented cells | 0.4–0.6 |
| pileups → homogeneous net substructure | 0.2–0.3 |
| disoriented cells → net substructure strips | 0.5–2.0 |
| disoriented cells → strips | 3.0 |
| strips → substructure with continuous disorientation | 3.0–4.0 |

case the list of governing parameters, like temperature and stress, can be extended to comprise the order parameters related to the dislocation substructures.

It appears that the significance of collective effects in the dislocation ensemble is enhanced; the dislocation structure is more directly affected by the current value of the dislocation density than by the current external stress.

Thus one is led to the assumption that the dislocation density is the independent variable on which the qualitative changes in the material structure and the mechanical responses depend. In addition, the hardening parameters are linked with the transformations in the substructures.

### 4.2.2. Microcrack (microshear) ensemble

Of all structural levels of subdislocation defects, the level of microcracks and microshear may be considered as the representative for the developed stage of plastic deformation and failure. Other defects (point defects, dislocations, dislocation pileups) have smaller values of intrinsic elastic fields and energies in comparison with microcracks and microshears. Moreover, the nucleation and growth of these defects (that are closest to the macroscopic level) are the final acts of the previous rearrangement of the dislocation substructures.

The density of these defects reaches $10^{12} - 10^{14}\mathrm{cm}^{-3}$, but each mesoscopic defect consists of a dislocation ensemble and exhibits the properties of this ensemble. Scenarios of the evolution of ensembles of these mesoscopic defects show features of nonequilibrium kinetic transitions, and experimental data obtained in a wide range of stress intensities and rates of strain confirm the universality of structural evolution and its effect on relaxation properties and failure.

The typical sizes and concentrations for the microcrack ensemble in different materials are represented in Table 2 [3].

The important features of the quasi-brittle fracture were established for the understanding of various stages of failure: damage, damage localization, crack nucleation and propagation. It was shown that microcracks have the dislocation nature and represent the hollow nuclei of the dislocation pileups. The model representation of microcrack as dislocation pileup [4] allowed the estimation of the microcrack

TABLE 2

| Material | $l(\mu m)$ | | $n(cm^{-3})$ | |
|---|---|---|---|---|
| | X-ray | Microscopy | X-ray | Microscopy |
| aluminum | 0.14 | 0.2 | $10^{11}$ | — |
| nickel | 0.08 | 0.1 | $10^{12}$ | $2 \times 10^{12}$ |
| gold, silver | — | 0.2 | — | $2 \times 10^{11}$ |
| copper, zinc | — | 0.25 | — | $5 \times 10^{11}$ |
| beryllium | 0.12 | — | $5 \times 10^{12}$ | — |
| steel 30CrMCN2A | — | 0.1 | — | — |
| NaCl | 2 | 1–3 | $10^8$–$10^9$ | $10^9$ |
| polyethylene | — | 0.015 | — | $6 \times 10^{15}$ |
| polypropylene | — | 0.02 | — | $7 \times 10^{14}$ |
| PMMA | — | 0.02 | — | $4 \times 10^{12}$ |

self-energy [3, 5]

$$E \approx \left[\frac{G}{V_0} \ln \frac{R}{r_0}\right] s^2, \tag{2}$$

where $\vec{B} = n\vec{b}$ is the total Burgers vector; $s = BS_D$ is the penny-shape microcrack volume; $S_D$ is the microcrack base; $V_0 = \frac{4}{3}r_0^3$ is the volume of the defect nuclei with $r_0$ the characteristic size of the defect nuclei and $R$ the characteristic scale of the elastic field produced by microcracks. The estimation given in [3, 4] showed that the power of the dislocation pileup is close to $n \approx 20$.

Two reasons are important for the dislocation representation of microcracks. The first one is the determination of the microcrack energy as the energy of the dislocation pileup. The second reason is the determination of microscopic parameters for the microcracks as the consequence of the symmetry change of displacement field due to the microcrack nucleation and growth. Study of the microcrack (microshear) size distribution for the different spatial scales revealed the self-similarity of the mesodefect pattern, Figure 1 [6]. The statistical self-similarity reflects the invariant form of the distribution function for the mesodefects of different structural levels. This fact has important consequences for the development of the statistical multifield theory of the evolution of the defect ensemble.

## 4.3. Order parameters of continuum with defects

### 4.3.1. Some results of gauge field theory

Currently, the gauge theory is used for the analysis of the structural and physical properties of materials with defects, being of great help in developing continuum models of such media for the following reasons. Nucleation and growth of defects change the diffeomorphic structure of displacement fields. Using the formalism of gauge field theory (the Yang–Mills formalism [7]), these changes can be introduced

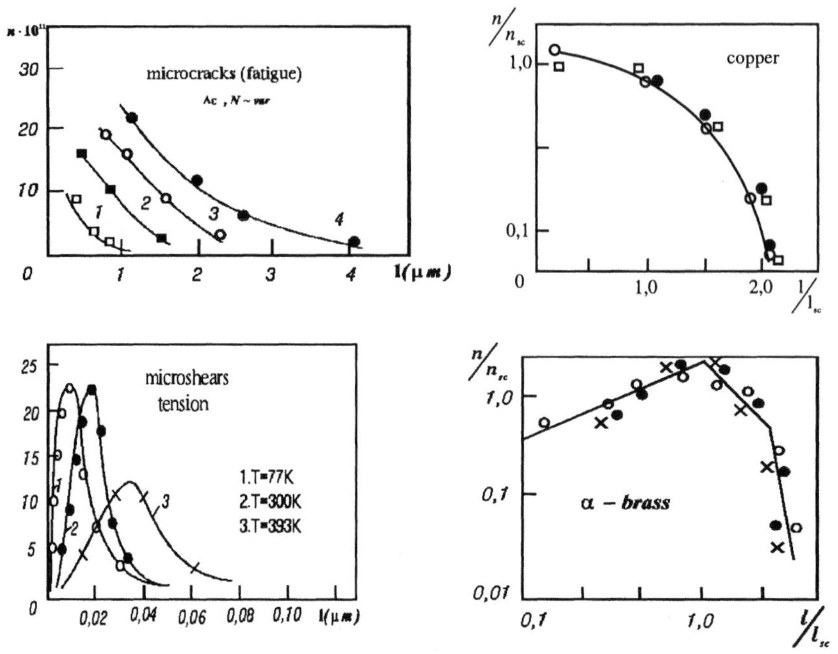


FIGURE 1. Distribution of microcracks and microshears in coordinates with and without dimension: $n$ is the microcrack concentration, $l$ is the characteristic size, and $n_{SC}$, $l_{SC}$ are scaling parameters [6].

as the localization of the corresponding symmetry group of the distortion tensor and considered as additional internal coordinates that are kinematically allowed.

The structure of the so-called gauge fields must correspond to the type of defects, and the basis for the application of gauge theory to our problem is the establishment of the internal symmetry group for the medium with defects. Most mechanical models of deformed continua are invariant with the respect to the uniform groups of translations, $T(3)$, and rotations, $SO(3)$. This means that the Lagrangian of the system is invariant under the transformation

$$x \to x' = g(\xi, t) + \tau(\xi, t),$$

where $x$ and $\xi$ are the current and initial coordinates, $g(\xi, t)$ and $\tau(\xi, t)$ are the rotation and translation operators, respectively.

The differentiation operator is invariant with the respect to the transformations $T(3)$ and $SO(3)$ and is given by the covariant derivative

$$D_\mu x = \partial_\mu x + \Gamma_\mu x + \beta_\mu,$$

where $\Gamma_\mu$ and $\beta_\mu$ are the differential operators related to rotation and translation [7].

For a medium with defects, the homogeneity of transformation $SO(3)$ and $T(3)$ is violated, the operators $\Gamma_\mu$ and $\beta_\mu$ correspond to the local rotations and translations related to the defects; the internal symmetry group corresponds to the so-called semidirect product $SO(3)T(3)$. The elements $\Gamma_\mu$ and $\beta_\mu$ of this group, which are functions of coordinates and time, are the gauge fields.

We can thus construct a Lagrangian of the medium with defects using the minimal expansion and write the Lagrangian of the system with two additional variables $\Gamma_\mu$ and $\beta_\mu$ [7]

$$L = -\frac{1}{2}C_1 g^{(1)}_{\mu\lambda} g^{(1)}_{\nu\chi} \alpha_{\mu\nu}\alpha_{\lambda\chi} - \frac{1}{2}C_2 g^{(2)}_{\mu\nu} g^{(2)}_{\nu\chi} \theta_{\mu\nu}\theta_{\lambda\chi}, \tag{3}$$

where

$$\theta_{\mu\nu} = \partial_\mu\Gamma_\nu - \partial_\nu\Gamma_\mu + \Gamma_\mu\Gamma_\nu - \Gamma_\nu\Gamma_\mu$$

$$\alpha_{\mu\nu} = \partial_\mu\beta_\mu - \partial_\nu\beta_\nu + \Gamma_\mu\beta_\nu - \Gamma_\nu\beta_\mu + \theta_{\mu\nu}x$$

are the so-called intensities of the gauge fields related to the local rotations and translations due to the presence of defects; $g_{ik}$ are the components of the metric tensor. The phenomenological parameters $C_1$ and $C_2$ in the Lagrangian $L$ are the constants of dislocation interaction. We will use expression (3) for the Lagrangian in our analysis of the statistical properties of the ensembles of defects.

### 4.3.2. Microscopic and macroscopic variables for microcracks (microshears) ensemble

The structural parameters associated with microcracks and microshears were introduced [3] as the derivative of the dislocation density tensor. These defects are described by symmetric tensors of the form

$$s_{ik} = s\nu_i\nu_k \tag{4}$$

in the case of microcracks and

$$s_{ik} = 1/2s(\nu_i l_k + l_i\nu_k) \tag{5}$$

for microshears. Here $\vec{\nu}$ is unit vector normal to the base of a microcrack or slip plane of a microscopic shear; $\vec{l}$ is a unit vector in the direction of shear; $s$ is the volume of a microcrack or the shear intensity for a microscopic shear.

The change of the diffeomorphic structure of the displacement field due to these defects also has important consequences from the point of view of the symmetry change of the system "solid with defects." This symmetry aspect can be used to model arbitrary defects both in crystalline and amorphous materials without the assumption concerning the dislocation nature of the defects that originally is the property of crystalline materials.

The average of the "microscopic" tensor $s_{ik}$ gives the macroscopic tensor of the microcrack or microshear density

$$p_{ik} = n\langle s_{ik}\rangle, \tag{6}$$

that coincides with the deformation caused by the defects, where $n$ is the defect concentration.

## 4.4. Statistical model of continuum with mesodefects

### 4.4.1. Effective field method

The effective field is frequently used as an auxiliary field (real or virtual) introduced into a theoretical model in order to take into account complex factors (like interparticle interactions) that are either too difficult to evaluate rigorously or not even clear in detail.

When we refer to "the effective field method" here, we mean that we use an auxiliary external multicomponent field constructed in such a way that, when added to the corresponding term in the Hamiltonian of the system, makes it a state of equilibrium at any given instant.

The effective field method was reintroduced [8] into statistical physics by Leontovich [9]. According to Leontovich, for an arbitrary nonequilibrium state of any thermally uniform system that is characterized by definite values of internal parameters, the transition into the equilibrium state with the same values of those internal parameters may be performed by introducing an additional force field. By definition, the entropy of this nonequilibrium state is equal to the entropy of the equilibrium (being that due the presence of the additional force field) state characterized by the same values of the considered material parameters.

The microscopic kinetics for the parameter $s_{ik}$ is determined by the Langevin equation

$$s_{ik} = K_{ik}(s_{lm}) - F_{ik} \tag{7}$$

where $K_{ik} = \frac{\partial E}{\partial s_{ik}}$; $E$ is the energy of the defect and $F_{ik}$ is a random part of the force field and satisfies the relations $\langle F_{ik}(t)\rangle = 0$ and $\langle F_{ik}(t')F_{ik}(t)\rangle = Q\delta(t-t')$. The parameter Q characterizes the mean value of the energy relief of the initial material structure (the energy of defect nuclei).

The statistical model of the microcrack (microshear) ensemble was developed in the terms of solutions of the Fokker–Planck equation in [2, 3]

$$\frac{\partial}{\partial t}W = -\frac{\partial}{\partial s_{ik}}K_{ik}W + \frac{1}{2}Q\frac{\partial^2}{\partial s_{ik}\partial s_{ik}}W. \tag{8}$$

According to the statistical self-similarity hypothesis the distribution function of defects can be represented in the form $W = Z^{-1}\exp(-\frac{E}{Q})$, where $Z$ is the normalization constant. By (8), the statistical properties of the defect ensemble can be described in terms of the defect energy $E$ and the dispersion properties of the system given by the value of $Q$.

In terms of the microscopic and macroscopic variables and according to the presentation of these mesodefects as the dislocation substructure, the energy of these defects (the Lagrangian) can be written in the form

$$E = E_0 - H_{ik}s_{ik} + \alpha s_{ik}^2, \tag{9}$$

where the quadratic term represents the self-energy of defects (2) and the term $H_{ik}s_{ik}$ describes the interaction of the defects with the external stress $\sigma_{ik}$ and with the ensemble of the defects in the effective field approximation:

$$H_{ik} = \sigma_{ik} + \lambda p_{ik} = \sigma_{ik} + \lambda n\langle s_{ik}\rangle, \tag{10}$$

where $\alpha$, $\lambda$ are material constants. The averaging procedure gives the self-consistency equation for the determination of the defect density tensor

$$p_{ik} = n\int s_{ik}W(s, \vec{\nu}, \vec{l})ds_{ik}. \tag{11}$$

For the dimensionless variables

$$\hat{p}_{ik} = \frac{1}{n}\sqrt{\frac{\alpha}{Q}}p_{ik}, \qquad \hat{s}_{ik} = \sqrt{\frac{\alpha}{Q}}, \qquad \hat{\sigma}_{ik} = \frac{\sigma_{ik}}{\sqrt{Q\alpha}},$$

the self-consistency equation has the form

$$\hat{p}_{ik} = \int \hat{s}_{ik}Z^{-1}\exp\left(\left(\hat{\sigma}_{ik} + \frac{1}{\delta}\hat{p}_{ik}\right)\hat{s}_{ik} - \hat{s}_{ik}^2\right)d\hat{s}_{ik}, \tag{12}$$

which includes the single dimensionless material parameter $\delta = \frac{\alpha}{\lambda n}$. Dimensional analysis allows us to estimate

$$\alpha \sim \frac{G}{V_0}, \qquad \lambda \sim G, \qquad n \sim R^{-3}.$$

Here $G$ is the elastic modulus, $V_0 \sim r_0^3$ is the mean volume of the defect nuclei, and $R$ is the distance between defects. Finally we obtain for $\delta$ the value $\delta \sim (\frac{R}{r_0})^3$ that corresponds with the hypothesis concerning the statistical self-similarity of the defect distribution on different structural levels. The solution of the self-consistency equation (12) was found for the case of the uniaxial tension and simple shear [2] (Figure 2).

The existence of characteristic nonlinear behavior of the defect ensemble in the corresponding ranges of $\delta(\delta > \delta_* \approx 1.3, \delta_c < \delta < \delta_*, \delta < \delta_c \approx 1)$ was established, where $\delta_c$ and $\delta_*$ are the bifurcation points. It was shown [3, 4] that the above ranges of $\delta$ are characteristic for the quasi-brittle ($\delta < \delta_c \approx 1$), ductile ($\delta_c < \delta < \delta_*$), and nanocrystalline ($\delta > \delta_* \approx 1.3$) responses of materials. Thus the behavior of the defect ensemble in the different ranges of $\delta$ is qualitatively different. The replacement of the stable material response for the fine grain materials with the metastable response for the ductile materials with the intermediate grain size occurs for the value of $\delta = \delta_* \approx 1.3$, when the interaction between the orientation modes of the defects has a more pronounced character. Therefore, the metastability has the nature of an orientation ordering in the defect ensemble. It will be shown in Section 4.8 that the continuous orientation transition due to the growth of the defect density (and, as a consequence, the decrease of $\delta$) provides the specific mechanism of the momentum transfer that is conventionally known as the plastic flow. As will be discussed in

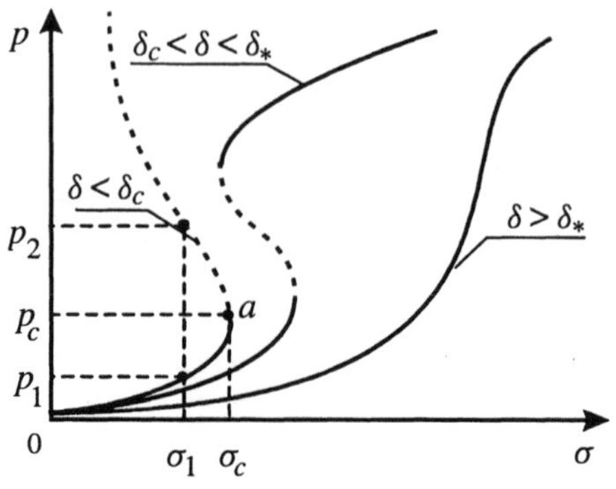


FIGURE 2. Characteristic responses of materials on defect growth.

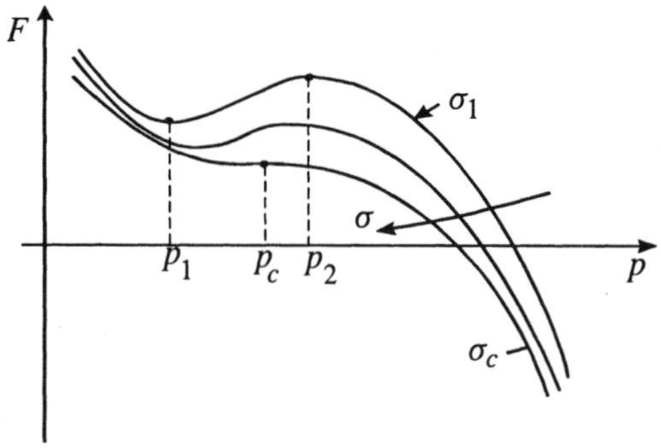


FIGURE 3. Free energy dependence on stress and defect density for $\delta < \delta_c \approx 1$.

Section 4.6, the range of $\delta < \delta_c \approx 1$ provides the specific nonlinear kinetics of defect evolution, which leads to the nucleation of failure hotspots.

## 4.5. Collective properties of ensembles of defects

### 4.5.1. Phenomenology of solids with defects. Free energy

The statistical description allows us to study the phenomenology of solids with defects on the basis of an appropriate choice of the free energy form $F$. As (11) is none other than $\frac{\partial F}{\delta p_{ik}} = 0$, the simple phenomenological form of the part of the free energy caused by defects (for the uniaxial case $p = p_{zz}, \sigma = \sigma_{zz}, \varepsilon = \varepsilon_{zz}$) is given by the six order expansion, similar to the well-known Ginzburg–Landau expansion [11]

$$F = \frac{1}{2}A\left(1 - \frac{\delta}{\delta_*}\right)p^2 - \frac{1}{4}Bp^4 + \frac{1}{6}C\left(1 - \frac{\delta}{\delta_c}\right)p^6 - D\sigma p + \chi(\nabla_l p)^2. \quad (13)$$

The bifurcation points $\delta_*, \delta_c$ play a role similar to that of the characteristic temperatures in the Ginzburg–Landau expansion in the phase transition theory. The gradient term in (13) describes the nonlocal interaction in the defect ensemble in the so-called long wave approximation; $A$, $B$, $C$, $D$, and $\chi$ are the phenomenological parameters.

The defect kinetics is determined by the evolution inequality [5]

$$\frac{\delta F}{\delta t} = \frac{\delta F}{\delta p}\frac{dp}{dt} \leq 0 \tag{14}$$

that leads to the kinetic equation for the defect density tensor

$$\frac{dp}{dt} = -\Gamma\left(A\left(1-\frac{\delta}{\delta_*}\right)p - Bp^3 + C\left(1-\frac{\delta}{\delta_c}\right)p^5 - D\sigma - \frac{\partial}{\partial x_l}\left(\chi\frac{\partial p}{\partial x_l}\right)\right), \tag{15}$$

where $\Gamma$ is the kinetic coefficient.

The kinetic equation (15) and the equation for the total deformation

$$\varepsilon = \hat{C}\sigma + p. \tag{16}$$

($\hat{C}$ is the component of the elastic compliance tensor) represent the system of the constitutive and evolution equations for a solid with the considered types of defects.

### 4.5.2. Collective properties of defect ensemble

As follows from the solution of (11), presented in Figure 2, transitions through the bifurcation points $\delta_c$ and $\delta_*$ lead to a sharp change in the symmetry of the distribution function as a result of the appearance of some orientationally pronounced macroscopic modes of the tensor $p_{ik}$. The effect of transitions on the evolution of the defect ensemble is determined by the type of bifurcation, i.e., by the group properties of the kinetic equation for the tensor $p_{ik}$ for different domains of $\delta(\delta > \delta_*$, $\delta_c < \delta < \delta_*$, $\delta < \delta_c)$. The qualitative relationships governing the changes in the behavior of the system are reflected in Figure 4 in the form of families of heteroclines, which are the solutions of the equation

$$A\left(1-\frac{\delta}{\delta_*}\right)p - Bp^3 + C\left(1-\frac{\delta}{\delta_c}\right)p^5 - D\sigma - \frac{\partial}{\partial x_l}\left(\chi\frac{\partial p}{\partial x_l}\right) = 0. \tag{17}$$

In the region $\delta > \delta_*$ this equation is of the elliptic type with periodic solutions with spatial scale $\Lambda$ and possesses $p$ anisotropy determined mainly by the applied stress. This distribution of $p$ gives rise to weak pulsations of the strain field. As $\delta \to \delta_*$, the solution of (17) passes the separatrix $S_2$, and the periodic solution transforms into a solitary-wave solution. This transition is accompanied by divergence of the inner scale $\Lambda : \Lambda \approx -\ln(\delta - \delta_*)$. In this case the solution has the form $p(\zeta) = p(x - Vt)$. The wave amplitude, velocity and the width of the wave front are determined by the parameters of nonequilibrium (orientation) transition:

$$p = \frac{1}{2}p_a[1 - \tanh(\zeta l^{-1})], \quad l = \frac{4}{p_a}\left(2\frac{\chi}{A}\right)^{\frac{1}{2}}. \tag{18}$$

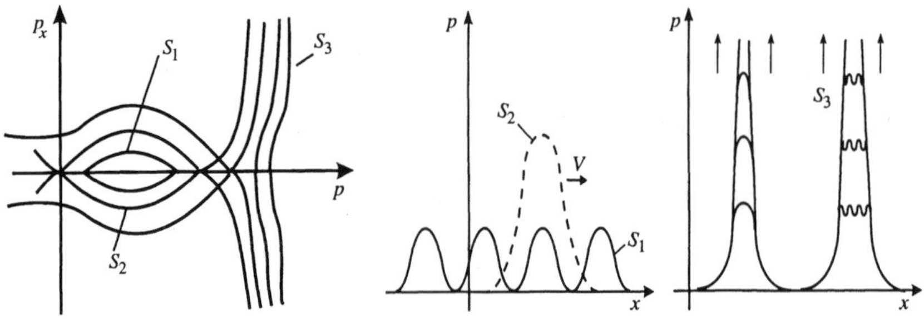


FIGURE 4. Types of heteroclines and the corresponding characteristic forms.

The velocity of the solitary wave is

$$V = \frac{\chi A(p_a - p_m)}{2\zeta^2},$$

where $(p_a - p_m)$ is the jump in $p$ in the course of an orientational transition. A transition through the bifurcation point $\delta_c$ (separatrix $S_3$) is accompanied by the appearance of spatio-temporal structures of a qualitatively new type characterized by explosive accumulation of defects as $t \to t_c$ in the spectrum of spatial scales ("blow-up" regime) [12, 13]. In this case the kinetics of $p$ is determined by the difference of the power of the terms in the expansion (13).

Assuming a power law dependence of the nonlocality parameter on $p$, the kinetic equation (15) can be written in the form

$$\frac{\partial p}{\partial t} \approx S(p_c)p^{\omega} + \frac{\partial}{\partial x}\left(\chi_0(p_c)p^{\beta}\frac{\partial p}{\partial x}\right), \tag{19}$$

where $\omega = 5/3$. It is shown in [12, 13] that for this type of equations the developed stage of the kinetics of $p$ in the limit $t \to t_c$ can be described by a self-similar solution

$$p(x,t) = \phi(t)f(\zeta), \quad \zeta = \frac{x}{L_c}, \quad \phi(t) = \Phi_0\left(1 - \frac{t}{t_c}\right)^{-m}, \tag{20}$$

where $m > 0$, $\Phi_0 > 0$ are the parameters related to the nonlinear form of (15); $L_c$ and $t_c$ are the scaling parameters. The function $f(\zeta)$ is determined by solving the corresponding eigenvalue problem. For example, for the case $\omega = \beta + 1$, the self-similar solution of (19) has the form

$$p(x,t) = [S(t - t_c)]^{-\frac{1}{\beta}}\left(\frac{2(\beta+1)}{\beta(\beta+2)}\sin^2\left(\frac{\pi x}{L_c} + \pi\theta\right)\right)^{\frac{1}{\beta}}, \tag{21}$$

where $\theta$ is a random value in the interval (0, 1). The scale $L_c$, the so-called fundamental length [13], has the meaning of a spatial period of the solution (20),

$$L_c = \frac{2\pi}{\beta}((\beta+1)\chi_0 S^{-1})^{\frac{1}{2}}.$$

The self-similar solution (20) describes the blow-up damage kinetics for $t \to t_c$ on the set of spatial scales $L_H = kL_c$, $k = 1, 2, \ldots, K$. The blow-up kinetics of damage localization allowed us to link the hotspots of failure with the above-mentioned self-similar structures.

## 4.6. Collective behavior of cracks and defects

### 4.6.1. Introductory remarks

The interaction of the main crack with the ensemble of defects is the subject of intensive experimental and theoretical studies that revealed some unresolved puzzles in the quasi-brittle failure. The longstanding problem is the limiting velocity of a crack. The linear elastic theory predicted that a crack should continuously accelerate up to the Rayleigh wave speed $V_R$; however, the experiments on a number of brittle materials [14] showed that speed of the crack will seldom reach even the half of this value. A portrait of the dissipation process was suggested in [5, 15] where the main role in the qualitative explanation of the mentioned effects was assigned to the collective modes of the defect ensemble and to the interaction of these modes with the moving crack. Below we will discuss the phenomena investigated intensively during the past decade within the nonlinear dynamics of crack propagation. These phenomena show the qualitative new features caused by the interaction of the crack with the ensemble of defects in the so-called process zone.

The renewed interest in the dynamic fracture during last decade is due to the variety of experimental results that cannot be explained within the classical fracture mechanics, within which only two steady-state velocities are predicted for a crack in an infinite plane specimen: zero and the Rayleigh speed [16]. The recent experimental study revealed the limiting steady-state crack velocity, a dynamical instability to microbranching [16, 17], the formation of a nonsmooth fracture surface [18], and the sudden variation of fracture energy (dissipative losses) with a crack velocity [19].

This renewed interest was the motivation for studying the interaction of defects at the crack tip area (process zone) with a moving crack. The still open problem of crack evolution is the condition of crack arrest, a condition related to the question of whether the crack velocity approaches zero smoothly as the load is decreased from a large value to the Griffith point [20]. There is also a problem at the low end of crack velocity: how a crack that is initially at rest might achieve its steady state.

### 4.6.2. Some results in crack mechanics

The subject of the rapid propagation of a crack in a brittle material is one whose roots go back to the classical works of Griffith and Mott [22, 23]. Since that time a great deal of both analytical and experimental work [24, 25, 26] has been expected to understand the phenomenon of fracture. To study the crack stability and the interaction of the main crack with the defect ensemble in the process zone we will consider first briefly the classical results in crack mechanics, where much analytical progress has been made in assuming that the medium behaves according to the equations of linear

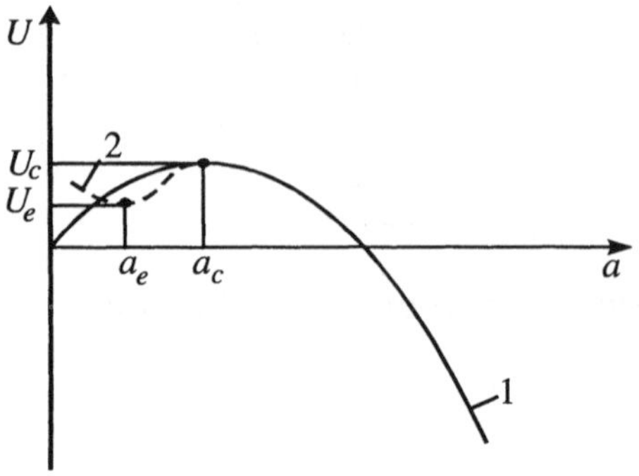


FIGURE 5. The Griffith (1) and Fraenkel (2) energy form of elastic solid with a crack.

elasticity. The important step to model the brittle failure was made by Griffith [22] when the additional characteristics of the crack resistance were introduced in the form of the energy of the development of the new surface at the crack tip. According to the Griffith theory the energy $U$ of elastic materials with a crack is represented in the form (Figure 5, curve 1)

$$U = -\frac{\sigma^2}{2G}\left(\frac{\pi a^2}{4}\right) + 2\gamma a, \tag{22}$$

where $\gamma$ is the surface energy; $\sigma$ is the applied stress; $a$ is the crack length; and $G$ is the elastic modulus.

Irwin [24] developed the Griffith conception and proposed the force version of the crack stability introducing the stress intensity factor

$$K_I = \sigma\sqrt{\pi a}. \tag{23}$$

Barenblatt [25] proposed a variant of the force version of the crack stability that reflects another view on the role of stress in the crack tip area. The effects of cohesion forces $G_B(s)$ were assumed to be relevant at the crack tip over the area $0 < s < d$. These forces have a molecular nature and alter the singularity of the stress field. Two hypotheses were assumed in the Barenblatt approach:

(i) small size of the crack tip zone in comparison with the crack length ($\frac{d}{a} \ll 1$);
(ii) autonomous behavior of crack tip, i.e., the self-similar evolution of the crack tip in the steady-state regime.

This hypothesis conforms with the properties that reveal quasi-brittle behavior. The self-similarity features of the crack tip evolution are the consequence of the small ratio of the applied stress $\sigma$ and the cohesion force $G_B$, i.e., $\frac{\sigma}{G_B} \ll 1$. This fact reflects the intermediate asymptotic character of quasi-brittle failure theories. A material parameter was introduced, the so-called the cohesive modulus, as an independent strength property of materials: $K_B = G_B\sqrt{d}$. Despite the similar form of the cohesive modulus $K_B$ and the stress intensity factor $K_I$ there is a difference between the Irwin criteria and those based on the cohesive modulus proposed by Barenblatt. The cohesive modulus determines the steady-state character of crack propagation but not the catastrophic one corresponding to the Griffith–Irwin approach. The qualitative difference between the two approaches is evidenced by the remarks by Fraenkel [27] in the critical analysis of the Griffith approach. Fraenkel wrote that the physically realistic form of the energy $U$ must contain the local minimum $U_e(a_e)$ (Figure 5, curve 2). The difference in the energy $\Delta U = U_c - U_e$ determines the work of the stress field at the crack tip under transition from the steady-state to the unstable regime of crack propagation. This work provides the means to surmount the energy barrier. It is natural to assume that the cohesive modulus is the force related to this barrier. We will show in the following that the metastable energy form, assumed by Fraenkel, is related to the collective behavior of the defect ensemble in the process zone and to the interaction of the defects with the main crack.

### 4.6.3. Origin of crack instability

The classic theory of fracture treats cracks as mathematical branch cuts that begin to move when an infinitesimal extension of the crack releases more energy than needed to create a fracture surface. This idea is successful in some cases in practice but conceptually incomplete. The experiments fail to confirm this idealized picture. The surface created by the crack is not necessarily smooth and flat. In a series of experiments on brittle fracture the simultaneous propagation of an ensemble of microcracks, instead of a single propagating crack, was observed [15]. The fracture process was then viewed as a coalescence of defects situated in the crack path; the mean acceleration drops, the crack velocity develops oscillations, and a structure is formed on the fracture surface [17, 18, 19, 20]. As the branches grow in size, they evolve into macroscopic large-scale crack branches.

A theoretical explanation of the limited steady-state crack velocity and the transition to the branching regime was proposed in [28, 29, 30] within a study of the collective behavior of the microcrack ensemble in the process zone. It was shown by the solution of the evolution equation for the defect density tensor that the kinetics of microcrack accumulation at the final damage stage includes the generation of spatial-temporal structures (dissipative structures with blow-up damage kinetics) that are precursors of the nucleation of the "daughter" cracks. The kinetics of the daughter crack generation is determined by two parameters that are given by the self-similar solution (20),

$$p(x,t) = \phi(t) f(\zeta), \quad \zeta = \frac{x}{L_c}, \quad \phi(t) = \Phi_0 \left(1 - \frac{t}{t_c}\right)^{-m}, \tag{24}$$

where $L_c$ and $t_c$ are the scaling parameters that can be found through the solution of the corresponding nonlinear eigenfunction problem [12, 13, 29]. These parameters are the spatial scales $L_C$ of the blow-up damage localization and the so-called "peak time" $t_c$, which is the time of damage localization in the self-similar blow-up regime. The velocity limit $V_C$ of the transition from the steady state to the irregular crack propagation is given by the ratio: $V_C \approx \frac{L_C}{t_c}$. As shown in Section 4.5, the set of spatial scales $L_H$ (daughter crack sizes) is proportional to $L_C$ and represents a new set of independent coordinates (collective modes of the defect ensemble) of the nonlinear system for $\sigma > \sigma_c$. These coordinates characterize the property of a second attractor that could determine the behavior of solutions of the nonlinear system. The first attractor corresponds to the well-known self-similar solution (23) for the stress distribution at the crack tip which is the background for the stress intensity factor conception. This solution is available in the presence of the metastability (local minimum) for $p$ in the range $\sigma < \sigma_c$.

The steady-state crack propagation occurs when the stress rise in the process zone provides the failure time $t_f > t_c = \frac{L_C}{V_C}$ for the creation of the daughter crack only in the straight crack path. The failure time $t_f$ follows from the kinetic equation (15) and represents the sum of the induction time $t_i$ (the time for the approach of the defect distribution to the self-similar profile on the $L_H = kL_c$ scales) and the peak time $t_c$: $t_f = t_i + t_c$. For velocities $V < V_C$ the induction time is $t_i \gg t_c$ and the daughter crack appears only along the initial main crack orientation. For the crack velocity $V \approx V_C$ there is a transient regime ($t_i \approx t_c$) of the creation of a number of localization scales (daughter cracks) in the main crack path. The crack velocity growth in the area $V > V_C$ leads to the sharp decrease of the induction time $t_i \to 0$, $t_f \to t_c$ that is accompanied by the extension of the process zone in both (tangential and longitudinal) directions where the multiple blow-up structures (daughter cracks) and, as a consequence, the main crack branching appear. The last situation is qualitatively similar to the resonance excitation of numerous mirror zones under the spall condition but in the nonuniform stress field in the range of angle with $\sigma > \sigma_c$.

### 4.6.4. Experimental study of nonlinear crack dynamics

EXPERIMENTAL SETUP Direct experimental study of crack dynamics in the preloaded PMMA plane specimen was carried out with the use of a high speed digital camera Remix REM 100-8 (time lag between pictures $10\mu s$) coupled with photoelasticity method (Figure 6) [29, 30, 31].

The pictures of stress distribution at the crack tip are shown in Figure 7 for slow ($V < V_C$) and fast ($V > V_C$) cracks. The experiment revealed that the path of the critical velocity $V_C$ is accompanied by the appearance of a stress wave pattern produced by the daughter crack growth in the process zone. Independent estimation of the critical velocity from the direct measurement of the crack tip coordinates and from the pronounced stress wave Doppler pattern gives a correspondence with the Fineberg data ($V_C \approx 0.4V_R$) [16].

CHARACTERISTIC CRACK VELOCITY The dependence of crack velocity on the initial stress is represented in Figure 8. Three portions with different slopes are shown. The existence of these portions determines three characteristic velocities: the velocity of the transition from the steady state to the nonmonotonic straight regime $V_S \approx 220$m/s, the transient velocity to the branching regime $V_C \approx 330$m/s, and the velocity $V_B \approx 600$m/s when the branches behave autonomously.

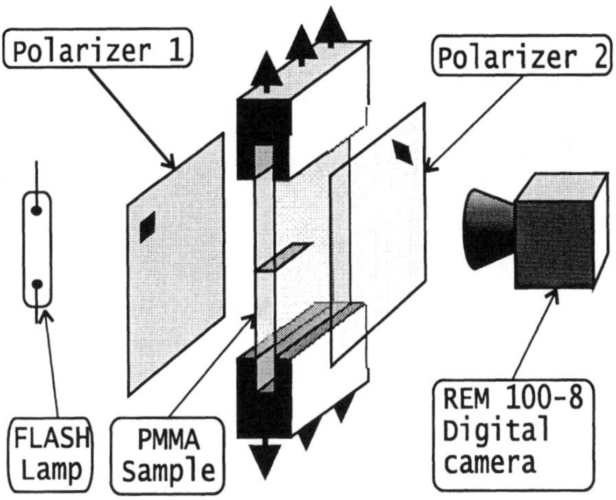


FIGURE 6. Scheme of the experiment.

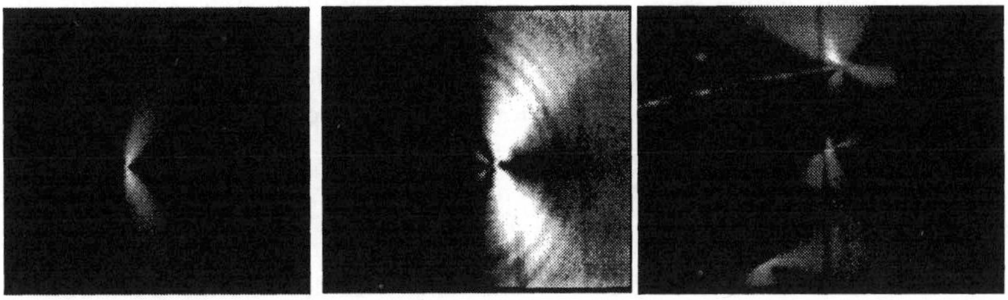

$V < V_C$ $V > V_C$ $V > V_B$

FIGURE 7. Different regimes of crack dynamics.

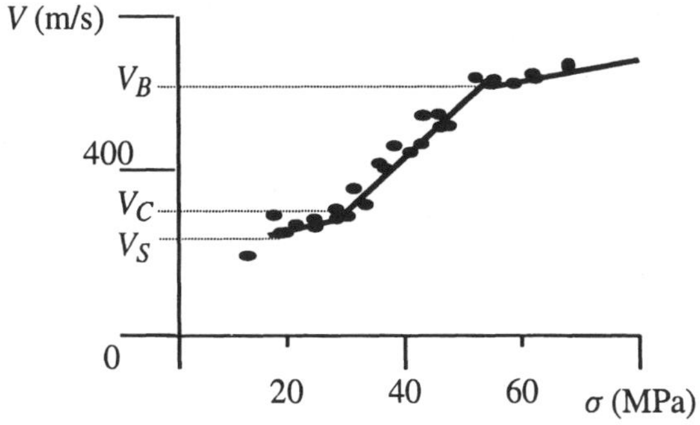


FIGURE 8. Crack velocity versus applied stress.

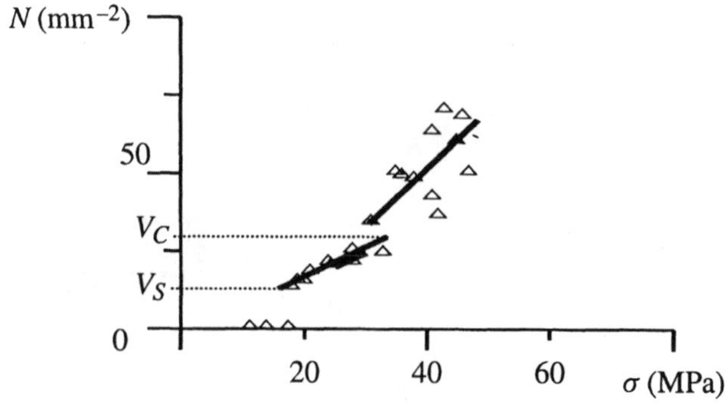


FIGURE 9. Mirror zone concentration versus applied stress.

The characteristic velocity $V_C \approx 330$m/s allowed us to estimate the peak time $t_c$ to measure the size of the mirror zone $L_C \approx 0.3$mm: $t_c = \frac{L_C}{L_V} \approx 1 \times 10^{-6}$s. This result led also to the explanation of the linear dependence of the branch length on the crack velocity [20]. Actually, since the failure time for $V > V_C$ is approximately constant ($t_f \approx t_c \approx 1\mu$s), there is a unique way to increase the crack velocity to extend the size of the process zone. The crack velocity $V$ is linked with the size of the process zone $L_{PZ}$ by the ratio $V = \frac{L_{PZ}}{t_c}$. In our experiments, the dependence of the density of the localized damage zone on the stress was observed (Figure 9). Since the branch length is limited by the size of the process zone, we obtain the linear dependence of branch length on the crack velocity. This fact explains the sharp dependence (quadratic law) of the energy dissipation on the crack velocity established in [20].

### 4.6.5. Scaling properties of failure

The scaling properties of failure stimulated great interest in the general problem of disordered media when the self-affinity of the failure surface was established in terms of the universality of the so-called roughness exponent. The self-affinity of the fracture surface was established first by Mandelbrot [32] as the existence of the power law of the distance $r$ measured within the horizontal plane for the points at which the heights $h(r)$ are measured. This defines the surface-roughness index $\zeta$ as $h(r) \propto r^{\zeta}$. Experimental data of the past decades on the measurement of the roughness exponent revealed the self-affinity of the fracture surface and it was established that the pattern of the fracture surface can be considered as scale invariant objects with the roughness index $\zeta \approx 0.8 \pm 0.05$ [33]. This fact allowed the determination of the length scales $r > r_0$ where the roughness exponent is the invariant. The range of scales $r > r_0$ for many materials shows the universal scaling properties (the roughness exponent $\zeta \approx 0.8$). For the scales $r < r_0$ the roughness exponent can change. The scale invariant property for $r > r_0$ means the transition from the roughness statistics caused by the initial structural heterogeneity (size of blocks, grains) to the statistics given by the collective properties of defects under transition from damage to fracture.

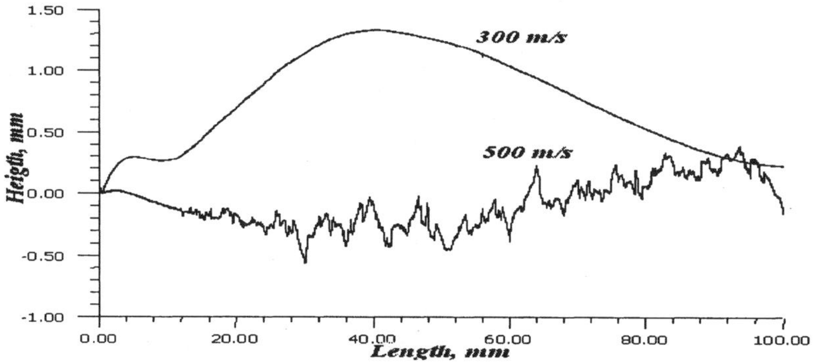


FIGURE 10. The roughness of the failure surface in PMMA.

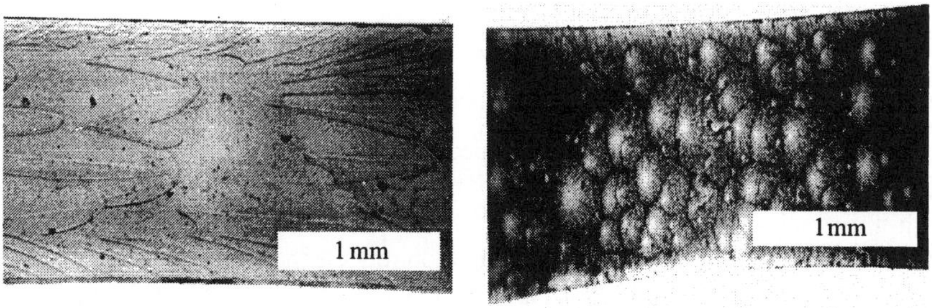


FIGURE 11. Failure surface for slow $V < V_C$ and fast $(V_B > V > V_C)$ cracks.

In our experiments the roughness profile was determined for the PMMA fracture surface (Figure 10) using the laser scanner system. Fracture surface analysis revealed the correspondence of the sharp change of the crack dynamics for the velocities $V_C$ and $V_B$ and the fractographic pattern. The fractographic image of the fracture surface was studied in the velocity range $V \sim 300 - 800$m/s when different regimes of the crack propagation were observed.

The first regime $V \sim 220-300$m/s is characterized by the mirror surface pattern (Figure 11). The increase of the crack velocity in this range leads to the characteristic pattern on the mirror surface in the form of the so-called conic markings [34]. The conic markings are the traces of the junctions of the main crack and the damage localization zones that nucleate in the process zone. These data reflect the influence of the damage kinetics on the characteristic size of the above mentioned failure structures on the surface.

The second regime appears in the velocity range 300–600m/s and the surface pattern includes numerous mirror zones.

The analysis of the roughness data in terms of the roughness exponent showed the dependence of the scaling properties on the regime of crack propagation. However,

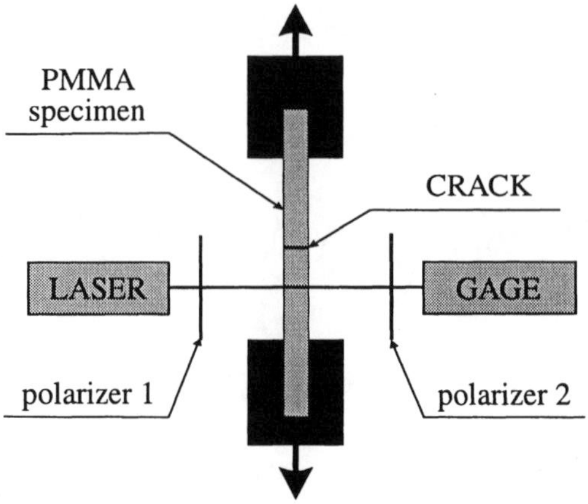


FIGURE 12. Scheme of stress phase portrait recording.

data have been collected on a group of specimens characterized by the exponent $\zeta \approx 0.8$. This fact allowed us to assume the existence of the regime of crack propagation with universal scaling index close to $\zeta \approx 0.8$. The existence of different scaling indexes for other regimes of crack propagation reflects the variety of the behavior of the investigated nonlinear system. As was shown, the crack dynamics in quasi-brittle materials is subject to two attractors. The first attractor is given by the intermediate asymptotic solution of the stress distribution at the crack tip. The self-similar solution (9) describes the blow-up damage kinetics on the set of spatial scales and determines the properties of the second attractor. This attractor controls the system behavior for $V > V_B$ when there is a range of angles with $\sigma > \sigma_c$. The universality of the roughness index can be considered also as a property of this attractor. In the transient regime $V_B > V > V_C$ the influence of two attractors can appear; here is a possible mechanism for the dispersion of experimentally measured roughness on the scale $r > r_0$.

The scaling properties of failure were studied also under the recording of the stress dynamics using the polarization scheme coupled with the laser system (Figure 12).

The temporal history of stress was measured at the marked point away from the main crack path at a fixed (4mm) distance. This allowed us to investigate the correlation property of the system using the stress phase portrait $\dot{\sigma} \sim \sigma$ for slow and fast cracks (Figure 13). These portraits display the periodic stress dynamics (Figure 13(a)) in correspondence with the local ellipticity of (15), for $\sigma < \sigma_c$ ($V < V_C$), and the stochastic dynamics (Figure 13(b)) for $V > V_C$, corresponding to the second type of the attractor. In the transient regime $V \approx V_C$ the coexistence of two attractors may appear, leading to the intermittency effect as the possible reason for the scaling index dispersion.

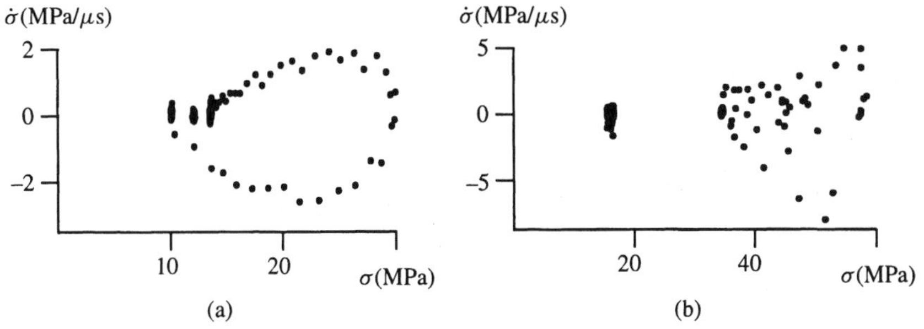


FIGURE 13. The Poincaré cross-section for the phase variables $\dot{\sigma} \sim \sigma$: (a) $V = 200$m/s. (b) $V = 615$m/s.

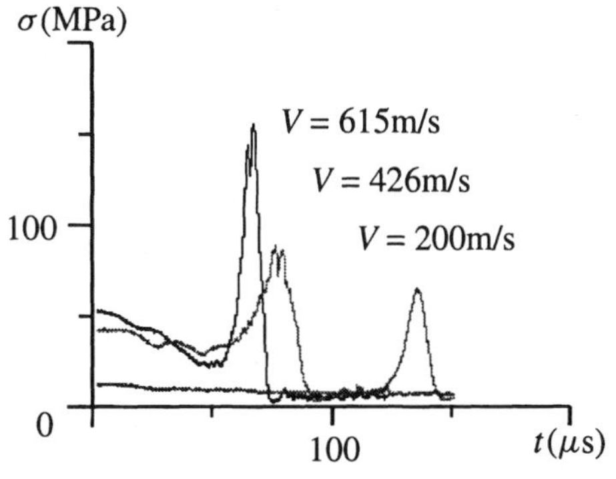


FIGURE 14. The stress history.

The recording of the temporal stress history at the marked point for $V > V_C$ revealed the appearance of finite amplitude stress fluctuations, which reflect the qualitative new structural changes in the process zone for the fast crack (Figure 14). The scaling properties, like the above attractor properties, were studied in terms of the correlation integral calculated from the stress phase pattern using the formula [35]

$$C(r) = \lim_{m\to\infty} \frac{1}{m^2} \sum_{i,j=1}^{m} H(r - |x_i - x_j|) \approx r^{\nu},$$

where $x_i$, $x_j$ are the coordinates of the points in the $\dot{\sigma} \sim \sigma$ space, and $H(\cdot)$ is the Heaviside function. The existence of the scales $r > r_0$ with the stable correlation index was established for the regimes $V < V_C$ and $V_B > V > V_C$ (Figure 15). The values of the correlation indexes in these regimes show the existence of two scaling regimes with the deterministic ($V = 200$m/s, $\nu \approx 0.8$) and stochastic ($V = 426, 613$m/s, $\nu \approx 0.4$) dynamics. The extension of the portions with constant indexes determines the scale of the process zone $L_{PZ}$. The length of the process zone increases with the growth of the crack velocity in the range $V_B > V > V_C$ with the preservation of the scaling property

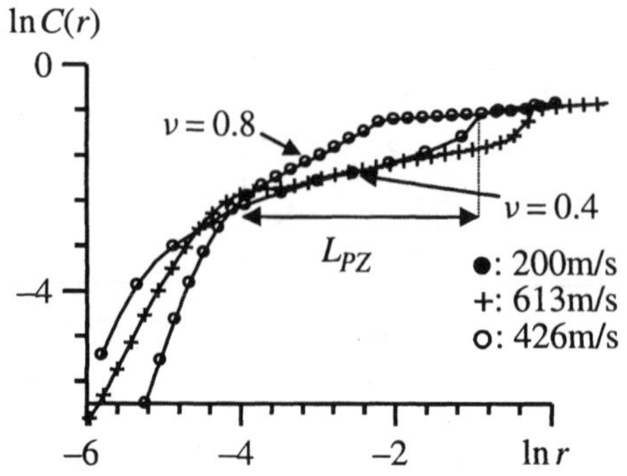


FIGURE 15. Correlation integral.

of the dynamic system. Numerical simulation of the damage kinetics in the process zone allowed us to conclude that this scaling is the consequence of the subordination of the failure kinetics to the blow-up self-similar solution that determines the collective behavior of the defect ensemble in the process zone [29, 30, 31].

### 4.6.6. Low velocity limit. Crack arrest

In this part we address the question of how a dynamic crack may approach zero velocity. This phenomenon was discussed in [21] and it was shown (considering a simplified version of the strip specimen with the radiation at the boundaries) that a steady-state velocity law with a square root behavior is expected as a function of the excess load over the Griffith load. This means that the steady-state velocity increases with an infinite slope near zero overload but in a smooth fashion otherwise. It was shown also that if the crack has no field inertia, the pass over the "trapping limit" will lead to crack movement. However, if the crack has a field inertia the crack velocity will exhibit a transient oscillation. A similar conclusion can be reached by comparing the estimation of the crack velocity given in [21],

$$V = w\sqrt{\frac{(E_e - 2\gamma)}{C(w)}},$$

where $w$ is the Barenblatt cohesive zone of the crack, $(E_e - 2\gamma)$ is the Griffith static term, and $C(w)$ is a material parameter, with the results predicted by the statistical model. Since $w$ is similar to $L_{PZ} \sim L_H$, the root term is the inverse characteristic time $t_C$ given by the self-similar solution. This fact allowed us to determine the range of the application of this generalized Griffith relation, where the crack can approach rest smoothly: $V < V_C$. For $V > V_C$, when the "wave part" of the energy increases with crack velocity, the crack arrest appears nonsmooth because the energy metastability provides "crack trapping." A similar view can be developed to analyze the crack overload above the Griffith value before any state motion.

## 4.7. Resonance excitation of failure

### 4.7.1. Delayed failure phenomenon. Failure waves

The phenomenon of failure wave in brittle materials has been the subject of intensive study during the last two decades. The term "failure wave" was introduced in [37] as the limit condition in damage evolution when the number of microshears is large enough for the determination of a front with a characteristic group velocity. This front separates the structured material from the failed area. The important feature of the failure wave phenomenon is that the velocity of failure wave doesn't depend on the velocity of propagation of the single crack having the theoretical limit equal to the Rayleigh wave velocity. The stored elastic energy in the material is the main factor, which provides the ability of a brittle solid to generate a failure wave. The high capacity of elastic energy in the material can be generated under bulk compression or under conditions of loading that lead to a state close to the bulk compression, for instance, under impact loading. High capacity of elastic energy can be realized by the removal of surface defects or structure homogenization. Brittle materials (glasses and ceramics) exhibit very high dynamic compressive strength, the Hugoniot elastic limits [38]. For glasses with high strength or high latent compressive stress self-keeping failure can be observed not only for compression but also for bending and tension.

Rasorenov et al. [39] were the first to observe the phenomenon of delayed failure behind an elastic wave in glass. Such a wave was introduced in [40] where the concept of a wave of fracture was discussed to explain the nature of the elastic limit. The existence of a failure wave was established by considering a small recompression signal in the VISAR record of the free surface velocity of K19 glass (similar to sodalime). This recompression signal resulted from a release returning after reflection of the shock at the glass rear surface reflecting again in compression at the lower impedance failure front. Recent studies have suggested that a wave of failure propagates behind the elastic wave in glass with a velocity in the range 1.5–2.5 km/s. Failure wave appeared in shocked brittle materials (glasses, ceramics) as a particular failure mode in which they lose strength behind a propagating front. Generally, interest in the failure wave phenomenon originates from the still open problem of the physical interpretation of traditionally used material characteristics such as the Hugoniot elastic limits, dynamic strength, and the relaxation mechanism of the elastic precursor.

Recent research has shown that glasses having open structures may fail in a characteristic way when loaded with plane shock waves. More filled materials such as sodalime glass have an elastic response to 6 or 7 GPa and plastic behavior beyond. The experiments in [41] confirmed the existence of these waves and extended measurements to glass states behind the shock (but ahead of the failure front) and behind the failure wave using manganin stress gauges. The tensile (spall) strength of the glass remained high behind the shock but dropped to zero behind the failure front. Additionally, the shear strength of the material dropped from a high value ahead, becoming lower behind the failure front. The experimental study established the sharp light opacity behind a front under high speed videocamera recording [42], reduction in acoustic impedance, and lowered sound speed [39]. The material states behind and ahead of the moving boundary (behind the shock) allowed the measurement of failure wave speed [43]. Comparison of these data with high speed framing in sodalime and borosilicate (Pyrex) glasses [38, 44] revealed the constant value of the failure wave speed.

Qualitative changes in silicate glasses behind the failure wave through the measure of the refractive index (refractive index increased) allowed Gibbons and Ahrens [45] to qualify this effect as a structural phase transformation. These results stimulated Clifton [46] to propose the phenomenological model in which the failure front was assumed to be a propagating phase boundary. According to this model the mechanism of failure wave nucleation and propagation results from local densification followed by shear failure around inhomogeneities triggered by the shock.

The description of the failure wave phenomenon as a consequence of the generation of collective burst modes of mesodefects was proposed in [47, 48, 49] in the course of a study of nonequilibrium transitions in defect ensembles. It was shown that a self-similar solution for the microshear density tensor exists that describes the qualitative changes in the microshear density kinetics during nonequilibrium transitions. The failure waves represent the specific dissipative structures (the "blow-up" dissipative structures) in the microshear ensemble that could be excited due to the transit of the elastic wave.

### 4.7.2. Self-similar solution

Equation (15) describes the characteristic stages of damage evolution. In the range of stress $\sigma < \sigma_c$ and defect density $p < p_c$ the damage kinetics is subject to the "thermodynamic branch" *oa* corresponding to the local minimum of the free energy (Figure 1). When the stress approaches the critical value $\sigma_c$ ($p \to p_c$) the properties of the kinetic equation (15) change qualitatively (from elliptic to parabolic) and the damage kinetics is subject to specific spatial-temporal structures that appear in the defect ensemble due to the interaction between defects [12]. These structures describe damage localization and nucleation of failure hotspots.

The spatial-temporal structures are given by the self-similar solution (20) of the kinetic equation (15) when passing the critical point $p_c$. The substitution of (20) into (14) leads to the equation for $f(\xi)$

$$-\frac{1}{\beta-1}f - \frac{\beta-(\omega+1)}{2(\beta-1)}\xi\frac{df}{d\xi} = \frac{d}{d\xi}\left[f^{\omega}\frac{df}{d\xi}\right] + f^{\beta}. \tag{25}$$

The conditions for $f$ at the front $\xi_f$ of the dissipative structure and the symmetry conditions at the center of the structure are

$$\xi = \xi_f: \quad f = 0, \quad f^{\omega}\frac{df}{d\xi} = 0, \tag{26}$$

$$\xi = 0: \quad f^{\omega}\frac{df}{d\xi} = 0. \tag{27}$$

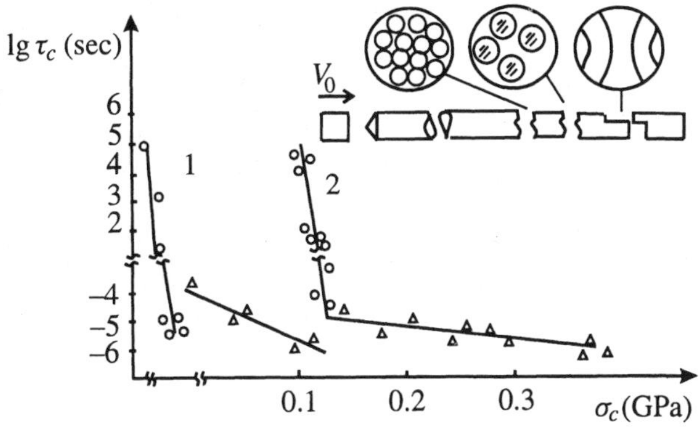


FIGURE 16. Fracture time $t_f$ for (1) a shocked rod of PMMA and (2) ultraporcelain versus stress amplitude $\sigma_a$. Inset: Surface pattern with mirror zones in different spall cross-sections.

We assume that the defect density in the dissipative structure exceeds essentially this density outside the failure hotspot. The difference in the number of conditions (26) and (27) and the differential power of (25) means that the solution exists only for specific values of $\xi_f = \bar{\xi}_f$, i.e., the eigenvalue problem for $\xi_f$ arises. A method for the solution of this problem was developed in [13], and it allowed the estimation of $\bar{\xi}_f$ and the definition of failure front propagation

$$x_f = \bar{\xi}_f \chi_0^{\frac{1}{2}} S^{-\frac{\omega}{2(\beta-1)}} t^{\frac{\beta-\omega+1}{2(\beta-1)}}. \tag{28}$$

Equation (28) determines three self-similar regimes depending on the relations between the parameters of the nonlinear medium. If the material properties and the stress levels provide the damage localization kinetics with parameters $\beta > \omega + 1$ the failure wave front will propagate with a group velocity given by the solution (28).

### 4.7.3. Some experimental results

The blow-up kinetics of damage localization allowed us to link the hotspots of failure with the above mentioned self-similar structures. The correspondence of these hotspots having the image of mirror zones to the above self-similar structures was confirmed experimentally in [50, 51]. The multiple mirror zones with equal sizes were excited on different spall cross sections in the shocked PMMA cylindrical rod when the stress wave amplitude exceeded some critical value corresponding to the transition to the so-called "dynamic branch" under spalling (Figure 16). The constant size of damage localization corresponds to the damage kinetics (free energy release rate) given by the following relationship between nonlinearity parameters $\beta \approx \omega + 1$.

The "dynamic branch" corresponds to the stress $\sigma > \sigma_c$, where the failure scenario is determined by the generation of the collective mode in the defect ensemble in the form of a dissipative structure with blow-up damage kinetics. The self-keeping features of failure corresponding to the conditions of failure waves were also observed

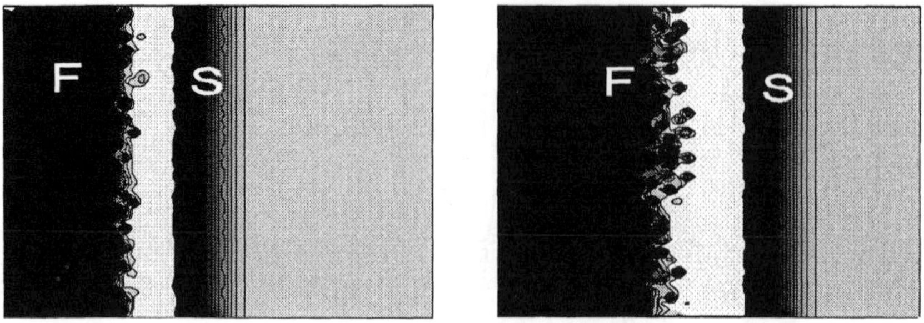


FIGURE 17. Propagation of stress (S) and failure (F) waves.

in the conditions of dynamic crack propagation. The framing of crack dynamics in the preloaded PMMA plane specimen established the existence of the transient velocity $V_B \approx 600$m/s, when the branches behave autonomously (Figure 8). The low rise of velocity for $\sigma > 60$MPa reflects the stress independent character of failure similar to the dynamic branch in spall fracture. The pictures of stress distribution at the crack tip is shown in Figure 11 for slow ($V < V_C$), fast ($V > V_C$), and branching ($V > V_B$) cracks.

### 4.7.4. Simulation of failure waves

The study of failure wave initiation and propagation was carried out on the basis of the constitutive equation (28) coupled with the momentum transfer equation. The defect density tensor in the compression stress wave represents the microshear density.

The system of equations was solved using an original finite element code. The simulation confirmed the delayed propagation of the failure front behind the stress wave (Figure 17). We observed also that the shear stress vanishes in the failure wave. The failure wave propagation leads to qualitative change in the transverse stress when this stress is approaching the longitudinal stress level (Figure 18).

### 4.7.5. Statistics of fragmentation

Qualitative new features in crack dynamics (transition from the steady-state branching regimes and the fragmentation) such as the phenomenon of delayed failure revealed specific features of dynamic failure that are properties of nonequilibrium systems with nonlinear and stochastic behavior caused by the collective effects in the defect ensemble. This fact allows us to develop a view on the dynamic fracture statistics (fragmentation problem) that links the nonlinear dynamic aspects of damage evolution with the change of symmetry properties in the system due to the generation of collective modes in the defect ensemble.

Some basic theories have emerged within the past 10 years for predicting the consequences of dynamic fragmentation induced by high velocity impact or explosive events. These theories have focused on the prediction of mean fragment size through energy and momentum balance principles [52, 55] and on statistical issues of fragment

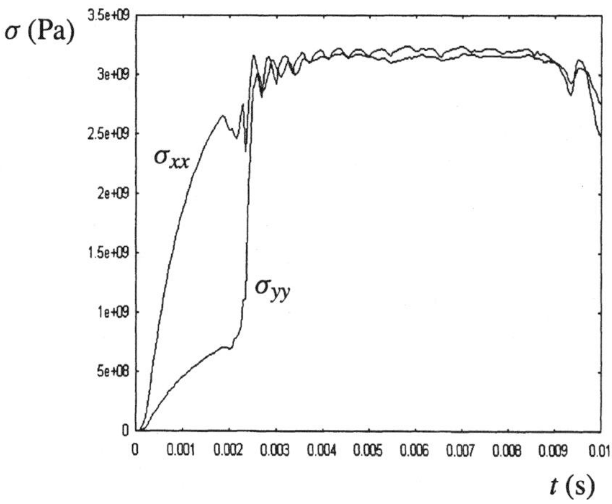


FIGURE 18. Longitudinal $\sigma_{xx}$ and transverse $\sigma_{yy}$ stress kinetics.

size distribution [56, 57]. However, there are a number of unresolved issues within the development of statistical energy balance theories of fragmentation.

Dynamic fragmentation was analyzed in [58] on the basis of local energy inequality and minimum fracture time requirement. This theory was later modified in [54] to include the kinetic energy available for fragmentation. However, the theories are not complete because the evolution process that leads to catastrophic failure of the materials was not considered.

Taking into account the dramatic changes in the scaling properties of the nonlinear system "solid with defects" due to the re-subjection of the system behavior to the collective modes of defects, the following scenario of the dynamic fragmentation can be discussed. The existence of characteristic stages of dynamic fracture under crack propagation (Figure 8) allowed us to establish the correlation of the dynamic fragmentation statistics with the energy density imparted on the material. The energy density $E < E_C$ ($E_C$ corresponds to the critical velocity $V_C$ of the steady-state branching transition) provides the stress intensity controlled failure scenario. The transient densities $E_B > E > E_C$ ($V_C < V < V_B$) lead to the fracture statistics that are sensitive to both self-similar solutions: the self-similar stress distribution at the crack tip and collective blow-up modes of damage localization. The intermittency effect, when the system reveals the complex statistics in the presence of two attractors, probably, has a phenomenological interpretation in the framework of Weibull statistics. Taking into account the theoretically predicted low limit of damage localization scale $L_C$, it can be assumed that a critical energy density exists that provides the limit size of fragmented structure close to $L_C$. Since these energy densities can be imparted in materials generally by shock, this homogeneous fragmentation can be realized because of failure wave excitation. The experiments in [59] with large specimens (optical glass K5 impacted by blunt steel cylinder) visualized the several fracture nucleations in the form of failure waves. Several spherical or nearly spherical waves were formed ahead of the main front.

## 4.8. Plasticity as metastable orientation transition in microshear ensemble

### 4.8.1. Defect induced mechanisms of momentum transfer

The physical particularity of momentum transfer under plastic flow is the motion of the deformation carriers (microshears in our case) in the conservative (elastic) field. This fact reflects the principal difference of irreversible deformation caused by the dislocation structure rearrangement from conventionally known mechanisms of viscous flow in liquids, which occurs due to momentum diffusion.

Much progress was achieved during the last decades in the understanding of the mechanisms of plastic flow and the phenomenology of plasticity for the large range of the stress and strain rate intensities. However, the attempt to use the structural aspects in the formulation of the plasticity constitutive equations did not allow the explanation of most main questions related to the specific nature of deformation carriers responsible for the plastic flow. Still unresolved problems are the regularities of the strain localization due to shear banding, the linkage of the structure evolution with deformation hardening, and the explanation of the structure of the plastic wave front in shocked materials.

The experimental data concerning the transitions in dislocation substructures under the plastic flow and the results of the statistical description of collective behavior of microshear ensemble allowed one to link plasticity rules with structure evolution. It was observed in Section 4.2 that the intermediate range of structural heterogeneity (for instance, grain size) corresponds to values of the structure sensitive parameter $\delta_c < \delta < \delta_*$ and provides the deformation response of solids caused by the generation of collective modes of the microshear ensemble with pronounced orientation. The formation of these modes is the consequence of the orientation instability in the microshear ensemble within the corresponding range of external stresses $\sigma_y^a < \sigma_y < \sigma_{\text{HEL}}$ for the given structural parameter $\delta$ (Figure 19). These stresses determine the range of the yield stresses $(\sigma_y^a, \sigma_{\text{HEL}})$, where the orientation metastability occurs. The yield stresses from this range represent the sum of the stresses $\sigma_y^a$ (where orientation metastability can appear). This stress has an athermal nature and determines the initial point of availability of the coherent microshear behavior as a qualitative new mechanism of stress relaxation. Part of the yield stress $(\sigma_y^m - \sigma_y^a)$ determines the additional stress, which provides the thermally activated transition $A \to D$ in the dislocation substructure in the more orientated state $D$.

The jump in the microshear density parameter $\Delta p_{AD}$ leads (according to (16)) to the sharp drop of elastic modulus, which precedes the plastic flow. The possible scenario of the microshear ensemble evolution at the point $D$ is shown in Figure 19 and can involve two opportunities. The first one, along the $Bd$ trajectory, here supposes high deformation hardening. The second scenario (along the trajectories $DHE$ or

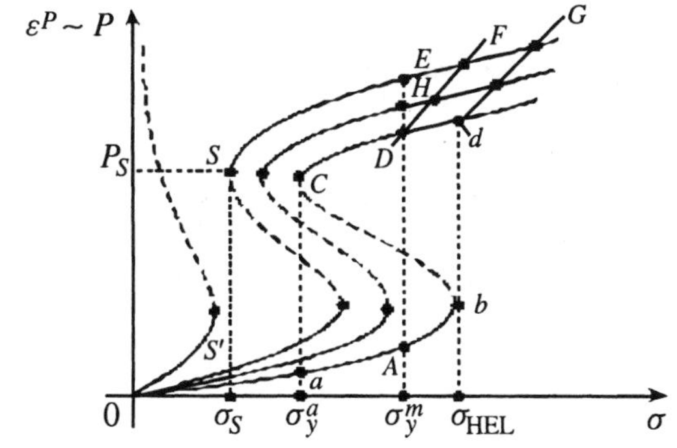


FIGURE 19. Metastable orientation transition in microshear ensemble.

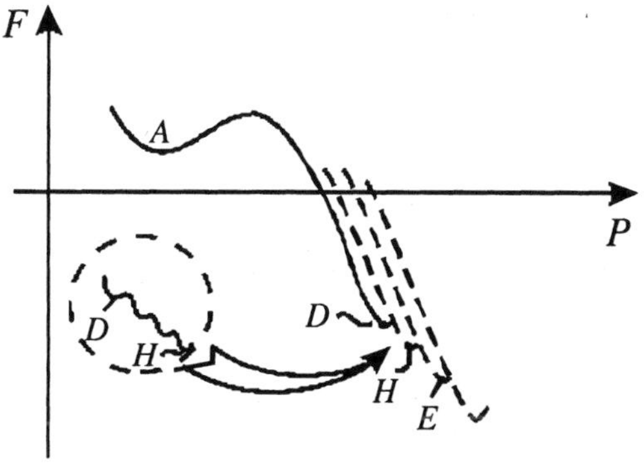


FIGURE 20. Free energy "transitions."

$DF$) presumes an increase of deformation caused by the defects in the continuous orientation ordering in the course of metastable transitions due to the change of value of the structure sensitive parameter $\delta$. Two realizations of this scenario ($DE$ and $DF$) correspond to the typical plastic deformation of solids that are conventionally known as perfect plasticity ($DE$) and plasticity with a hardening ($DF$).

This type of plastic flow is described in the framework of the incremental plasticity theory, when the increment of plastic deformation $d\varepsilon_{ik}^p (= dp_{ik})$ is determined by the so-called normality rule

$$d\varepsilon_{ik}^p = d\lambda \frac{\partial \Phi}{\partial \sigma_{ik}}, \tag{29}$$

where $\Phi$ is the plastic potential and $\lambda$ is the multiplier. The incremental law of plastic flow determines the plastic strain decrement as the result of the "expansion" of the yield surface, given by the plastic potential $\Phi$.

The consideration of the plastic flow as nonequilibrium orientation transition led to the interpretation of the phenomenological background of the incremental plasticity. According to this result the driving force of the plastic deformation is the minimization of the free energy $F(\sigma, p, \delta)$ in the current system state, for instance, $(\sigma_D, p_D, \delta_D)$, under the continuous ordering of defects $(D \to H \to \cdots)$ and the formation of different dislocation substructures. The description of these transitions can be obtained as the generalization of (14) considering $\delta$ as the independent variable

$$\delta F = \frac{\partial F}{\partial p} dp + \frac{\partial F}{\partial \delta} d\delta \leq 0. \tag{30}$$

It is evident that the increment $d\delta$ plays the role of the increment $d\lambda$ and provides the change of the plastic flow potential in term of $\delta$. For the quasi-static loading the relations (30) allowed the expression of the increment $d\delta$ in the term of current values of stress and defect induced plastic strain $p$. The unloading, for instance, from point $F$ occurs along the trajectory $FES$ in the "partly unloaded state."

Kinetics of the defect density and the scaling parameter can be described with the usage of the evolution inequality in the form

$$\frac{\delta F}{\delta t} = \frac{\partial F}{\partial p}\frac{\partial p}{\partial t} + \frac{\partial F}{\partial \delta}\frac{\partial \delta}{\partial t} \leq 0. \tag{31}$$

The analogue of $\sigma_y^a$ under unloading is the threshold stress $\sigma_S$. Taking into account that the plastic flow realizes as continuous orientation ordering of dislocation substructures with the scaling parameter $\delta$ related to the size of the dislocation substructure $l_n$ and the distance between ones $l_c$, the real scenario of the plastic straining depends on the loading conditions, for instance, the stress or the strain rate.

The strain rate sensitivity of plastic flow is observed typically at strain rates exceeding $10^3\text{s}^{-1}$. It is important to use in this range of strain rate $\dot{\varepsilon} > 10^3\text{s}^{-1}$ the description of independent kinetics of defect induced plastic straining and the structure evolution in the form that follows from (31),

$$\frac{dp}{dt} = -L_p \frac{\partial F}{\partial p}, \tag{32}$$

$$\frac{d\delta}{dt} = -L_\delta \frac{\partial F}{\partial \delta}, \tag{33}$$

where $L_p$ and $L_\delta$ are the kinetic coefficients. The relations between the kinetics of the external loading (for instance, imposed stress or strain rate), the kinetics of transition in the $p$ metastability area, and the kinetics of structure sensitive parameter $\delta$ determine the variety of the material behavior for high strain rate loading and the response of shocked materials.

The kinetics of $\delta$-parameter determines the hardening law due to the formation of new dislocation substructures responsible for the momentum transfer in more high structural level that is realized as the rescaling of initial structural scales $l_n \to L_n(p)$, $l_c \to L_c(p)$.

Experimental data concerning the variation of the flow stress with a strain rate established the dramatic increase of the flow stress at the strain rate $\dot{\varepsilon} > 10^3 - 10^4\text{s}^{-1}$.

The description of the plastic flow as a metastable orientation transition in the defect ensemble allowed us to propose the explanation of the anomalous hardening. Equation (34) can be represented in the form

$$\frac{d\delta}{dt} = -\frac{1}{\tau_\delta}\frac{\partial \tilde{F}}{\partial \delta}, \tag{34}$$

where $\bar{F} = \frac{F}{A}$, $\tau_\delta = (L_\delta A)^{-1}$ is the characteristic time of structure rearrangement in terms of $\delta$. The anomalous hardening for the strain rate $\dot{\varepsilon} > 10^3 - 10^4 s^{-1}$ is the consequence of limiting rate ability of materials to the dislocation substructure rearrangement with characteristic time $\tau_\delta \approx 10^{-5}$s. For strain rates exceeding $\dot{\varepsilon} > 10^4 s^{-1}$, the plasticity, as a metastable transition to the dislocation substructure with more pronounced defect orientation on large spatial scales, could not be realized and defect ensemble evolution could occur along the branch $CDD'd$ with a constant value of $\delta$ and maximal hardening.

## 4.9. Kinetics of metastable transitions and mechanical responses of shocked condensed matter

### 4.9.1. Structure of stress wave in solid under impact loading

Complicate material responses to the increase of the strain rate are observed in the nonlinear behavior of deformation and in the changes of yield stresses of plasticity and strength. The attempts were undertaken in [47, 50, 51] to establish a link of the mesodefect evolution (microcracks, microshears) with relaxation properties and failure kinetics. A statistical approach allowed us to establish specific features of the defect ensemble evolution caused by the initial solid state (structural heterogeneity in terms of $\delta$) and the interaction between defects. The scenario of defect ensemble evolution has the form of nonequilibrium kinetic transitions, which appear as a specific form of self-similarity. This self-similarity is particularly clear at plastic instability and damage localization under dynamic loading. The self-similarity in the behavior of a solid loaded dynamically is caused by the excitation of spatial-time structures in the defect ensemble. The appearance of these structures is accompanied by the qualitative change of solid responses to dynamic loading.

The established correspondence of the nonlinear behavior of the defect ensemble and the structure of the shock wave profile leads to the explanation of some regularity of shocked material responses.

One-dimensional plane shock waves have been widely used to study the mechanical response of solids to high velocity deformation. For many materials under a certain range of impact pressures, there exist two-wave structures of which the first wave, the so-called elastic precursor, travels at the velocity of sound waves, while the second wave, the plastic shock wave, travels at a slower speed that increases with impact pressure. While the full two-wave structure is thus not steady (does not propagate without change of form), each component wave may be treated as steady after sufficient propagation distance. The steady-wave profile propagates without change in shape.

The stress wave structure was examined in [60] for the impact loading of an aluminum plate loaded by the quartz disk at the rate of 400m/s. In the case of the plane wave propagation in the $z$ direction the system of constitutive equations coupled with conservation laws of mass and impulse are written as

$$\frac{\partial p'_{zz}}{\partial t} = -L'_p \frac{\partial F'}{\partial p'_{zz}}, \qquad \frac{\partial p_0}{\partial t} = -L_p^0 \frac{\partial F^0}{\partial p_0}, \tag{35}$$

$$\dot{\varepsilon}_{zz} = \dot{\varepsilon}^e_{zz} + \dot{p}_{zz}, \qquad \dot{\varepsilon}^e_{zz} = \frac{\dot{\sigma}_{zz}}{E}, \tag{36}$$

$$\rho \frac{\partial^2 u_z}{\partial t^2} = \frac{\partial \sigma_{zz}}{\partial z}, \tag{37}$$

$$\varepsilon_{zz} = \frac{\partial u_z}{\partial z}, \tag{38}$$

where $p'_{zz}$ and $p_0$ are deviatoric and isotropic parts of $p_{zz}$, $F = F'(p'_{zz}, \sigma'_{zz}) + F^0(p_0, \sigma_0)$, $\rho$ is the material density, and $L'_p$ and $L^0_p$ are the kinetic coefficients. Only the $p'_{zz}$ component is nonnull in compression waves.

The solution of system (35)–(38) has to satisfy the boundary and initial conditions

$$\sigma_{zz}(0,t) = \sigma_0(t), \qquad \sigma_{zz}(h,t) = 0,$$
$$v_z(z,0) = \sigma(z,0) = p_{zz}(z,0) = 0, \qquad \rho(z,0) = \rho_0,$$

where $h$ is the plate thickness, and$\sigma_0(t)$ was determined on the basis of the solution of the collision problem. Material parameters have been determined from quasi-static data of aluminum testing (uniaxial tension) using direct methods of the registration of the microcrack accumulation. The parameters were taken to be $\rho = 2.71 \times 10^3$kg/m$^3$, $G = 109.7$GPa, $\tau_p = L_p/G = 2.1 \times 10^{-6}$s, $\tau_l = h/C_l = 1.96 \times 10^{-6}$s$(C_l = (G/\rho)^{1/2})$. Results of a numerical simulation of the stress wave propagation and the time dependence of stress and microcrack density parameter in the spall section are presented in Figure 21. In the stress area corresponding approximately to the dynamic yield stress the orientation kinetic transition for the parameter $p_{zz}$ occurs and results in the abrupt increase in stress relaxation tempo, a change in the plastic wave profile and the separation of the elastic precursor (bold parts of the curves in Figure 21).

A sharp transition to the highly ordered structure due to the metastable transition may lead to the behavior that has been commonly referred to as dynamic failure due to plastic shear instability. Spatial scales of the orientation area (shear localization zone) are determined by the parameters of the solitary wave of shear instability given by the solution (18).

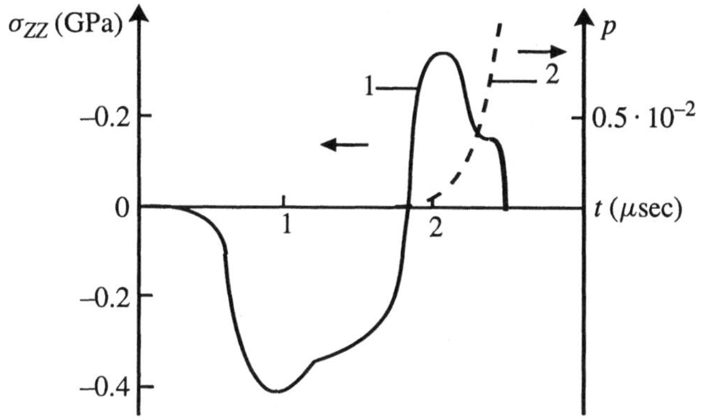


FIGURE 21. Structure of stress waves and damage kinetics in spall cross section of aluminum target.

### 4.9.2. Strain rate universality at the steady-state plastic wave

Experimental evidence of viscous-like effects in solids was lacking until the results of Sakharov's group [61] when the measure of shock viscosity in solids was inferred using an indirect method. The existence of steady-wave shocks as significant features of stress waves in solids is treated in [62, 63]. With the development of laser interferometry techniques the first direct measurements of high resolution wave profiles were provided in [64] for aluminum. The observation of the shock wave rise showed the very rapid increase in steady-wave strain rate with increasing peak stress. This increase is substantially greater than would be predicted by a simple Newtonian proportionality between viscous stress and strain rate. The existence of steady-wave solutions for the Navier–Stokes equations was recognized first by Rayleigh and Taylor [65, 66]. Since then a large literature on this subject in the field of fluids has emerged [67, 68].

The unique feature of the large amplitude wave profile is the steady-wave plastic shock profile. The steady-wave profile propagates without change in form and, as pointed out in [69], is a consequence of a stable balance between the competing processes of stress–strain nonlinearity and dissipative or viscous material behavior.

According to the results presented in Section 4.5, the steady-wave profile for different shock amplitudes appears as that of the self-similar auto-solitary wave induced by the collective orientation mode in the microshear ensemble. The rescaling of current values of $\delta$ ($DF$-path, Figure 19) provides the self-similar auto-solitary response for the variation of the stress amplitude.

The rate of the transition $\dot{p}_{zz}$ from the lower to upper branches reaches the maximum under the deepest penetration into the metastability area (point $b$ in Figure 19). This is assumed to be a main reason for the universality of the plastic strain rate dependence on the stress amplitude $\dot{p}_{zz} \approx A\sigma_{\text{amp}}^4$ established by Barker and Grady for a wide class of materials for strain rates $\dot{\varepsilon} > 10^5\text{s}^{-1}$. This relation follows from the self-similar solution (18), when the driving force of the transition under the $\delta$ rescaling can be represented in terms of the stress as the difference in the free energy for the current level along the $dG$-path (Figure 19) and the energy for metastability limits (the analogue point $b$, Figure 19) for the curves corresponding to the current values of $\delta$.

This result reflects the importance of the collective effects (orientation transition) in the system of defects providing the intriguing material independent feature in the steady-shock process and reveals the specific character of the solid viscous response. On the other hand the universality of this fourth-power dependence for a large class of materials leads to the assumption that in the intensive shock environment the complex structural processes (which are involved due to the plastic flow) become simpler. The study of these general mechanisms seems to be important for the understanding of the collective effects in the defect ensembles responsible for the irreversible deformation in solids.

### 4.9.3. Relaxation properties and defect-induced instabilities in shocked liquids

The problem of the true structure of liquids can not at present be considered as finally explained. A dependence of mechanical properties on the relation between the relaxation time of the medium and the characteristic time of the loading is not only a property of solids but is also valid for liquids. All theories of viscous liquid flow start out to some extent with an assumption about the nature of the structure of the liquid. For this reason investigation of the relaxation properties of a liquid under high strain rate loading produced by shock waves, is of great scientific and applied interest. The viscosity of condensed matter plays the principal role in effecting the irreversible shock compression. It is well known that, if account is not taken of viscosity, one can not produce a continuous distribution of all thermodynamic quantities in the shock wave front (SWF).

The method of investigating the viscosity of materials behind the SWF, proposed first by Sakharov [61], is based on the experimental study of small perturbations on the SWF and then by Barker [64] with the use of the Doppler interferometry (VISAR) technique.

A still unresolved puzzle arose in Sakharov's experiments when the shocked liquid (water and mercury) and solid (aluminum and lead) revealed the same viscosity $\eta = 10^4$Pz at approximately the same values of the strain rate $\dot{\varepsilon} = 10^5\mathrm{s}^{-1}$ at the SWF. It is appropriate to emphasize that these substances are very different under normal conditions. Nevertheless, for the pressure $P \approx$ 80–100KBar water and mercury are characterized by the same dynamic viscosity as that of an unmelted solid (pressure range $P \approx$ 400–500KBar) behind SWF. One is led to the conclusion that the mentioned liquids in the pressure range $P \approx$ 80–100KBar have a relaxation time $\tau > \dot{\varepsilon}^{-1} \sim 10^{-5}$s differing by six orders of magnitude from the molecular (diffusion) relaxation times estimated by the Einstein formula as $\tau_D = \frac{\Delta^2}{6D_{sd}} \sim 10^{-11}$s, where $\Delta$ is the distance between the particles and $D_{sd}$ is the self-diffusion coefficient. It means that a liquid behaves in the quoted experiments effectively like a solid.

The physical mechanisms leading to the development of instabilities in condensed media indicate the possibility of describing instabilities in liquids on the basis of an analysis of the kinetics of fluctuations, if the latter are viewed as defects in the structure of liquids [12, 70]. In the case of liquids the mesoscopic defects, being by their nature similar to the fluctuation of the displacement field in solids, can be also regarded as real structural defects that are produced during collective motion of groups of molecules. This mechanism of motion does not correspond to the conventional diffusion mechanism of momentum transfer in simple liquids.

The attempt to explain the noncontradictory nature of the viscosity of shock compressed liquid and solid for the strain rate $\dot{\varepsilon} = 10^5 s^{-1}$ was undertaken in [61] using the Fraenkel's idea concerning the role of defects ("holes") produced by the propagating SWF. The validity of this view was qualitatively confirmed by the measure of the electric conductivity of water behind the SWF, when a dramatic change of the resistivity was found.

Apparently, Fraenkel [71] was the first to call attention to the analogy in the mechanisms of flow of solids and liquids, noting that "... x-ray diffraction pictures of liquids are similar to those of microcrystalline solids, and it would be possible to interpret their general features on the basis of the idea that a liquid consists of a large number of randomly oriented submicroscopic-size crystals" and "... the widely held view that the fluidity of liquids is caused by the absence of shear elasticity, i.e., by a zero shear modulus ... is incorrect (except, possibly, for the case of liquid helium II)." These statements are confirmed in [72] by measurements of relaxation spectra in a shear simple liquid flow accompanying the superposition of shear oscillations in a liquid at frequencies $10^5$Hz when the existence of shear elasticity was established. The explanation of this long relaxation time anomaly is linked in [72] with a coordinated displacement and reorientation of groups of molecules which involves longer times. A coordinated displacement of groups of molecules (similarly to the relative slipping of blocks or grains in solids) can be realized during nucleation of mesoscopic defects arising between these groups of molecules.

The dissipative function for a medium relaxing by flow and as a result of the development of defects has the form [12]

$$TP_s = -\frac{1}{T} q_k \nabla_k T + \sigma_{ik} e_{ik}^v - \frac{\delta F}{\delta p_{ik}} \dot{p}_{ik} \geq 0, \tag{39}$$

where T is the temperature, $q_k$ is heat flux, $\delta F/\delta p_{ik}$ is the thermodynamic force acting on a system when $p_{ik}$ is different from its equilibrium value, and $e_{ik}^v = e_{ik} - \dot{p}_{ik}$ is the "viscous" component of the strain rate tensor. The condition that the dissipative function be positive definite leads to a system of equations for the tensor variables:

$$\sigma_{ik} = \eta e_{ik}^v + \chi \dot{p}_{ik}, \tag{40}$$

$$-\frac{\delta F}{\delta p_{ik}} = -\chi e_{ik}^v + \varsigma \dot{p}_{ik}, \tag{41}$$

where $\eta$, $\chi$, and $\varsigma$ are kinetic coefficients. For the case of simple shear, from (40) follows an expression for the effective viscosity

$$\eta_{im} = \frac{\sigma_{xz}}{e_{xz}} = \eta - (\eta - \chi)\frac{\dot{p}_{xz}}{\dot{\varepsilon}_{xz}}. \tag{42}$$

In the interval $\delta_c < \delta < \delta_*$, the strain rate fluctuations "conform" to the spectrum of solitary waves, meaning that $\dot{\varepsilon}_{ik} \approx \dot{p}_{ik}$, and (42) gives the asymptotic viscosity $\eta = \chi$. The independence of viscosity of condensed matter $\eta \approx 10^4$Pz for the strain rates $\dot{\varepsilon} \approx 10^4 - 10^6 s^{-1}$ is the consequence of the dependence of the deformation response on the strain rates induced by the defect kinetics in the form of finite amplitude solitary wave disturbances.

The following scenario of the development of instabilities in a liquid is possible in accordance with the characteristic features of the nonlinear behavior of ensembles of defects.

The weak periodic pulsations of the velocity caused by defect-induced fluctuations of the strain rates appear in the region $\delta > \delta_*$. The flow intensification can lead to the transition through the bifurcation point $\delta_*$ (the region $\delta_c < \delta < \delta_*$) and the nucleation of the cascade of solitary waves as a result of orientation transition in an ensemble of microscopic shears. The inertial cascade of the solitary waves is generated under the rescaling of $\delta$ due to the appearance of new spatial scales linked with the solitary wave front. The second inertial interval appears during the transition for $\delta = \delta_c$, when dissipative structures with blow-up kinetics of the defect-induced strain rate are excited.

It is interesting to compare the real picture of turbulent flow with the scenario given by the dynamics of $p_{ik}$. The dynamics of the development of turbulence has been investigated in [73] in an analysis of the evolution of a turbulent spot in a Poiseuille flow with Reynolds number $R \sim 840$–$1500$. It is observed that the turbulent spot has the form of a triangular wing, and the turbulent motion arises inside a region on the boundary of which disturbances in the form of solitary waves arise. These waves, generated on the boundary of the spot, propagate into the laminar zone and transform into quasi-periodic damped pulsations of the velocity. The motion of solitary waves in the direction of the spot results in secondary instability and growth of the spot. This experimentally observed scenario agrees with the nonlinear dynamics of macroscopic defect-induced strain rate fluctuations. The laws established for the development of instabilities in condensed media suggest the possible existence of a bifurcation tree, leading to turbulence [74], that can be described on the basis of a kinetic approach [75].

## 4.10. Summary

A multifield statistical approach is presented and developed for the study of the multiscale nature of plastic deformation and failure. The tensor order parameter, the defect density tensor, is introduced for typical mesoscopic defects, microcracks and microshears, which are the structural images of dislocation substructures at different material scales.

This statistical model leads to the existence of a dimensionless structural parameter of scaling, which provides the link between the continuous growth of the defect density under deformation and the rescaling of the structural level involved in the successive mechanism of momentum transfer due to the collective motion of defects. This parameter represents the ratio of the current scale of dislocation substructure and the distance between substructures. The statistical model reveals the variety in behavior of ensembles of mesoscopic defects caused by the interaction of defects among themselves and with an external field. The collective properties of defects due to their interaction play a most important role in the formation of large-scale mesodefect substructures and finally in the mechanisms of plastic flow and failure.

Large-scale mesodefect substructures are generated as orientation transitions in the low scale mesodefect substructure. This transition occurs during the excitation in the material of the auto-solitary strain waves with a front length corresponding to the orientation transition area. The important consequence of this result is that the plastic flow develops as the continuous orientation ordering of dislocation substructures. The driving force of this transition is the free energy release under conditions of growth of the defect density and the rescaling of the structural level of the material, which provides the optimal mechanism of momentum transfer.

The orientation transitions in mesodefect ensembles and the nucleation of spatial areas with pronounced orientation of defects can be considered as mechanisms of adiabatic shear bands and of strain localization. The spatial distribution of the initial susceptibility in terms of the above scaling parameter can lead to the multiple generation of the set of collective modes (auto-solitary strain waves) in the bulk of the specimen. These modes can be considered as new phase variables that determine the material behavior and decrease the system symmetry. The irregular character of plastic deformation (Henky flow) is the consequence of the stochastic dynamics due to the generation and the interaction of auto-solitary collective modes in the mesodefect system.

The important result of statistical multifield theory is the determination of the type of bifurcation that leads to qualitative change for the evolution equation of the defect density tensor during transition from plastic mechanism of structural relaxation to the damage localization and failure. This bifurcation type arises for a critical value of scaling parameter and leads to the generation of the collective modes, i.e., the dissipative structures with blow-up kinetics of the defect density growth on the corresponding spatial scales. The set of these collective modes is bounded with the coordinates of the second attractor, which defines another multiscale nature of damage-failure transition.

The collective mesoscopic modes of dislocation substructures (auto-solitary orientation waves and blow-up dissipative structures) have the meaning of self-similar solutions–eigenfunction spectrum of corresponding nonlinear problems. The loading condition can provide the resonance excitation of this type of collective modes, which may lead to the anomalous deformation responses in the form of specific types of strain rate localization, failure waves, and self-similar structures of wave fronts in shocked condensed matter.

**Acknowledgments**

This work was supported in part by Russian Foundation of Basic Research project 02-01-0736 and ISTC projects 1181 and 2146. The author thanks Sergey Uvarov, Mikhail Sokovikov, Vladimir Leont'ev, Oleg Plekhov, Vladimir Barannikov, and Marina Davydova for collaboration and useful discussions.

## References

[1] N. A. Koneva, S. P. Lychagin, L. T. Trishkona, and E. V. Kozlov, in *Strength of Metals and Alloys: Proceedings of the 7th International Conference, Montreal, Canada*, Vol. 1, 1985, 21.

[2] N. Hansen and D. Kuhlmann-Wilsdorf, *Materials Sci. Engrg.*, **81** (1986), 141.

[3] V. I. Betechtin, O. B. Naimark, and V. V. Silbershmidt, in *Proceedings of the International Conference on Fracture (ICF* 7), Vol. 6, 1989, 38.

[4] V. I. Betechtin and V. I. Vladimirov, in S. N. Zhurkov, ed., *Problems of Strength and Plasticity of Solids*, Nauka, Leningrad, 1979, 142.

[5] O. B. Naimark, in J. R. Willis, ed., *Proceedings of the IUTAM Symposium on Nonlinear Analysis of Fracture*, Kluver Academic Publishers, Dordrecht, The Netherlands, 1997, 285–298.

[6] G. I. Barenblatt and L. R. Botvina, *Izv. An. SSSR Mech. Tv. Tela*, **4** (1983), 161 (in Russian).

[7] A. Kadic and G. B. Edelen, Lecture Notes in Physics 174, Springer-Verlag, Berlin, 1983.

[8] Yu. L. Raikher and M. I. Shliomis, W. Coffey, ed., in *Relaxation Phenomena in Condensed Matter*, Advances in Chemical Physics LXXXVI, John Wiley, New York, 1994, 595.

[9] M. A. Leontovich, Introduction to thermodynamics, in *Statistical Physics*, Nauka, Moscow, 1983, Chapter 3.

[10] O. B. Naimark and V. V. Silbershmidt, *Europ. J. Mech. Ser.* A *Solids*, **10** (1991), 607.

[11] L. D. Landau and E. M. Lifshitz, *Course of Theoretical Physics: Statistical Physics*, Pergamon Press, Oxford, UK, 1980.

[12] O. B. Naimark, *JETP Lett.*, **67**-9 (1998), 751.

[13] S. P. Kurdjumov, in *Dissipative Structures and Chaos in Non-Linear Space*, Vol. 1, Utopia, Singapore, 1988, 431.

[14] J. Fineberg, S. P. Gross, and E. Sharon, in J. R. Willis, ed., *Proceedings of the IUTAM Symposium on Nonlinear Analysis of Fracture*, Kluver Academic Publishers, Dordrecht, The Netherlands, 1997, 177.

[15] K. Ravi-Chandar and W. G. Knauss, *Internat. J. Fracture*, **26** (1982), 65.

[16] L. B. Freund, *Dynamic Fracture Mechanics*, Cambridge University Press, Cambridge, UK, 1990.

[17] J. Fineberg, S. Gross, M. Marder, and H. Swinney, *Phys. Rev. Lett.*, **67** (1991), 457.

[18] E. Sharon, S. P. Gross, and J. Fineberg, *Phys. Rev. Lett.*, **74** (1995), 5096.

[19] J. F. Boudet, S. Ciliberto, and V. Steinberg, *J. Physique*, **6** (1993), 1493.

[20] E. Sharon, S. P. Gross, and F. Fineberg, *Phys. Rev. Lett.*, **76** (1996), 2117.

[21] B. L. Holian and R. Thomson, *Phys. Rev.* E, **56**-1 (1997), 1071.

[22] A. A. Griffith, *Philos. Trans. Roy. Soc. London Ser.* A, **221** (1921), 163.

[23] N. F. Mott, *Engineering*, **165** (1948), 16.

[24] G. R. Irwin, *J. Appl. Mech.*, **24** (1957), 361.

[25] G. I. Barenblatt, *Adv. Appl. Mech.*, **7** (1962), 55.

[26] J. R. Rice, *J. Appl. Mech.*, **35** (1968), 37.

[27] Ya. I. Frenkel, *J. Tech. Phys.*, **22**-11 (1952), 1857.

[28] O. B. Naimark, M. M. Davydova, and O. A. Plekhov, in G. Frantziskonis, ed., *Proceedings of the NATO Workshop "Probamat: 21st Century"*, Kluwer Academic Publishers, Dordrecht, The Netherlands, 1998, 127.

[29] O. B. Naimark, M. M. Davydova, O. A. Plekhov, and S. V. Uvarov, *Phys. Mesomech.*, **2–3** (1999), 47.

[30] O. B. Naimark, M. M. Davydova, and O. A. Plekhov, *Comput. Struct.*, **76** (2000), 67.

[31] O. B. Naimark, Plasticity and damage (plenary lecture), in D. Miannay, P. Costa, D. Francois, and A. Pineau, eds., *Proceedings of EUROMAT* 2000: *Advances in Mechanical Behavior*, Vol. 1, Elsevier, Amsterdam, 2000, 15–28.

[32] B. B. Mandelbrot, D. E. Passoja, and A. J. Paullay, *Nature*, **308** (1984), 721.

[33] E. Bouchaud, in A. Carpinteri, ed., *Proceedings of the IUTAM Symposium "Size-Scale Effects in Failure Mechanisms of Materials and Structures"*, Kluwer Academic Publishers, Dordrecht, The Netherlands, 1996, 121.

[34] J. J. Mecholsky, in C. R. Kurkjian, ed., *Strength of Inorganic Materials*, Plenum Press, New York, 1995, 569.

[35] J. Feder, *Fractals*, Plenum Press, New York, London, 1988.

[36] O. B. Naimark and M. M. Davydova, *J. Physique* III, **6** (1996), 259.

[37] L. A. Galin and G. P. Cherepanov, *Soviet Phys. Dokl.*, **167** (1966), 543.

[38] N. Bourne, J. Millett, Z. Rosenberg, and N. Murray, *J. Mech. Phys. Solids*, **46** (1998), 1887.

[39] S. V. Rasorenov, G. J. Kanel, V. E. Fortov, and M. M. Abasenov, *High Pressure Res.*, **6** (1991), 225.

[40] V. N. Nikolaevskii, *Internat. J. Engrg. Sci.*, **19** (1981), 41.

[41] N. K. Brar and S. J. Bless, *High Pressure Res.*, **10** (1992), 773.

[42] N. Bourne, Z. Rosenberg, and J. E. Field, *J. Appl. Phys.*, **78** (1995), 3736.

[43] D. P. Dandekar and P. A. Beaulieu, in L. E. Murr, K. P. Staudhammer, and M. A. Meyers, eds., *Metallurgical and Materials Applications of Shock-Wave and High-Strain-Rate Phenomena*, Elsevier Science, Amsterdam, 1995, 211.

[44] N. Bourne, Z. Rosenberg, J. E. Field and I. G. Crouch, *J. Phys.* IV *Colloq.* C, **8** (1994), 635.

[45] R. V. Gibbons and T. J. Ahrens, *J. Geophys. Res.*, **76** (1971), 5489.

[46] R. J. Clifton, *Appl. Mech. Rev.*, **46** (1993), 540.

[47] O. B. Naimark, Keynote lecture, in B. Karihaloo, ed., *Proceedings of the* IX*th International Conference of Fracture*, Vol. 6, Sydney, 1997, 2795.

[48] O. B. Naimark, F. Collombet, and J.-L. Lataillade, *J. Phys.* IV *Colloq.* C, **7** (1998), 773.

[49] O. A. Plekhov, D. N. Eremeev, and O. B. Naimark, J*J. Phys.* IV *Colloq.* C, **10** (2000), 811.

[50] V. V. Beljaev and O. B. Naimark, *Soviet Phys. Dokl.*, **312**-2 (1990), 289.

[51] E. Bellendir, V. V. Beljaev, and O. B. Naimark, *Soviet Tech. Phys. Lett.*, **15**-3 (1989), 90.

[52] D. E. Grady, *J. Appl. Phys.*, **53** (1982), 322.

[53] M. E. Kipp and D. E. Grady, *J. Mech. Phys. Solids*, **33** (1986), 399.

[54] L. A. Glenn and A. Chudnovsky, *J. Appl. Phys.*, **59** (1986), 1379.

[55] D. E. Grady, *J. Mech. Phys. Solids*, **36** (1988), 353.

[56] D. E. Grady and M. E. Kipp, *J. Appl. Phys.*, **58**-3 (1985), 1210.

[57] D. E. Grady, *J. Appl. Phys.*, **68** (1990), 6099.

[58] D. E. Grady, Fragmentation by blasting, in W. L. Fourney, R. R. Boade, and L. S. Costin, eds., *Experimental Mechanics*, Society for Experimental Mechanics, Bethel, CT, 1985, 63.

[59] H. Senf, E. Strauburger, and H. Rothenhausler, in L. E. Murr, K. P. Staudhammer, and M. A. Meyers, eds., *Metallurgical and Materials Applications of Shock-Wave and High-Strain-Rate Phenomena*, Elsevier Science, Amsterdam, 1995, 163.

[60] O. B. Naimark and V. V. Belayev, *Phys. Combust. Explosion*, **25** (1989), 115.

[61] A. D. Sakharov, R. M. Zaidel, V. N. Mineev, and A. G. Oleinik, *Soviet Phys. Dokl.*, **9** (1965), 1091; V. N. Mineev and E. N. Savinov, *Soviet Phys. JETP*, **25** (1967), 411; V. N. Mineev and R. M. Zaidel, *Sov. Phys. JETP*, **27** (1968), 874.

[62] W. Band and G. E. Duval, *Amer. J. Phys.*, **29** (1961), 780.

[63] D. C. Wallace, *Phys. Rev.* B, **24** (1981), 5597 and 5607.

[64] L. M. Barker, *Behavior of Dense Media Under High Pressures*, Gordon and Breach, New York, 1968.

[65] Lord Rayleigh, *Proc. Roy. Soc. London*, **84** (1910), 247.

[66] G. I. Taylor, *Proc. Roy. Soc. London*, **84** (1910), 371.

[67] R. Von Mises, *J. Aero. Sci.*, **17** (1950), 551.

[68] M. J. Lighthill, *Surveys in Mechanics*, Cambridge University, Cambridge, UK, 1965.

[69] J. W. Swegle and D. E. Grady, *J. Appl. Phys.*, **58**-2 (1985).

[70] O. B. Naimark, *Soviet Tech. Phys. Lett.*, **23**-7 (1997), 529.

[71] Ja. I. Frenkel, *Kinetic Theory of Liquids*, Clarendon Press, Oxford, UK, 1946.

[72] B. V. Derjagin et al., *Polymer*, **30** (1989), 1.

[73] D. R. Carlson, S. E. Widnall, and M. F. Peeters, *J. Fluid Mech.*, **121** (1982), 487.

[74] G. M. Zaslavskii and R. Z. Sagdeev, *Introduction to Nonlinear Physics*, Nauka, Moscow, 1988 (in Russian).

[75] Yu. L. Klimontovich, *Phys.* B, **229** (1996), 51.

# 5. Extended Thermodynamics: A Multifield Theory Par Exellence

I. Müller*

**Abstract.** A monatomic ideal gas, upon close inspection, reveals itself as a complex material whose proper description requires the use of hundreds, or even thousands of fields, at least in the rarefied state.

## 5.1. Introduction

Nothing should be simpler than a monatomic ideal gas, or so one might think, and that intuitive feeling is largely confirmed by thermodynamic calculations in a dense gas. Indeed in such a gas the equations of Navier–Stokes–Fourier represent the constitutive properties of the gas quite well. However, for low pressures—i.e., at great heights—the gases are rarefied and the constitutive equations of Navier–Stokes–Fourier do not hold anymore.

That is the range of extended thermodynamics (see [1]), a multifield theory that employs hundreds or thousands of moments to describe the state of the gas. It is shown here that only a multifield theory of that type is capable of describing the observed properties of rarefied gases satisfactorily. The phenomena considered are

- light scattering in a gas,
- shock structure,
- the development of a shock in a shock tube.

## 5.2. Field and field equations

### 5.2.1. General

The fields of extended thermodynamics of monatomic ideal gases are the moments

$$F_{i_1 i_2 \ldots i_k} = m \int c_{i_1} c_{i_2} \cdots c_{i_k} f d\mathbf{c} \quad (k = 0, 1, \ldots, N) \tag{1}$$

of the distribution function which represents the number density of atoms with a certain velocity $\mathbf{c}$. The moments are symmetric tensors of up to order $N$ and, because

*Faculty of Process Engineering, Thermodynamics, Technical University of Berlin, 10623 Berlin, Germany, im@thermodynamik.tu-berlin.de.

of the symmetry, the number of independent variables equal

$$n = \frac{1}{2}\sum_{\alpha=0}^{N}(\alpha+1)(\alpha+2) = \frac{1}{6}(N+1)(N+2)(N+3). \tag{2}$$

The field equations are of balance type. They relate the rate of change of the densities $F_{i_1 i_2 \ldots i_k}$ to the divergence of the fluxes $F_{i i_1 i_2 \ldots i_k}$ and the productions $\Pi_{i_1 i_2 \ldots i_k}$, viz.

$$\frac{\partial F_{i_1 i_2 \ldots i_k}}{\partial t} + \frac{\partial F_{i i_1 i_2 \ldots i_k}}{\partial x_i} = \Pi_{i_1 i_2 \ldots i_k}. \tag{3}$$

We observe that the fluxes $F_{i i_1 i_2 \ldots i_k}$ are themselves moments of rank $k+1$. This is typical for monatomic ideal gases.

In order to close the system we need constitutive relations for the production $\Pi_{i_1 i_2 \ldots i_k}$ and for the last flux $F_{i i_1 \ldots i_N}$. In extended thermodynamics we achieve closure by use of the entropy principle, which determines the distribution function appropriate to given variables $F_{i i_1 i_2 \ldots i_k}$ $(k = 1, 2, \ldots, N)$. The distribution function has the form

$$f = y e^{-\frac{1}{k}\chi} \quad \text{with } \chi = \sum_{k=1}^{N}\sum_{i_1=1}^{3}\cdots\sum_{i_k=1}^{3} \Lambda_{i_1 i_2 \ldots i_k} c_{i_1} c_{i_2} \cdots c_{i_k}, \tag{4}$$

where $k$ is the Boltzmann constant and $y$ determines the smallest element $d\mathbf{x}d\mathbf{c}$ of phase space. It is related to the Planck constant. The $\Lambda$'s are Lagrange multipliers which are introduced in the course of exploiting the entropy inequality. In the end they must be eliminated by use of the relations

$$F_{i_1 i_2 \ldots i_k} = m \int c_{i_1} c_{i_2} \cdots c_{i_k} y e^{-\frac{1}{k}\sum_{k=1}^{N}\sum_{i_1=1}^{3}\cdots\sum_{i_k=1}^{3}\Lambda_{i_1 i_2 \ldots i_k} c_{i_1} c_{i_2} \cdots c_{i_k}}. \tag{5}$$

(In the usual procedure, the integrand in (5) is linearized in nonequilibrium quantities. After that the calculation of $\Lambda_{i_1 i_2 \ldots i_k}$ in terms of $F_{i_1 i_2 \ldots i_k}$ is reduced to the solution of a linear system of equations.)

Because of the entropy principle the system of balance equations is symmetric hyperbolic—at least if it is written in terms of the Lagrange multipliers $\Lambda_{i_1 i_2 \ldots i_k}$—and therefore all characteristic speeds are finite. It seems appropriate to mention this fact, since some of the commonly used field equations for gases are parabolic and predict infinite wave speeds. Such is the case for the Navier–Stokes–Fourier equations and this phenomenon of infinite wave speeds was called a paradox by Cattaneo [2].

One of the motifs for the development of extended thermodynamics was the desire to remove that paradox.

### 5.2.2. The special case $N = 3$

For $N = 3$, we have 20 independent equations, which are represented in Figure 1. In the figure, the synthetic notation used hitherto has been abandoned and we have represented the moments—as far as possible—in their canonical notation: $\rho$ for the density, $v_i$ for velocity, $T$ for temperature, $t_{ij}$ for stress, and $q_i$ for heat flux. These are 13 moments; the remaining six, viz. $\Delta$ and $\rho_{\langle ijk \rangle}$, have no canonical notation,

$$
\begin{aligned}
&\frac{\partial \rho}{\partial t}+\bar{\rho}\frac{\partial v_i}{\partial x_i}=0\\
&\frac{\partial v_i}{\partial t}+\frac{\left(\frac{k}{m}\bar{T}\right)}{\bar{\rho}}\frac{\partial \rho}{\partial x_i}+\frac{\partial}{\partial x_i}\left(\frac{k}{m}T\right)-\frac{1}{\bar{\rho}}\frac{\partial t_{\langle ij\rangle}}{\partial x_j}=0\\
&\frac{\partial\left(\frac{k}{m}T\right)}{\partial t}+\frac{2}{3}\left(\frac{k}{m}\bar{T}\right)\frac{\partial v_k}{\partial x_k}+\frac{2}{3}\frac{1}{\bar{\rho}}\frac{\partial q_k}{\partial x_k}=0.\\
&\frac{\partial t_{\langle ij\rangle}}{\partial t}-\frac{4}{5}\frac{\partial q_{\langle i}}{\partial x_{j\rangle}}-2\bar{\rho}\left(\frac{k}{m}\bar{T}\right)\frac{\partial v_{\langle i}}{\partial x_{j\rangle}}-\frac{\partial \rho_{\langle ijk\rangle}}{\partial x_k}=\sigma t_{\langle ij\rangle}\\
&\frac{\partial q_i}{\partial t}-\left(\frac{k}{m}\bar{T}\right)\frac{\partial t_{\langle ik\rangle}}{\partial x_k}+\frac{5}{2}\bar{\rho}\left(\frac{k}{m}\bar{T}\right)\frac{\partial \frac{k}{m}T}{\partial x_i}=\frac{2}{3}\sigma q_i\\
&\frac{\partial \rho_{\langle ijk\rangle}}{\partial t}-3\frac{k}{m}\bar{T}\left\{\frac{\partial t_{\langle\langle ij\rangle}}{\partial x_{k\rangle}}-\frac{2}{5}\frac{\partial t_{\langle r\langle j\rangle}}{\partial x_r}\delta_{jk\rangle}\right\}=\frac{3}{2}\sigma\rho_{\langle ijk\rangle}
\end{aligned}
$$

$$
\begin{aligned}
&\frac{\partial \rho}{\partial t}+\bar{\rho}\frac{\partial v_i}{\partial x_i}=0\\
&\frac{\partial v_i}{\partial t}+\frac{\left(\frac{k}{m}\bar{T}\right)}{\bar{\rho}}\frac{\partial \rho}{\partial x_i}+\frac{\partial}{\partial x_i}\left(\frac{k}{m}T\right)-\frac{1}{\bar{\rho}}\frac{\partial t_{\langle ij\rangle}}{\partial x_j}=0\\
&\frac{\partial\left(\frac{k}{m}T\right)}{\partial t}+\frac{2}{3}\left(\frac{k}{m}\bar{T}\right)\frac{\partial v_k}{\partial x_k}+\frac{2}{3}\frac{1}{\bar{\rho}}\frac{\partial q_k}{\partial x_k}=0.\\
&\frac{\partial t_{\langle ij\rangle}}{\partial t}-\frac{4}{5}\frac{\partial q_{\langle i}}{\partial x_{j\rangle}}-2\bar{\rho}\left(\frac{k}{m}\bar{T}\right)\frac{\partial v_{\langle i}}{\partial x_{j\rangle}}-\frac{\partial \rho_{\langle ijk\rangle}}{\partial x_k}=\sigma t_{\langle ij\rangle}\\
&\frac{\partial q_i}{\partial t}-\left(\frac{k}{m}\bar{T}\right)\frac{\partial t_{\langle ik\rangle}}{\partial x_k}+\frac{5}{2}\bar{\rho}\left(\frac{k}{m}\bar{T}\right)\frac{\partial \frac{k}{m}T}{\partial x_i}=\frac{2}{3}\sigma q_i\\
&\frac{\partial \rho_{\langle ijk\rangle}}{\partial t}-3\frac{k}{m}\bar{T}\left\{\frac{\partial t_{\langle\langle ij\rangle}}{\partial x_{k\rangle}}-\frac{2}{5}\frac{\partial t_{\langle r\langle j\rangle}}{\partial x_r}\delta_{jk\rangle}\right\}=\frac{3}{2}\sigma\rho_{\langle ijk\rangle}
\end{aligned}
$$

FIGURE 1. The system of 20 moment equations. Left: Euler equations within the frame. Right: Navier–Stokes–Fourier equations within the frame.

nor specific names, other than that they are moments. However, they must satisfy the explicit and specific system of partial differential equations.

Indeed, the left-hand side of the system is entirely specific and the right-hand side, here calculated for Maxwellian molecules, contains one single parameter $\sigma$, which can be determined from a single measurement of viscosity or thermal conductivity.

We stop here a moment in order to appreciate the system of 20 equations shown in Figure 1. If we ignore all terms except those in the frame of Figure 1(left), we recognize the Euler equations which ignore all dissipation. More realistic are the equations of Navier–Stokes–Fourier, which we find within the frame of Figure 1(right). We recognize that, according to that frame, the stress is proportional to the temperature gradient. We also recognize, however, that we must leave out entirely specific terms in order to obtain the Navier–Stokes–Fourier equations. Such terms are rates and gradients of stress and heat flux.

Therefore we conclude that the theory of Navier–Stokes–Fourier may suffice, if rates and gradients are small. On the other hand, extended thermodynamics is needed, if rates are fast and gradients are steep. And, since in a gas "fast" and "steep" are measured in terms of mean times of free flight and mean free paths, we conclude that extended thermodynamics is needed for rarefied gases where flight times and free paths are long, while in a dense gas Navier–Stokes–Fourier may be sufficient.

Wolf Weiss [3] has the explicit moment equations available at the touch of a button and he has determined the characteristic speeds predicted by them, i.e., the speeds of acceleration waves. The biggest one of them, the pulse speed, is listed in Figure 2(left): For $n = 20$, i.e., the system of Figure 1, the pulse speed is $\mathrm{Ma} = 1.81$, while for $n = 2300$, it is $\mathrm{Ma} = 6.46$, and for $n = 15180$, it is $\mathrm{Ma} = 9.36$. In Figure 2(right), we see from the line identified by crosses that the pulse speed seems to increase monotonically, and this was confirmed by Boillat and Ruggeri [4], who

| No. of moments | $V_{max}/c_0$ |
|---|---|
| 4 | 0.77459667 |
| 10 | 1.34164079 |
| 20 | 1.80822948 |
| 33 | 2.21299946 |
| 56 | 2.57495874 |
| 84 | 2.90507811 |
| 120 | 3.21035245 |
| 165 | 3.49555791 |
| 220 | 3.76412372 |
| 286 | 4.01860847 |
| 364 | 4.26098014 |
| 455 | 4.49279023 |
| 560 | 4.71528716 |
| 680 | 4.92949284 |
| 816 | 5.13625617 |
| 969 | 5.3362913 |
| 1140 | 5.53020569 |
| 1330 | 5.71852112 |
| 1540 | 5.90168962 |
| 1771 | 6.08010585 |
| 2021 | 6.25411673 |
| 2300 | 6.42402919 |
| 2600 | 6.59011627 |
| 2925 | 6.75262213 |
| 3276 | 6.91176615 |
| 3654 | 7.06774631 |
| 4060 | 7.22074198 |
| 4495 | 7.37091629 |
| 4960 | 7.51841807 |
| 5456 | 7.66338362 |
| 5984 | 7.80593804 |
| 6545 | 7.94619654 |
| 7140 | 8.08426549 |
| 7770 | 8.22024331 |
| 8436 | 8.35422129 |
| 9139 | 8.48628432 |
| 9860 | 8.61651144 |
| 10660 | 8.74497644 |
| 11480 | 8.87174833 |
| 12341 | 8.99680171 |
| 13244 | 9.12046722 |
| 14190 | 9.24253184 |
| 15180 | 9.36313918 |

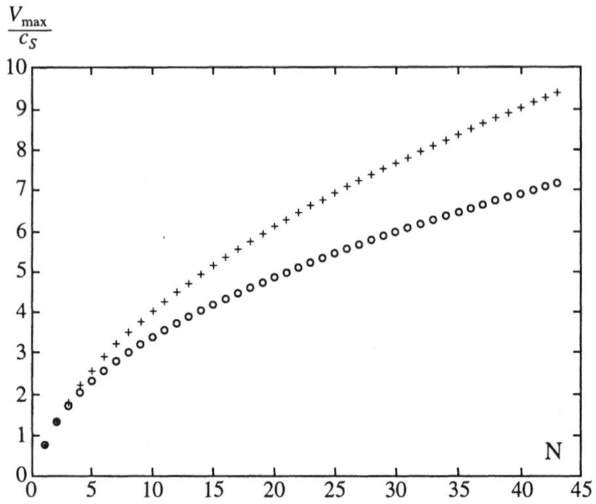


FIGURE 2. Characteristic speeds vs. number of moments. Crosses: Exact values by Weiss [3]. Circles: Lower bound by Boillat and Ruggeri [4].

calculated a lower bound for the pulse speed, viz.

$$\frac{V_{\max}}{c_0} \geq \sqrt{\frac{6}{5}\left(N - \frac{1}{2}\right)}. \tag{6}$$

Thus for infinitely many moments the pulse speed tends to infinity.

In some way, this result is an anticlimax when we recall that the original motif of extended thermodynamics was the desire to obtain finite speeds. We have a theory now that predicts a finite speed for all finite numbers of moments and has an infinite speed only for infinitely many moments. That seems satisfactory, but for those who like the anticlimax aspect better, let me say that by now extended thermodynamics has outgrown its former exclusive concern for speeds; it is now an explicit and definite theory for the description of rarefied gases and nowhere is this more evident than in its treatment of light scattering to which we shall proceed in the next chapter.

Before that, however, two remarks seem to be in order. Firstly, in a relativistic theory, the lower bound (6) is not valid. It is replaced by another estimate, also due to Boillat and Ruggeri [5], which implies that the largest pulse speed is the speed of light and that may occur only for infinitely many moments. Secondly, it is clear that, apart from the pulse speed, a high moment theory predicts many longitudinal speeds; indeed, hundreds of them, if the number of moments is in the thousands. Weiss [6] has calculated them all and has plotted them (see Figure 3). Some are below the speed of sound but most are above and the spectrum of speeds becomes nearly continuous. The question is why do we not see them, or, better, why do we not hear them? We shall discuss that question when we deal with shock structures in Section 5.4. The short answer is that these propagation modes are quickly damped out.

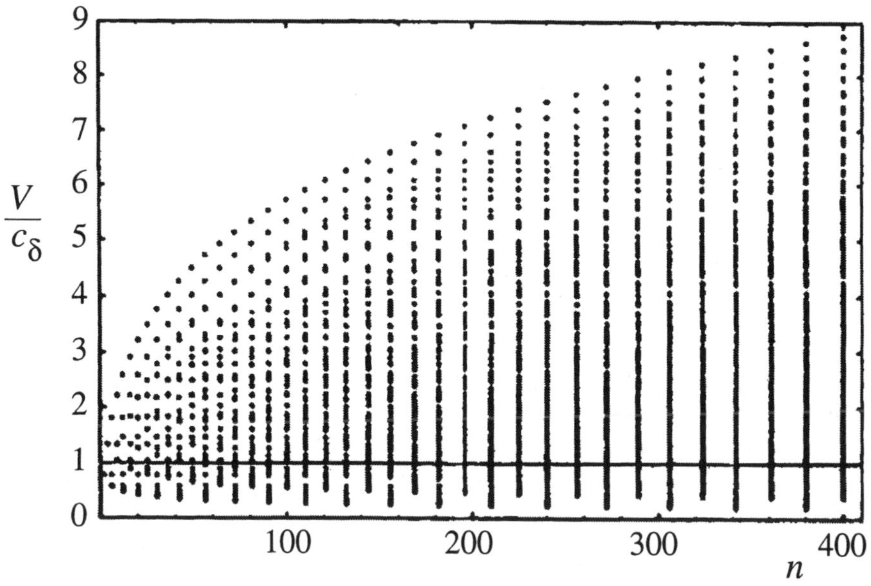


FIGURE 3. Pulse speeds.

## 5.3. Light scattering

Light scattering occurs when a light wave with frequency $\omega$ falls on a gas in equilibrium. It is scattered by the density fluctuations of the gas and the scattered light may be investigated in a detector. It turns out that the scattered light consists mostly of light of the incident frequency $\omega$, but neighbouring frequencies are also represented and in a dense gas the spectrum of the scattered gas is represented by a graph with three peaks, the central one at the incident frequency (see Figure 4).

The relative heights of the peaks, their widths and distances permit us to read off thermodynamic constitutive properties of the gas, like specific heats, compressibilities, thermal conductivity, and viscosity. It is now appropriate to ask why it is

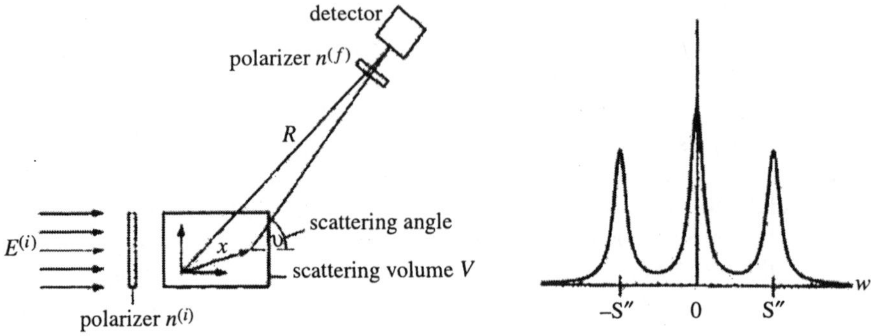


FIGURE 4. Light scattering and the spectrum of scattered light in a dense gas.

that a microscopic density fluctuation through the light which it scatters can carry information about a macroscopic property of the gas, like its viscosity. And this is where the Onsager hypothesis comes in.

Onsager has postulated—and that postulate has been convincingly confirmed by experiment—that the mean regression of a microscopic fluctuation may be calculated from the macroscopic field equations.

Once this postulate is accepted, it is easily exploited. What we have to do with our explicit set of field equations is this:

- Fourier transform spatially,
- Laplace transform in time,
- solve algebraically,
- average over initial data.

In this manner, according to Onsager, we should obtain the scattering spectrum. And indeed, for a dense gas, where we know the Navier–Stokes–Fourier equations to be good enough, we obtain the three-peaked graph of the experiment.

However, for even a moderate degree of rarefaction the measured spectrum cannot be described by a Navier–Stokes–Fourier theory; in that case we need extended thermodynamics. In Figure 5(left), we see the experimental dots, measured by Clark [7] and the predictions of extended thermodynamics of 20, 35, 56, and 84 moments. None of them agrees with the experiments. So now the reader may think that it is time for an adjustment of parameters in order to fit the curves. However, extended thermodynamics has no adjustable parameters. It is a completely specific *theory of theories* of increasing moment numbers. And the only fitting parameter is the number of moments.

So we increase the number of moments. And indeed for $n = 120$ we obtain a satisfactory agreement with experiment. Moreover, the predicted spectrum of ET 120 does not change when we increase the number of moments further. Thus in Figure 5(right), we see that the curves of ET 120, ET 165, ET 220, and ET 286 all lie on

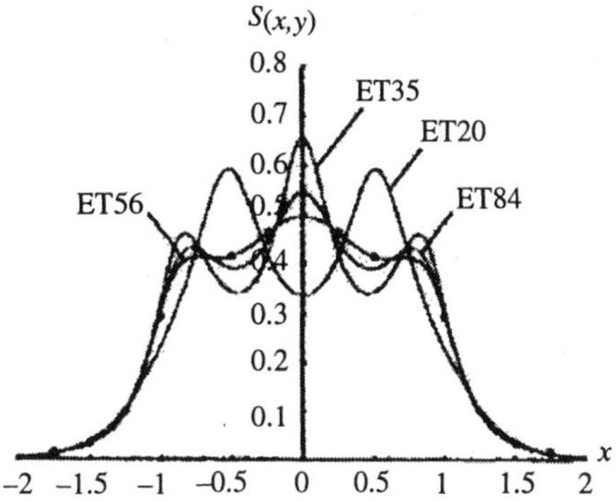


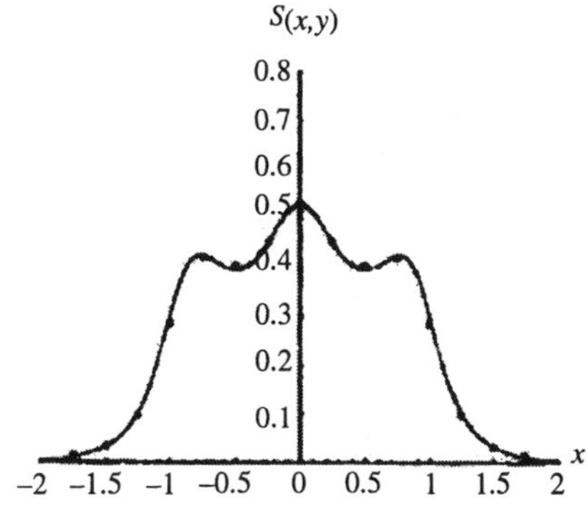


FIGURE 5. Scattering spectra for different moment numbers. Left: ET 20, ET 35, ET 56, ET 84. Right: ET 120, ET 165, ET 220, ET 286.

top of each other—and on the experimental dots! We conclude that there is a convergence and that convergence allows us to determine the range of validity of the theory. As soon as in our theory of theories with increasing numbers of moments one theory predicts the same result as the previous one, the latter one was good enough.

These results are both satisfactory and disappointing. They are satisfactory because now for the first time we have succeeded in describing a scattering spectrum in a good manner. However, there is also disappointment: To be sure, in view of the experimental findings, nobody expected Navier–Stokes–Fourier to be good enough, but one might have hoped that ET 13 or ET 14 would suffice. Instead we need hundreds of moments. In fact, the numbers between 200 and 300 that were good in Figure 5 for a moderate degree of rarefaction become insufficient in a more rarefied gas as shown in Figure 6. It is true that the disagreement in that case is restricted to a plait-like region, but nevertheless it is disagreement. Incidentally, the plait winds around a Gaussian curve which reflects the Maxwellian distribution of atomic velocities which the scattering spectrum approaches for an increasing degree of rarefaction.

## 5.4. Shocks

### 5.4.1. Shock structure

Given that extended thermodynamics is a theory for rapid changes and steep gradients it is only natural that shock waves should be investigated by the theory. Also, it is well known that the shock structure is badly represented by the Navier–Stokes–Fourier theory. Figure 7 shows both the measured shock thickness and the one calculated from the Navier–Stokes–Fourier theory as functions of the Mach number. The disagreement is obvious.

This seems to be the occasion for trying out extended thermodynamics. The first one to do this was Grad [8], who tried his 13-moment theory. (To be sure, Grad did

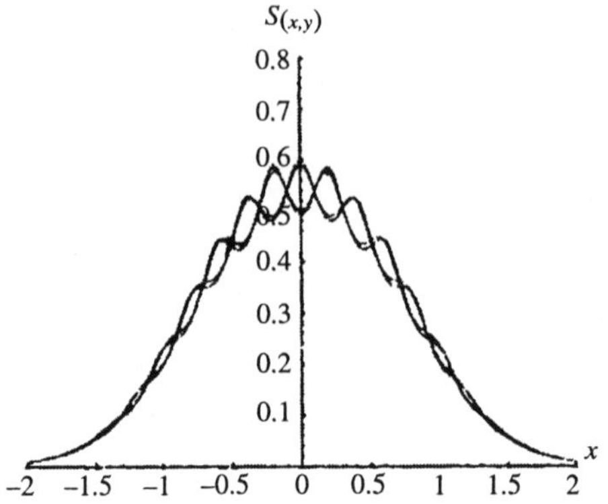


FIGURE 6. Calculated scattering spectra for ET 210, ET 225, ET 240, and ET 256.

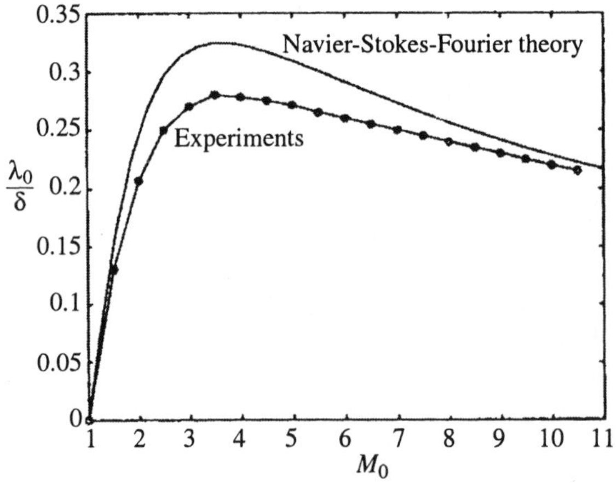


FIGURE 7. Shock thickness: Experiment and Navier–Stokes–Fourier theory.

not know extended thermodynamics. However, his 13-moment approximation in the kinetic theory of gases is essentially equivalent to ET 13.) He found that the theory, far from being better, was dramatically worse. Indeed for Ma $>$ 1.65 he found no smooth shock at all but rather a subshock (see Figure 8).

It is not certain whether Grad realized the significance of the number Ma $= 1.65$. This speed is the pulse speed of the 13-moment theory. Nothing in the 13-moment theory can be faster, and therefore, if the shock moves faster, the information about its arrival cannot make itself felt in the upstream region before the shock. Facetiously we may say that Ma $>$ 1.65 is the "true" supersonic range for ET 13.

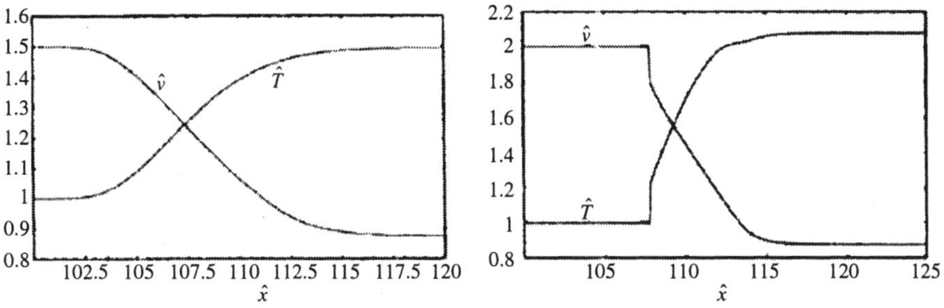


FIGURE 8. Left: ET 13. Smooth shock structure at Ma $= 1.5$. Right: ET 13. Subshock for $M = 2$.

However, this is no problem in extended thermodynamics, at least not in principle. If the pulse speed of ET 13 is too small, we choose a theory with more variables from our theory of theories. And we already know that the pulse speed goes up with the number of moments (see Figure 2). Therefore, Au [9] has calculated the shock structures for higher moment theories. The task is infinitely more difficult than the corresponding one for light scattering because the nonlinear equations must be found and implemented. Yet Au was able to go up to ET 560 and to obtain shock structures in excellent agreement with experiments—but only up to Ma $= 1.8$. Clearly, one should go higher with the number of moments, if one wishes to obtain good results for higher Mach numbers. At this time, the computing power proved to be insufficient for the task.

### 5.4.2. Shock tube experiment

A related problem to the shock structure calculation is the question about how the stationary shock structure forms to begin with. Shocks are usually investigated in a shock tube: a long tube initially divided by a membrane into a high pressure region and a low pressure one. When the membrane is removed, according to the Euler theory of gas dynamics, a shock and a contact discontinuity move into the low pressure region, while a continuous but nonsmooth rarefaction "fan" moves into the high pressure region. See the dashed lines in Figure 9, which shows columns of graphs for density $\rho$, pressure $p$, and velocity $v$ at different times after the removal of the membrane.

The solid lines in Figure 9 show the calculated results for ET 13. Let us consider them: For long times—actually, no longer than 313 mean times of free flight—the Euler result is closely approximated, except that

- the Euler shock is replaced by a continuous and smooth shock structure;
- the Euler contact discontinuity is replaced by a continuous and smooth structure;
- the nonsmooth rarefaction fan is approximated by a smooth function.

For short times, however, we see the effect of the two characteristic speeds of ET 13, one with Ma $= 1.65$ and the other one with Ma $= 0.63$. Accordingly, there are two

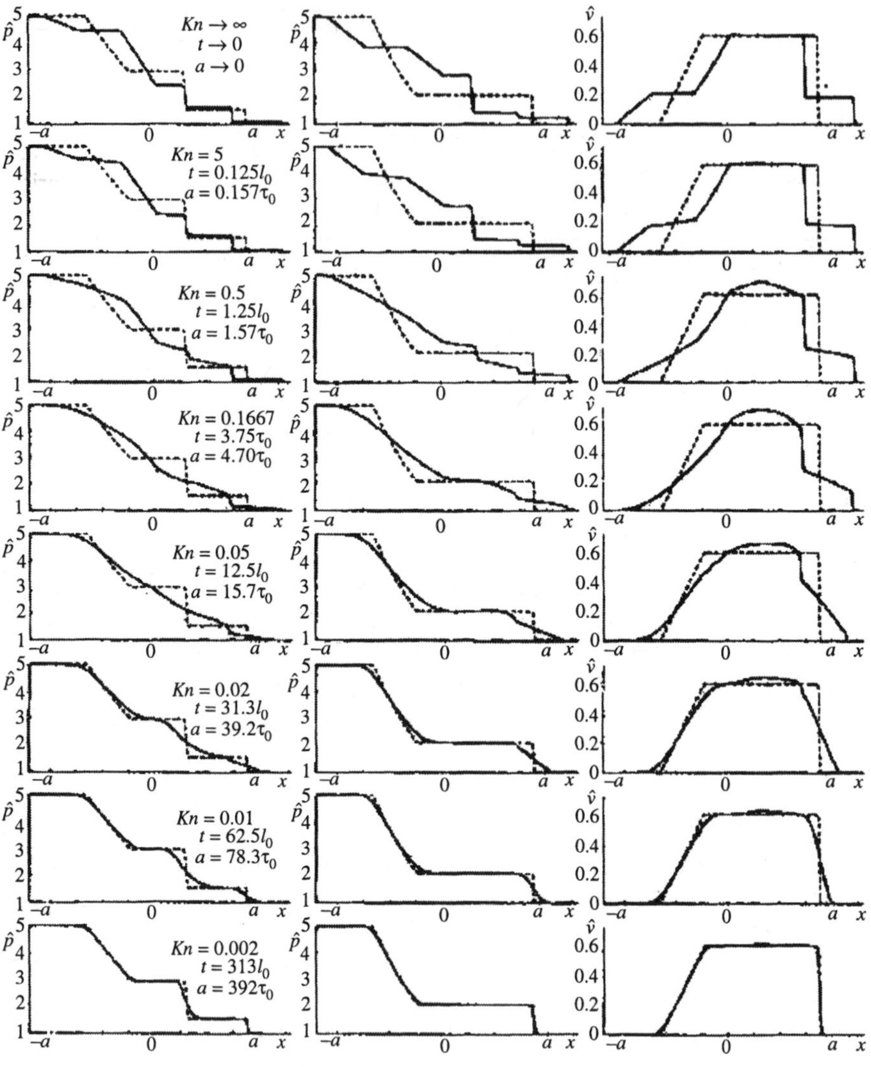


FIGURE 9. The emergence of the shock structure in ET 13 and according to the Euler equations (dashed lines).

shocks, a fast and a slow one, propagating to the right and two rarefaction fans, again a fast and a slow one.

The tableau of graphs of Figure 9 shows instructively how the initial situation develops, by dispersion and absorption, into the final one. While at the beginning we still see the individual waves, we see that, at the end, the shock structure propagates with the Mach number appropriate to the Euler shock and that that Ma number was not present initially.

The situation is perhaps made clearer by Figure 10, which shows calculations immediately after the removal of the membrane and for a large number of moments. In that case we can also solve the Boltzmann equation and obtain the dashed line. The predicted fields of extended thermodynamics show the many waves appropriate to a high number of moments, and those fields zigzag around the solution of the Boltzmann equation. The zigzag line becomes tighter and tighter as the number of moments increases and approximates the exact solution in this manner.

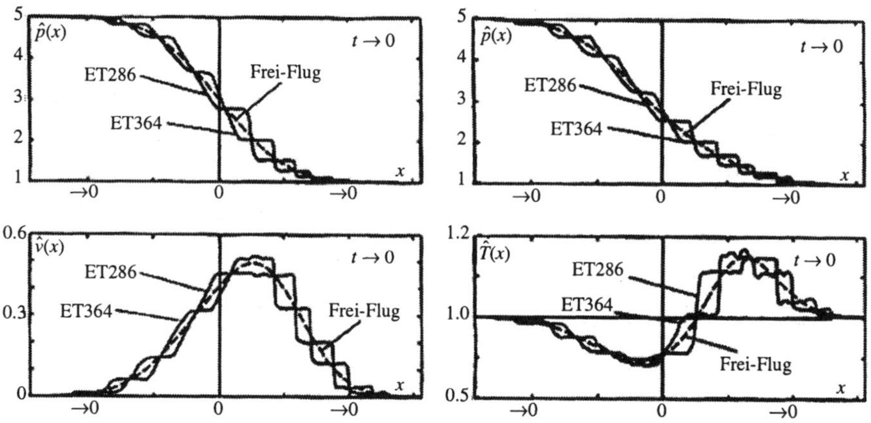


FIGURE 10. The shock tube fields immediately after removal of the membrane [10].

We must realize from the graphs of Figures 9 and 10 that the highest characteristic speeds are quickly damped. That is the reason why we do not hear them.

## References

[1] I. Müller and T. Ruggeri, *Extended Thermodynamics* (1st ed.) and *Rational Extended Thermodynamics* (2nd ed.), Springer Tracts of Natural Philosophy, Springer-Verlag, New York, 1993 and 1998.

[2] C. Cattaneo, Sulla conduzione del calore, *Atti Sem. Mat. Fis. Univ. Modena*, **3** (1948), 83–101.

[3] W. Weiss, *Zur Hierarchie der erweiterten Thermodynamik*, dissertation, Technical University of Berlin, Berlin, 1990.

[4] G. Boillat and T. Ruggeri, Moment equations in the kinetic theory of gases and wave velocities, *Cont. Mech. Thermodynam.*, **9** (1997), 205–212.

[5] G. Boillat and T. Ruggeri, Maximum wave velocity in the moment system of a relativistic gas, *Cont. Mech. Thermodynam.*, **11** (1999), 107–111.

[6] W. Weiss, *Die Berechnung von kontinuierlichen Stoßstrukturen in der kinetischen Gastheorie*, Habilitation thesis, Technical University of Berlin, Berlin, 1996.

[7] N. A. Clark, Inelastic light scattering from density fluctuations in dilute gases: The kinetic-hydrodynamic transition in a monatomic gas, *Phys. Rev.* A, **12** (1975), 232–244.

[8] H. Grad, *The Profile of a Steady Plane Shock Wave*, Communications in Pure and Applied Mathematics 5, John Wiley, New York, 1952.

[9] J. D. Au, *Lösung nichtlinearer Probleme in der erweiterten Thermodynamik*, dissertation, Technical University of Berlin, Berlin, 2001.

[10] J. D. Au, M. Torrilhon, and W. Weiss, The shock tube study in extended thermodynamics, *Phys. Fluids*, **13** (2001), 2423–2432.

# 6. Microstructure and Turbulence in Dilute Polymer Solutions

E. De Angelis*, C. M. Casciola*, P. M. Mariano†, and R. Piva†

**Abstract.** An appropriate picture of the interaction of polymers chains and turbulence structure is crucial to grasp the drag-reducing mechanisms of dilute polymers solutions. In most models the physically small diffusion is normally neglected. However, in the presence of a continuous spectrum of length and time scales, like in turbulence, the introduction of a diffusion term, however small, is crucial to enforce a cutoff at large wave number. Such a term can also be regarded as a natural consequence of a detailed picture of the substructural interactions between the polymeric chains and the fluid. The results obtained through numerical simulations are used in the appropriate thermodynamic framework to extract valuable information concerning the interaction between turbulence and microstructure. A general multifield formulation is finally employed to explore possible additional interaction mechanisms between neighboring populations of polymers that may play a role in accounting for slightly nonlocal interactions between polymer macromolecules in the solvent.

## 6.1. Introduction

Flows of dilute polymer solutions are highly complex due to the interaction between phenomena occurring at the level of the microstructure, i.e., at the level of the macromolecules, with those taking place at hydrodynamical scales. Modeling is already not simple for laminar flows. In presence of turbulence the scenario gets worse since the smallest hydrodynamical scale shrinks. In fact, even for the largest molecules, the interaction between the two levels does not occur in terms of spatial scales, the polymer being, almost in general, smaller than the smallest scale of turbulence. Temporal scales, instead, may easily become comparable when the most striking macroscopic effect of the polymers on turbulence occurs in the form of a substantial reduction of drag with the flow still remaining turbulent.

There are typically two different (but strictly connected) approaches to the construction of a model for polymeric liquids. On the one hand, a micromechanical model of the elementary polymer chain may be considered as a basic block to develop a statistical description of the polymers inseminated in the solvent. Clearly, the

*Dipartimento di Meccanica e Aeronautica, Università di Roma La Sapienza, Rome, Italy.
†Dipartimento di Ingegneria Strutturale e Geotecnica, Università di Roma La Sapienza, Rome, Italy.

achieved level of detail largely affects the range of macroscopic phenomena that can be addressed. However, the microrheological complexity should be kept to a minimum, since easily, the overall description of the solution becomes too cumbersome for investigations of turbulence. On the other hand, one may wish to minimize explicit assumptions concerning the microdynamics, preserving the capability of the model to capture the phenomenology of interest. Since the microstructure plays a crucial role, appropriate descriptors of the microstructure must be introduced and one is naturally led to a multifield description. In the present paper both these approaches are considered. The microrheological description is reviewed first to address effects related to the diffusion of the macromolecules. Afterwards, a multifield formalism is presented with the physical interpretation of the descriptors naturally emerging from the comparison with the microrheological formulation. The purpose is to establish the general form of equations suitable to model microstructural interactions of a dissipative nature between neighboring material elements.

Even in the most elementary model of polymer solutions, the thermodynamics involved includes irreversible terms that account for the dissipative features of the system consistently with the viscoelastic nature of such materials. However, at variance with the Newtonian component of the dissipation, the peculiar dissipation of the polymers principally acts on a purely local basis.

For instance, in the simplest dumbbell model, where a polymer chain is modeled as a spring connecting two massless beads, the dissipation is localized on the beads and is due to the friction in the relative motion of the beads and the supporting fluid. Physically, elementary models of this class assume that the spatial scale of variation of the microstructural descriptors is so large that no nonequilibrium effects can arise due to spatial nonuniformity.

Neglecting diffusion-like effects due to spatial variations of the microstructure is inherently inconsistent and this simplifying procedure can be safely adopted only on the basis of an a posteri verification. In fact, in turbulence, it always happens that the solution of the system of PDEs, where polymer diffusion processes are neglected, blows up via an ultraviolet instability. This makes the simplified system impossible to solve numerically, unless stabilization is artificially achieved through the numerical scheme. Thus, cross-element interactions are crucial for computational applications of polymer modeling in turbulent flows. The present paper is organized as follows. In Section 6.2 a brief overview of the elastic dumbbell model is presented. Section 6.4 is devoted to the description of the effects of turbulence on polymers dynamics by discussing a few numerical results. Finally, in Section 6.5, the multifield approach is addressed to hint at possible weakly nonlocal interaction mechanisms.

## 6.2. The microrheological model

Approaching the simulation of dilute polymers in a turbulent flow, one has to find a correct compromise between the need of a realistic description and the feasibility of the simulation. In fact, many degrees of freedom are generally required by turbulence itself and the presence of polymers increases drastically this demand. Specifically, we address here the elastic dumbbell model under the assumption of isothermal conditions, which is one of the simplest reasonable descriptions for the dilute polymer solutions. The relevant mechanical model accounts for the basic physical mechanisms that characterize the behavior of polymer molecules. Each molecule of polymer can be described as a chain of linear elements. In equilibrium conditions, due to thermal agitation, the chains assume the aspect of statistically spherical random coils. When a substantial deformation rate is present, the coils elongate but naturally tend to return to their equilibrium configuration. The restoring force is of an entropic origin. If the coils are abruptly released they move back to the equilibrium configuration and their dynamics is characterized by a spectrum of relaxation times.

### 6.2.1. Langevin and Fokker–Planck equations

In the rheological model used here, only one relaxation time is retained and each molecule is described as a dumbbell, i.e., two massless beads connected by a spring. The spring accounts for the "entropic elasticity" of the polymer molecules while the viscous forces due to the interaction with the Newtonian solvent are modeled by the friction on the beads. The dumbbells are also forced by a Brownian noise that accounts for the thermal motion [2]. In order to present the relevant equations as concisely as possible, a linear spring is considered.

The model provides each point in space with a collection of dumbbells, which introduces a component of the stress in addition to that of the solvent, the so-called extra stress.

The evolution is controlled by the Langevin equations for the two beads or, equivalently, by the Fokker Planck equation for the corresponding probability density function (PDF).

Denoted by $\mathbf{x}_1$ and $\mathbf{x}_2$ the placements of the two beads, the balance equations for the beads read

$$\begin{aligned} \zeta(\dot{\mathbf{x}}_1 - \mathbf{u}_1) &= -H(\mathbf{x}_1 - \mathbf{x}_2) + \boldsymbol{\xi}_1, \\ \zeta(\dot{\mathbf{x}}_2 - \mathbf{u}_2) &= -H(\mathbf{x}_2 - \mathbf{x}_1) + \boldsymbol{\xi}_2, \end{aligned} \tag{1}$$

where $\zeta$ is the friction coefficient, $\mathbf{u}_1$ and $\mathbf{u}_2$ is the macroscopic velocity field evaluated at the bead positions, and $\boldsymbol{\xi}_1$ and $\boldsymbol{\xi}_2$ are independent Gaussian white noise processes with identical variances $\sigma^2 = 2k_B\Theta\zeta$. $H$ is the spring coefficient, which is typically expressed as $H = 3k_B\Theta/nb^2$, $k_B$ is the Boltzmann constant, $\Theta$ is the absolute temperature, and $n$ is the number of rigid segments of length $b$ used to model the macromolecule as a Kuhn chain. Introducing the end-to-end vector $\mathbf{r} = \mathbf{x}_2 - \mathbf{x}_1$ and the geometric center $\mathbf{x}_c = (\mathbf{x}_1 + \mathbf{x}_2)/2$, equations (1) are recast in the form

$$\begin{aligned} \zeta(\dot{\mathbf{r}} - \mathbf{w}_r) &= -2H\mathbf{r} + \boldsymbol{\xi}_r, \\ \zeta(\dot{\mathbf{x}}_c - \mathbf{w}_c) &= \boldsymbol{\xi}_c, \end{aligned} \tag{2}$$

where $\mathbf{w}_r = \mathbf{u}_2 - \mathbf{u}_1$, $\mathbf{w}_c = (\mathbf{u}_1 + \mathbf{u}_2)/2$, $\boldsymbol{\xi}_r = \boldsymbol{\xi}_2 - \boldsymbol{\xi}_1$, and $\boldsymbol{\xi}_c = (\boldsymbol{\xi}_1 + \boldsymbol{\xi}_2)/2$ and the corresponding variances are $\sigma_r^2 = 2\sigma^2$ and $\sigma_c^2 = \sigma^2/2$. Denoted by

$\psi(\mathbf{x}_c, \mathbf{r}, t) d^3\mathbf{x}_c d^3\mathbf{r}$ the probability to find a molecule having extension in the neighborhood of $\mathbf{r}$ with geometric center in the neighborhood of $\mathbf{x}_c$, the Fokker–Planck equation for $\psi(\mathbf{x}_c, \mathbf{r}, t)$ corresponding to the Langevin equations (2) may be written as [6]

$$\begin{aligned} \frac{\partial \psi}{\partial t} &= \frac{\partial}{\partial \mathbf{x}_c}\left[D_{\text{tr}}\frac{\partial \psi}{\partial \mathbf{x}_c} - \mathbf{w}_c \psi\right] \\ &+ \frac{\partial}{\partial \mathbf{r}}\left[D_{\text{rot}}\frac{\partial \psi}{\partial \mathbf{r}} + \frac{2}{\zeta}\psi\frac{\partial}{\partial \mathbf{r}}(V)\right] - \frac{\partial}{\partial \mathbf{r}}(\psi \mathbf{w}_r), \end{aligned} \tag{3}$$

where $\mathbf{w}_c$ and $\mathbf{w}_r$ depend by definition both on $\mathbf{x}_c$ and $\mathbf{r}$. In equation (3) the elastic force has been rewritten in terms of its potential $V$, and the constants $D_{\text{tr}}$ and $D_{\text{rot}}$, called *translational* and *rotational diffusivity*, respectively, are related to the friction coefficient $\zeta$ and the variances of the stochastic processes $\xi_1, \xi_2$, by the equations

$$D_{\text{rot}} = \frac{1}{2}\frac{\sigma_r^2}{\zeta^2} = \frac{2k\Theta}{\zeta}, \qquad D_{\text{tr}} = D_{\text{rot}}/4. \tag{4}$$

Equation (3) provides the evolution equation for the moments of the PDF and, in particular, for the following observables: the number density $\rho_p$ of the dumbbells with a given geometric center $\mathbf{x}_c$, defined as

$$\rho_p(\mathbf{x}_c, t) = \int_{\mathbb{R}^3} \psi, \tag{5}$$

the average dumbbell extention at $\mathbf{x}_c$,

$$\mathbf{r}_m(\mathbf{x}_c, t) = \int_{\mathbb{R}^3} \psi \mathbf{r}, \tag{6}$$

and the covariance matrix, or conformation tensor,

$$\mathcal{R}(\mathbf{x}_c, t) = \int_{\mathbb{R}^3} \psi \mathbf{r} \otimes \mathbf{r}. \tag{7}$$

The evolution equation for the number density given a divergence-free velocity field $\mathbf{u}$ then follows as

$$\frac{\partial \rho_p}{\partial t} + u_k \frac{\partial \rho_p}{\partial x_k} = \frac{\partial}{\partial x_k}\left(D_{\text{tr}}\frac{\partial \rho_p}{\partial x_k}\right) + \frac{\partial j'_k}{\partial x_k}, \tag{8}$$

where

$$\mathbf{j}' = \int_{\mathbb{R}^3} \mathbf{w}'_c \psi, \tag{9}$$

and $\mathbf{w}'_c = \mathbf{w}_c - \mathbf{u}_c$. $\mathbf{u}_c$ is the velocity of the fluid at the location of the center of mass. Due to the arbitrariness of the numbering of the beads, $\psi$ and $\mathbf{w}_c$ should be invariant under the transformation $\mathbf{r}$ in $-\mathbf{r}$, while $\mathbf{w}_r$ is changed into $-\mathbf{w}_r$. This implies that

$\mathbf{r}_m = 0$. Concerning the conformation tensor it follows that

$$\frac{\partial \mathcal{R}_{ij}}{\partial t} + u_k \frac{\partial \mathcal{R}_{ij}}{\partial x_k}$$
$$= 2D_{\rm rot}\rho_p\delta_{ij} - \frac{4H}{\zeta}\mathcal{R}_{ij} + K_{ir}\mathcal{R}_{rj} + \mathcal{R}_{ir}K_{jr} + \frac{\partial}{\partial x_k}\left(D_{\rm tr}\frac{\partial \mathcal{R}_{ij}}{\partial x_k}\right) + \mathcal{L}'_{ij}, \tag{10}$$

where

$$\mathcal{L}'_{ij} = -\partial/\partial x_k \int_{\mathbb{R}^3} (w'_c)_k \psi r_i r_j + \int_{\mathbb{R}^3} (w'_r)_i \psi r_j + \int_{\mathbb{R}^3} (w'_r)_j \psi r_i \tag{11}$$

with $\mathbf{w}'_r = \mathbf{w}_r - \mathbf{K}\mathbf{r}$ and $\mathbf{K} = \nabla\mathbf{u}$. If the typical length scale associated with spatial velocity variations is large compared to the dumbbell dimensions, the velocity difference between the beads can be expressed as $\mathbf{w}_r = \mathbf{K}\mathbf{r}$, which is the first term in the Taylor series, and $\mathbf{w}_c = \mathbf{u}_c$. In this case, the terms $\mathbf{j}'$ and $\mathcal{L}'_{ij}$ vanish. It is worth noticing that, in this approximation, equation (8) for the number density reduces to a convection-diffusion equation. Thus, if the initial concentration is homogeneous, homogeneity persists in time and equation (8) becomes redundant. On the contrary, when the velocity field at dumbbell scale cannot be approximated by a linear distribution, migration of dumbbells occurs due to the number density flux $\mathbf{j}$, and the concentration is in general not uniform [1].

For long chain polymers, the Brownian forcing is typically very small compared to the other terms in the Langevin equations, and one may be tempted to neglected it in the equation for the geometric center. However, it must be retained, at least in principle, in the equation for the end-to-end vector, since otherwise the beads may eventually collapse one upon the other. Under these assumptions, all terms multiplied by $D_{\rm tr}$ drop from the system and a deterministic equation for the trajectory of the geometric center of the dumbbell is obtained. Consistently the PDF does depend on the position only through the velocity field and the associated Fokker–Planck equation simplifies to

$$\frac{\partial \psi}{\partial t} = D_{\rm rot}\frac{\partial}{\partial \mathbf{r}}\left[\frac{\partial \psi}{\partial \mathbf{r}} + \psi\frac{\partial}{\partial \mathbf{r}}\left(\frac{V}{k\Theta}\right)\right] - \frac{\partial}{\partial \mathbf{r}}(\psi \mathbf{K}\mathbf{r}), \tag{12}$$

where the dependency from the geometric center has disappeared and the velocity difference is linearly approximated. The corresponding equation for the conformation tensor follows from (10) by neglecting $\mathcal{L}$ and assuming $\rho_p = \text{const}$ and $D_{\rm tr} = 0$.

Typically, as already anticipated, the translational diffusivity is so small that the corresponding diffusive terms are normally safely neglected in laminar flows. In turbulence these diffusion-like terms are instead crucial to prevent ultraviolet instabilities. Actually, turbulent flows are characterized by a continuous flux of energy from the low towards the high wave numbers. This effect is related to the meandering of vortex lines due to stretching and folding of material lines. The same process operates on the conformation tensor, via the term $K\mathcal{R} + \mathcal{R}K^T$, and leads to the formation of smaller and smaller scales. In fact, the tiny translational diffusivity is a singular perturbation, and eventually, a sufficiently fine scale develops where the cutoff finally starts operating. In numerical simulations, however, the resolution needed to reach the dissipative cutoff is never available for physically meaningful choices of the parameters, and the diffusion term is usually artificially enhanced as a tool to control the high wave number instability [7].

As easily verified by direct substitution, for a linear spring, hence a quadratic potential $V$, the solution of equation (3) is a Gaussian distribution and the PDF is fully described by its covariance tensor $\mathcal{R}$. In the finite elongation nonlinear elastic (FENE) model, where the potential is chosen as

$$V = -\frac{1}{2} H (r_{\max}^2 - r_0^2) \log \left[ \frac{r_{\max}^2 - r^2}{r_{\max}^2 - r_0^2} \right] \tag{13}$$

with $r$ the modulus of $\mathbf{r}$, the Gaussian ansatz does not hold and the solution of the full equation for the PDF is required. This procedure is too cumbersome to be used in a direct numerical simulation (DNS) of wall turbulence. Instead a closure assumption, known as Peterlin's approximation, is adopted. This allows one to account for finite extensibility without substantially increasing the complexity of the model with respect to the case of linear elasticity.

### 6.2.2. Evolution equation for the conformation tensor

Peterlin's hypothesis, i.e., the spring coefficient of each dumbbell depends only on the local average elongation, allows for the introduction of a Gaussian approximation for the PDF (FENE-P model [17], where "P" stands for Peterlin), and reduces the complexity of the problem to the equation for the covariance matrix $\mathcal{R}_{ij}$, which in dimensionless form becomes

$$\frac{D\mathcal{R}_{ij}}{Dt} = -\frac{1}{\mathrm{De}} f(\mathcal{R})\mathcal{R}_{ij} + K_{ir}\mathcal{R}_{rj} + \mathcal{R}_{ir}K_{jr} + \frac{1}{\mathrm{De}}\delta_{ij} + \frac{1}{\mathrm{Pe}}\nabla^2\mathcal{R}_{ij}, \tag{14}$$

where $D/Dt$ denotes the material time derivative and

$$f(\mathcal{R}) = \frac{R_{\max}^2 - R_0^2}{R_{\max}^2 - \operatorname{tr}\mathcal{R}}. \tag{15}$$

The conformation tensor is directly linked to the extra stress $T_{ij}^p$ caused by the dumbbells

$$T_{ij}^p = \frac{\eta_p}{\mathrm{De}}(f(\mathcal{R})\mathcal{R}_{ij} - \delta_{ij}), \tag{16}$$

which provides the back reaction of the polymers on the macroscopic velocity field assumed to be governed by the standard incompressible Navier–Stokes equations with the additional terms given by the divergence of the extra stress.

The dimensionless parameters appearing in the previous expressions are the following: the Deborah number De, which is the ratio between the polymer relaxation time and the fluid dynamic time scale; $\eta_p$, which is the ratio between the polymer contribution to the viscosity and the viscosity of the solvent; $R_{\max}^2$, which is the polymer maximum allowed elongation. Finally, Pe is the Peclet number, i.e., the ratio of the convective to the diffusive time scale of the polymers.

## 6.3. Thermodynamic aspects of the polymer solution

In order to grasp the mechanism of interaction between turbulence and microstructure, it is helpful to present a few considerations on the thermodynamic behavior of dilute polymer solutions. The basic idea is that the dissipation of the polymers, i.e., the friction between the beads and the solvent, principally acts on a purely local basis, an aspect captured also by neglecting the diffusivity. In this context the thermodynamics of the system may be described in terms of the free energy of the ensemble to successively recover the additional contributions coming from the diffusivity.

### 6.3.1. Equilibrium distribution and free energy

The first term on the left-hand side of the simplified kinetic equation (12), after rearranging the terms in brackets, becomes

$$\begin{aligned} D_{\text{rot}} \frac{\partial}{\partial \mathbf{r}} \left[ \frac{\partial \psi}{\partial \mathbf{r}} + \psi \frac{\partial}{\partial \mathbf{r}} \left( \frac{V}{k\Theta} \right) \right] \\ = \frac{1}{\zeta} \frac{\partial}{\partial \mathbf{r}} \left[ \psi \left( k_B \Theta \frac{\partial}{\partial \mathbf{r}} \log \psi - k_B \Theta \frac{\partial}{\partial \mathbf{r}} \log \Psi_{\text{eq}} \right) \right], \end{aligned} \tag{17}$$

where the expression for the equilibrium PDF is

$$\Psi_{\text{eq}}(\mathbf{r}) = c e^{-\frac{V}{k_B \Theta}}, \quad V = \frac{3 k_B \Theta}{2 n b^2} \mathbf{r}^2. \tag{18}$$

(12) may be rewritten as

$$\frac{\partial \psi}{\partial t} = \frac{1}{\zeta} \frac{\partial}{\partial \mathbf{r}} \left[ \psi \left( k_B \Theta \frac{\partial}{\partial \mathbf{r}} \log \frac{\psi}{\Psi_{\text{eq}}} \right) \right] - \frac{\partial}{\partial \mathbf{r}} [\psi \mathbf{K} \mathbf{r}]. \tag{19}$$

The term in round brackets is the time-dependent potential related to the Helmholtz free energy of the solution (see [12])

$$A^p = n_p k_B \Theta \int \psi(\mathbf{r}) \log \frac{\psi}{\Psi_{\text{eq}}}. \tag{20}$$

Using the explicit expressions for the two PDFs, where the end-to-end vector is made dimensionless with the root-mean-square value at the equilibrium,

$$\Psi_{\text{eq}} = \left( \frac{1}{2\pi} \right)^{\frac{3}{2}} e^{-\frac{1}{2} R_i R_i}, \qquad \psi = \left( \frac{1}{2\pi} \right)^{\frac{3}{2}} \left( \frac{1}{\det \mathcal{R}} \right)^{\frac{1}{2}} e^{-\frac{1}{2} R_i \mathcal{R}_{ij}^{-1} R_j}, \tag{21}$$

the dimensionless free energy is

$$A^p = +\frac{1}{2} \frac{\eta_p}{\text{De}} [\text{tr}(\mathcal{R} - \delta) - \log(\det \mathcal{R})]. \tag{22}$$

For the FENE-P model the free energy reads (see [2])

$$A^p = -\frac{1}{2} \frac{\eta_p}{\text{De}} \left[ (R_{\max}^2 - 3) \log \left( \frac{R_{\max}^2 - \text{tr}\, \mathcal{R}}{R_{\max}^2 - 3} \right) + \log(\det \mathcal{R}) \right]. \tag{23}$$

### 6.3.2. Entropy production

The evolution equation for the free energy gives the relation between the rate of change of the free energy, the polymer stress power $T^p_{ij} E_{ij}$ with $E_{ij}$ the strain tensor, and the dissipation due to the polymers. Following the chain rule, the rate of change of the free energy is

$$\frac{DA^p}{Dt} = \frac{\partial A^p}{\partial \mathcal{R}_{ij}} \frac{D\mathcal{R}_{ij}}{Dt}, \tag{24}$$

where from (23),

$$\begin{aligned}\frac{\partial A^p}{\partial \mathcal{R}_{ij}} &= \frac{1}{2}\frac{\eta_p}{\mathrm{De}}\left[f(\mathcal{R}_{ii})\delta_{ij} - \frac{1}{\det \mathcal{R}}\frac{\partial(\det \mathcal{R})}{\partial \mathcal{R}_{ij}}\right] \\ &= \frac{1}{2}\frac{\eta_p}{\mathrm{De}}[f(\mathcal{R}_{ii})\delta_{ij} - \mathcal{R}^{-1}_{ij}].\end{aligned} \tag{25}$$

Hence, neglecting diffusion in equation (14), the material derivative in (24) follows as

$$\begin{aligned}\frac{DA^p}{Dt} &= \frac{\eta_p}{\mathrm{De}}[f(\mathcal{R}_{ii})\mathcal{R}_{ij}E_{ij} - K_{ii}] \\ &\quad - \frac{1}{2}\frac{\eta_p}{\mathrm{De}^2}[f(\mathcal{R}_{ii})^2\mathcal{R}_{ii} - 6f(\mathcal{R}_{ii}) + \mathcal{R}^{-1}_{ii}].\end{aligned} \tag{26}$$

For incompressible flows $K_{ii} = 0$ and the first square brackets reduces to the polymer stress power (16). Finally, the last two terms in (14) can be written as

$$\left(-\frac{1}{\mathrm{De}} f(\mathcal{R})\mathcal{R}_{ij} + \frac{1}{\mathrm{De}}\delta_{ij}\right) = -\frac{2}{\eta_p}\frac{\partial A^p}{\partial \mathcal{R}_{ik}}\mathcal{R}_{kj}. \tag{27}$$

The second square bracket in (26) reduces to

$$-\frac{2}{\eta_p}\frac{\partial A^p}{\partial \mathcal{R}_{ik}}\mathcal{R}_{kj}\frac{\partial A^p}{\partial \mathcal{R}_{ij}} = -\Phi, \tag{28}$$

i.e.,

$$\frac{\eta_p}{2}\Phi = \frac{\partial A^p}{\partial \mathcal{R}_{ik}}\mathcal{R}^{\frac{1}{2}}_{kl}\mathcal{R}^{\frac{1}{2}}_{lj}\frac{\partial A^p}{\partial \mathcal{R}_{ij}} = \mathcal{G}_{jl}\mathcal{G}^T_{lj} = \mathcal{G}_{jl}\mathcal{G}_{jl}, \tag{29}$$

which, as a positive definite quadratic form, implies that $\Phi$ cannot be negative. Consequently, the stress power

$$\mathbf{T}^p \cdot \mathbf{E} = \frac{DA^p}{Dt} + \Phi \tag{30}$$

is expressed as the sum of two contributions, a dissipation term $\Phi$, and the rate of change of free energy accumulated or released. Since the free energy of the polymer chains is directly related to their entropy,

$$A^p = -\Theta S, \tag{31}$$

in isothermal conditions the configurational entropy $S^p$ of the polymers is given by

$$\frac{DS^p}{Dt} = -\frac{1}{\Theta}\mathbf{T}^p \cdot \mathbf{E} + \frac{\Phi}{\Theta}. \tag{32}$$

To recover the effect of diffusion, it is sufficient to retain the corresponding term in the derivation of equation (26) to yield

$$\mathbf{T}^p \cdot \mathbf{E} = \frac{DA^p}{Dt} + \Phi + \Phi_D. \tag{33}$$

### 6.3.3. Energy balance

The rate of change of the total energy $U_T = U + q$, with $U$ the internal energy and $q = 1/2|\mathbf{u}|^2$, can be expressed as

$$\rho\frac{DU_T}{Dt} = \rho u_i f_i + \frac{\partial u_i T_{ij}}{\partial x_j} + \rho Q, \tag{34}$$

where $T_{ij}$ is the total stress and $Q$ enforces isothermal conditions. Given the balance of mechanical energy

$$\left(\rho\frac{Du_i}{Dt} - \rho f_i + \frac{\partial T_{ij}}{\partial x_j}\right) u_i = 0, \tag{35}$$

the equation for the internal energy reads

$$\frac{DU}{Dt} = -p\frac{\partial u_i}{\partial x_i} + \sigma_{ij}E_{ij} + T^p_{ij}E_{ij} + \rho Q, \tag{36}$$

where the stress tensor is decomposed into a Newtonian part, the pressure $p$ and viscous term $\sigma_{ij}$, and a non-Newtonian part, $T^p_{ij}$. Since in incompressible flows the pressure term is zero, the free energy of the solution is entirely given by (23). Accounting for isothermal conditions,

$$\frac{DA}{Dt} = \frac{DU}{Dt} - \Theta\frac{DS}{Dt}, \tag{37}$$

where $S$ is the entropy of the solution, whose rate is given by

$$\frac{DS}{Dt} = \frac{1}{\Theta}\sigma_{ij}\frac{\partial u_i}{\partial x_j} + \frac{\Phi}{\Theta} + \frac{Q}{\Theta}. \tag{38}$$

In the entropy, two contributions are naturally identified, one associated with the polymers (32) and the other given by

$$\frac{DS^s}{Dt} = \frac{1}{\Theta}\sigma_{ij}E_{ij} + \frac{1}{\Theta}T^p_{ij}E_{ij} + \frac{Q}{\Theta}. \tag{39}$$

In the above decomposition, the stress power normalized by the temperature corresponds to the entropy exchange between the two components. Typically, when the polymers are elongated, their entropy decreases, and consequently, the stress power tends to increase the other contribution by a corresponding amount. The previous remarks play a crucial role in understanding the interactions between turbulence and microstructure. The equation for the kinetic energy $q$ is

$$\begin{aligned}\frac{Dq}{Dt} &= \frac{\partial}{\partial x_j}\left(-pu_j + \frac{2}{\mathrm{Re}}u_iE_{ij} + \frac{1}{\mathrm{Re}}u_iT^p_{ij}\right)\\ &\quad - \frac{2}{\mathrm{Re}}E_{ij}E_{ij} - \frac{1}{\mathrm{Re}}E_{ij}T^p_{ij},\end{aligned} \tag{40}$$

where the term in parentheses can only spatially redistribute energy. The other terms are the energy dissipated through molecular viscosity of the Newtonian solvent and the stress power. The latter is responsible for the energy drain and supply due to the polymers. The polymers introduce an additional dissipation term, and equation (33) provides the link between energy exchange, extra dissipation, and polymer free energy. This means that kinetic energy may be, at least partially, converted into free energy and vice versa and that spatial energy transfer could, in principle, occur via convection of polymer molecules, which extract energy from the macroscopic field at a certain point and might give it back elsewhere.

## 6.4. Polymer dynamics and turbulence

In this section, a few properties of turbulent flows are recalled to discuss the effect of turbulence on polymers. Numerical results are used to discuss the alteration of turbulence by polymers.

### 6.4.1. The scales of turbulence

Turbulence is associated with chaotic behavior in both space and time. Often several symmetries are present in the system, such as stationarity and homogeneity in one or more directions. For example, a turbulent flow in a channel may be assumed to be statistically stationary and invariant under translations in the directions parallel to the channel walls. The typical turbulent field presents a characteristic scale, the integral scale of turbulence $L$, above which the correlation tensor, defined by

$$R_{ij}(\mathbf{x}-\mathbf{x}_0) := \langle\langle u_i(\mathbf{x},t)u_j(\mathbf{x}_0,t)\rangle\rangle, \tag{41}$$

where the double angular brackets denote an ensemble average with respect to turbulent realizations, rapidly falls to zero. The correlation tensor may be Fourier transformed in the directions of homogeneity to yield the energy spectrum. For the channel geometry, the spectrum is given by

$$E(y, y_0; \mathbf{k}) := \frac{1}{(2\pi)^2}\int_{\mathbb{R}^2} R_{ij}(\mathbf{x}-\mathbf{x}_0)e^{-\jmath\mathbf{k}\cdot(\mathbf{x}-\mathbf{x}_0)}d(x-x_0)d(z-z_0), \tag{42}$$

where $x$ and $z$ are the two coordinates parallel to the walls and $y$ is normal. The typical spectrum exhibits a viscous cutoff above a certain wave number $k_K$ that defines the smallest characteristic scale of the turbulent field $\eta = 2\pi/k_K$, called the Kolmogorov scale. The presence of the cutoff implies that the kinetic energy content in the Fourier modes greater than $k_K$ is negligible. This behavior is well known in Newtonian flows at large Reynolds number, where the small viscosity coefficient makes dissipation of energy efficient only at scales comparable with the Kolmogorov scale. Actually, the Kolmogorov description of turbulence provides a link between the scale $\eta$ and the local average dissipation rate $\epsilon$

$$\eta = \left(\frac{\nu^3}{\epsilon}\right)^{1/4} \tag{43}$$

and predicts the existence of a range of scales much smaller than $L$ and much larger than $\eta$, the inertial subrange, where the direct effect of viscosity is negligible and where the turbulence is unaffected by the specific energy injection process. In this range, characterized by a self-similar spectrum of the form $E(k) \propto k^{-5/3}$, energy is transferred at a constant rate from small to large wave numbers. This cascade process continuously moves energy from the larger scales, where power injection occurs due to flow instabilities or external forcing, towards the smaller scales, where dissipation takes place. The physical mechanism behind energy transfer is provided by the advection term in the Navier–Stokes equation via stretching and tilting of material lines. Actually, in inviscid flows the evolution of vorticity, $\boldsymbol{\zeta}$, parallels that of material line elements, as implied by the transport equation

$$\frac{D\boldsymbol{\zeta}}{Dt} - \boldsymbol{\zeta}\mathbf{K} = 0. \tag{44}$$

To grasp the nature of the effects of turbulence on polymers it is instrumental to comment on the dynamics described by an equation of the form

$$\frac{D\mathcal{R}}{Dt} - (\mathbf{K}\mathcal{R} + \mathcal{R}\mathbf{K}^T) = 0, \tag{45}$$

which involves the crucial nonlinear operator appearing in the equation for the conformation tensor, equation (10). According to equation (45), $\mathcal{R}$ is frozen in the fluid, exactly as for a vector governed by equation (44), since again $\mathcal{R}$ is entirely reconstructed from its initial condition and the deformation gradient $\mathbf{F}$.

The analogy implies that the same mechanism of stretching and folding of material lines produces a similar cascade process in equations of the form (10). In fact both the equations for vorticity and conformation tensor contain additional terms. To stop the cascade, the crucial terms are those related to diffusion. As already recalled, the process of meandering of vortex lines stops at the level of the Kolmogorov scale. The corresponding process on $\mathcal{R}$ may instead continue indefinitely, unless diffusion-like terms are retained in the equation. The picture is essentially the same found in the mixing of strictly nondiffusing tracers: In the stretching and folding process, material elements with different concentrations are brought into contact, implying the possibility of unboundedly large concentration gradients. Concerning polymers, the situation is worse since they react back on the velocity field through the divergence of the extra stress, which is directly related to $\mathcal{R}$, equation (16). The process develops high wave number modes, and due to the small diffusivity of the macromolecules, the physical cutoff is moved to such large wave numbers as to make a well-resolved numerical solution unaffordable, unless numerical dissipation is introduced in the equations.

As shown in the next section, the artificial diffusivity approach makes interesting simulations available for the investigation of polymers in a turbulent environment. However, in principle, at very small diffusivities, intense spatial gradients of the descriptors of the polymers are produced. This raises the question, addressed in Section 6.5, of possible additional microstructural interactions, beyond diffusion and migration, between neighboring material elements.

### 6.4.2. The alteration of wall turbulence

The model described in Section 6.2.1 has been successfully applied to DNS only recently [13, 5, 14]. In the channel flow simulation discussed here, turbulence is maintained through a pressure gradient that varies in order to provide a constant mass flux. Figure 1 compares a portion of the time history of $\mathrm{Re}_\tau$ with the Newtonian simulation with the same mass flux. Here $\mathrm{Re}_\tau = u_\tau h/\nu$, with $u_\tau = \sqrt{\tau_w/\rho}$, $\tau_w$ being the average tangential stress at the wall. A reduction in the average value of $\mathrm{Re}_\tau$ from 175 for the Newtonian to 150 for the viscoelastic case is apparent, giving a drag reduction of 36%. It is worth recalling here that the chaotic behavior of turbulence is always associated with the random local occurrence of certain typical patterns in the relevant fields, which are known in the literature as coherent structures (see, e.g., [11, 16]). The presence of regions of more or less locally ordered motion is actually the origin of the intermittency of the energy dissipation, which is found to occur in spots immersed in regions where energy is mainly spatially advected.

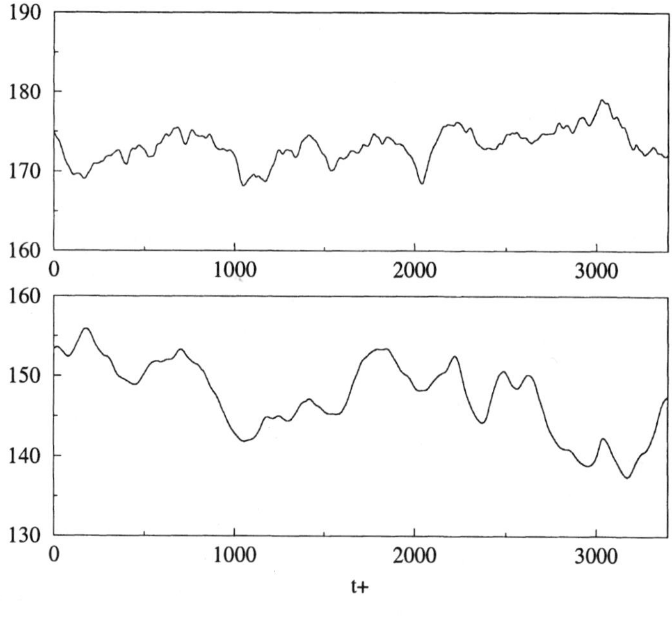


FIGURE 1. Friction Reynolds number $\mathrm{Re}_\tau$ vs. time for the Newtonian (upper plot) and the viscoelastic flow (lower plot). The simulations have been performed on a domain $2\pi h \times 2h \times 1.2\pi h$ with a computational grid of $96 \times 129 \times 96$ at a nominal Reynolds number of 5000. The specific parameters of the viscoelastic simulation are $\mathrm{De} = 10$, $\eta_p = 0.1$, $R^2_{\max} = 1000$, and $\mathrm{Pe} = 1700$.

In fact intermittency was a most unexpected feature found as soon as anemometers with sufficient spatial and temporal resolution became available. The coherent structures are particularly significant in shear flows where they provide the main source of turbulent kinetic energy, being associated with the relatively large scale fluctuations responsible for the Reynolds stresses, hence for energy draining from the mean field. Numerical simulations have been able to isolate the elemental phenomena underlying birth, evolution, disappearance, and rebirth of the coherent structures, the so-called regeneration cycle of wall turbulence. This is done by considering a small portion of the flow where only a few coherent patterns can be found. In such conditions all spatially averaged quantities show characteristic fluctuations that are the footprints of the regeneration process. This behavior is recognized in the time history of the average wall friction, as expressed by the friction Reynolds number shown in Figure 1. There the random fluctuations associated with the chaotic dynamics of the wall structures display time scales that are substantially increased by drag-reducing (bottom plot) with respect to classical Newtonian cases (top plot) [15]. The dimensions of the structures involved in the process may be estimated by addressing the correlation tensor, equation (41). Significant examples are reported in Figure 2, where, with respect to Newtonian flows, the transverse correlation lengths are substantially increased [18].

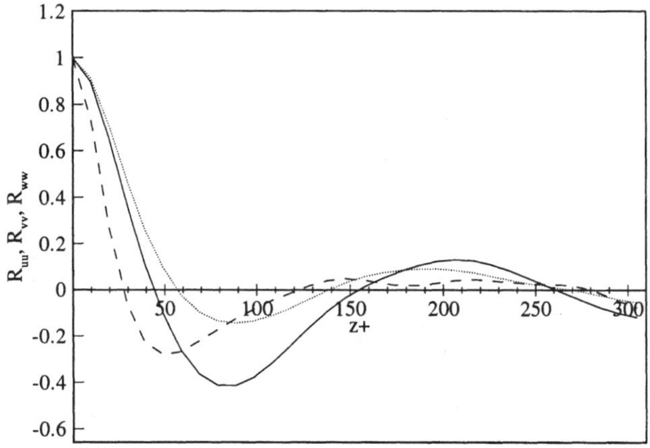


FIGURE 2. Spanwise correlation at $y^+ := u_\tau y/\nu = 10$ of the velocity components for the viscoelastic flow: $R_{uu}$ (solid line), $R_{vv}$ (dashed line), $R_{ww}$ (dotted line) vs. $z^+ = u_\tau z/\nu$.

Qualitatively, Figure 3 offers the explanation of this effect by showing a well-organized pattern in the streamwise velocity field near the wall. This substantial level of organization is inherited by the viscoelastic reaction, the divergence of the extra stress, equation (15). The organization of the polymers passes through a process of time scale selection, with the polymers most affected by fluctuations at frequencies comparable with their natural time scale, defined as the principal relaxation time. It is worth emphasizing that these scales, belonging to the intermediate hydrodynamical frequency range, carry a dynamics that is largely affected by convection, implying that the correct framework for physical interpretation is the Lagrangian one. The DNS also makes available all the thermodynamic quantities discussed in Section 6.3, namely, the free energy, the dissipation, and the polymer stress powers, as well as the different contribution to the entropy. As an example, in Figure 4, the rate of change of the polymer free energy is superimposed to the isolines of the streamwise perturbation velocity. From the figure, the correlation between the elementary regeneration events and free energy accumulation are apparent.

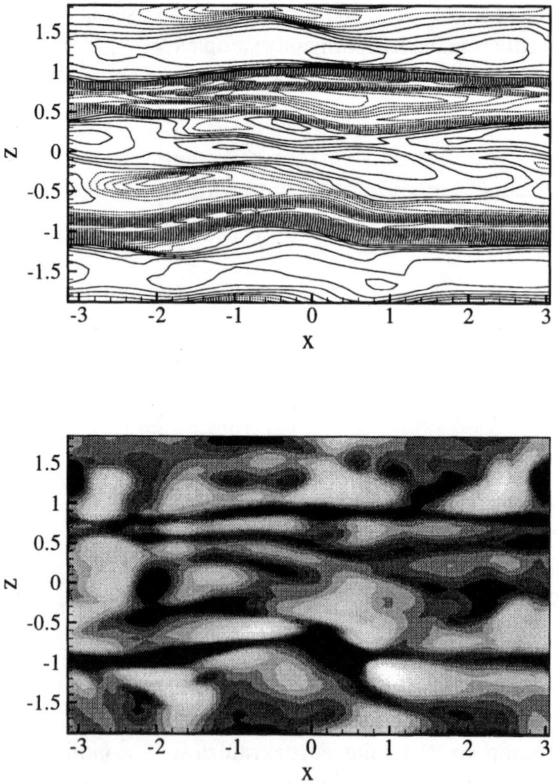


FIGURE 3. Plane section parallel to the walls at $y^+ = 30$. Top: Contour lines of the streamwise component of perturbation velocity. Bottom: Streamwise component of the viscoelastic reaction, defined as the divergence of the extra stress.

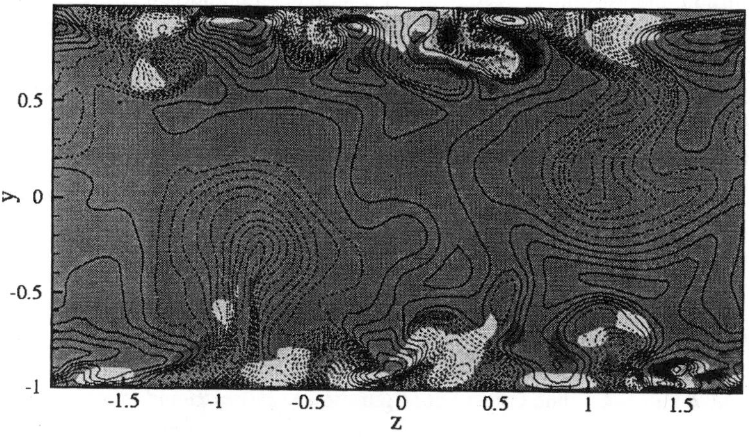


FIGURE 4. Cross-flow section: Isocontours of the streamwise component of perturbation velocity and of the rate of change of the polymer free energy, lines, and flood, respectively.

As a summary, the numerical results obtained by the FENE-P model with diffusion show the capability of the model to shed light on the rather involved dynamics of polymers in a turbulent environment. A point remains open to discussion concerning the possible existence of mechanisms of interaction of nearby populations of polymers at small diffusivities, a topic addressed from a general perspective in the next section.

## 6.5. A multifield description of substructural interaction

The microrheological models presented in the previous sections describe the statistical behavior of the polymers limiting possible interactions between neighboring elements to diffusion or migration effects. As already mentioned at the end of Section 6.2.1, diffusion terms are crucial to control the ultraviolet instability that may take place in a turbulent environment. A general investigation of the possible cross-element interactions is thus worthwhile. However, the adopted point of view requires a clear knowledge of the microdynamics, which may be lacking in more general conditions. To overcome this difficulty it is useful to adopt a general framework to generate appropriate balance equations for standard and substructural interactions, independently of any special microrheological mechanism. Specific phenomenological constitutive equations may then account for the different substructural mechanisms to be considered. In any case, such a general framework should recover as special case the equation obtained from specific microrheological models able to fit experimental data for polymeric liquids. To this purpose the appropriate setting is given by the multifield theories introduced to study continua with microstructure [3, 9], where proper descriptors are used to derive the balance equations for the interactions. Here, conforming to the previous approach, the conformation tensor $\mathcal{R}$ is selected as the appropriate descriptor for the population of the polymeric chains. The requirements of invariance with respect to changes of observers are then applied to obtain the balances of standard and substructural interactions.

### 6.5.1. Placements and descriptors

The regular region of the three-dimensional Euclidean point space $\mathcal{E}^3$ occupied by the fluid in its reference configuration is denoted by $\mathcal{B}_0$. Points of $\mathcal{B}_0$ are indicated with $\mathbf{X}$. The standard deformation of $\mathcal{B}_0$ is described by orientation-preserving diffeomorphisms $\hat{\mathbf{x}}(\cdot){:}\mathcal{B}_0 \to \mathcal{E}^3$. The current configuration $\hat{\mathbf{x}}(\mathcal{B}_0) = \mathcal{B}$ and is regular, too. Any regular subset $\mathfrak{b}$ of $\mathcal{B}$ is called *part* of the fluid in the following. The picture of the configuration of the fluid just summarized is not sufficient because no geometrical information is given on the substructure generated by the polymeric chains. Here each material element $P$ placed at $\mathbf{X}$ in the reference configuration is endowed with a population of chains. Problems related to possible entanglement of polymeric chains are presently excluded. Following the microrheological approach, each chain is described by an end-to-end stretchable vector $\mathbf{r}$ and at each point $\mathbf{X}$ of $\mathcal{B}_0$ a distribution function $\psi_{P(\mathbf{X})}(\mathbf{r})$ represents the population of the chains within the material element $P$ placed at $\mathbf{X}$. The requirement of indifference to the transformation $\mathbf{r} \to -\mathbf{r}$ suggests that the minimal complexity in the description is achieved by using the second-order tensor

$$\mathcal{R}^{\#}(\mathbf{X}) = \int_{\mathbb{R}^3} \psi_{P(\mathbf{X})}(\mathbf{r})\mathbf{r} \otimes \mathbf{r} \tag{46}$$

to provide a rough coarse-grained geometrical descriptor of the population itself (see (7)). $\mathcal{R}^{\#}$ is used as the order parameter in the following, where the star denotes the material representation of the field.

The descriptor $\mathcal{R}^{\#}$ of the substructure is also associated with each material element. A *spatial* representation $\mathcal{R}$ of $\mathcal{R}^{\#}$ is clearly given by $\mathcal{R} = \hat{\mathcal{R}}(\mathbf{x}(\mathbf{X}))$. The current placement and the value of the order parameter at the instant $t$ of the material element at $\mathbf{X}$ are given by $\mathbf{x}(\mathbf{X}, t)$ and $\mathcal{R}^{\#}(\mathbf{X}, t)$. Rates in the material representation are given by $\dot{\mathbf{x}}(\mathbf{X}, t)$ and $\dot{\mathcal{R}}^{\#}(\mathbf{X}, t)$ while in the spatial representation they are given by $\mathbf{u}(\mathbf{x}, t)$ and $\dot{\mathcal{R}}(\mathbf{x}, t) = D\mathcal{R}(\mathbf{x}, t)/Dt$. Clearly, $\mathbf{x} \doteq \mathbf{u}$ and by the chain rule $\dot{\mathcal{R}} = \partial_t \mathcal{R} + (\nabla \mathcal{R})\mathbf{u}$.

As a crucial point of the modeling in the present context, the order parameter is considered an *observable* object; i.e., it should sense a change of observer. Two external spatial observers (e.g., $\mathcal{O}$ and $\mathcal{O}'$), seeing the same configuration, evaluate two different pairs of rates. The relevant rules of change of observers related by rigid-body motions, i.e., by proper orthogonal tensors $\mathbf{Q}$, read

$$\mathbf{u}^* = \mathbf{u} + \mathbf{c}(t) + \dot{\mathbf{q}}(t) \times (\mathbf{x} - \mathbf{x}_0), \tag{47}$$

$$\dot{\mathcal{R}}^* = \dot{\mathcal{R}} + \mathcal{A}\dot{\mathbf{q}}. \tag{48}$$

In (47), $\mathbf{c}(t)$ is the translational velocity, $\dot{\mathbf{q}}$ the rotational velocity (i.e., the characteristic vector of $\mathbf{Q}^T\dot{\mathbf{Q}}$), and $\mathbf{x}_0$ a fixed point in space, while in (3) the linear operator $\mathcal{A}$ is a third-order tensor whose covariant components are given by $\mathcal{A}_{ijk} = \mathsf{e}_{ilj}\mathcal{R}_{lk} - \mathcal{R}_{il}\mathsf{e}_{jlk}$, where $\mathsf{e}$ is Ricci's tensor (see, e.g., [9] for a complete description of changes of observer).

### 6.5.2. Microstresses and balance equations

Since the order parameter is an observable object, the interactions developing power in the rate of change of the order parameter have to be accounted for. In the following only contact interactions exchanged by neighboring material elements are considered as a consequence of the rearrangement of the polymeric chains in each element. These mechanisms, though as yet unidentified in microrheological terms, are to be thought of as additional to the diffusion discussed in Section 6.1, which is not readdressed here under the multifield viewpoint. For the sake of simplicity bulk interactions on the polymeric chains are not considered. These bulk interactions could have a twofold nature: (i) they could account for an inertial contribution of the agitation of the chains on the macroscopic motion; (ii) they could be of electrical nature in the case of polyelectrolyte polymers (see, e.g., [8]). However, the inertial contribution of the chains is basically negligible and the case of applied external electric fields are not considered.

Given any arbitrary part $\mathfrak{b}$ of $\mathcal{B}$, $\mathcal{P}_{\mathfrak{b}}^{\text{ext}}$ is the power developed on it by all the external standard and substructural interactions as a linear functional on the space of the rates. In this way the general framework of multifield theories describing material substructures is followed as developed in [3, 4, 9] and the substructural interactions between neighboring polymeric chains is represented accordingly. With these premises, the expression of $\mathcal{P}_{\mathfrak{b}}^{\text{ext}}$ is given by

$$\mathcal{P}_{\mathfrak{b}}^{\text{ext}}(\mathbf{u}, \dot{\mathcal{R}}) = \int_{\mathfrak{b}} \mathbf{b} \cdot \mathbf{u} + \int_{\partial\mathfrak{b}} (\mathbf{T}\mathbf{n} \cdot \mathbf{u} + \mathcal{S}\mathbf{n} \cdot \dot{\mathcal{R}}), \tag{49}$$

where $\mathbf{b}$ is the vector of standard bulk forces, $\mathbf{T}$ the Cauchy stress, $\mathbf{n}$ the outward unit normal to the boundary $\partial\mathfrak{b}$ of $\mathfrak{b}$, and $\mathcal{S}$ is a third-order tensor called the *microstress*. It represents the interactions exchanged by neighboring material elements across the surface $\partial\mathfrak{b}$ as a consequence of the rearrangement of the polymeric chains in at least one of them.

The basic requirement is then

> $\mathcal{P}_{\mathfrak{b}}^{\text{ext}}$ *is invariant under changes of observers ruled by* (47) *and* (48) *for any* $\mathfrak{b}$, i.e.,
>
> $$\mathcal{P}_{\mathfrak{b}}^{\text{ext}}(\mathbf{u}^*, \dot{\mathcal{R}}^*) = \mathcal{P}_{\mathfrak{b}}^{\text{ext}}(\mathbf{u}, \dot{\mathcal{R}}) \tag{50}$$
>
> *for any choice of* $\mathbf{c}(t)$, $\dot{\mathbf{q}}(t)$, *and* $\mathfrak{b}$.

As a consequence of the requirement of invariance, the equation

$$\begin{aligned} &\mathbf{c} \cdot \left( \int_{\mathfrak{b}} \mathbf{b} + \int_{\partial \mathfrak{b}} \mathbf{T}\mathbf{n} \right) \\ &+ \dot{\mathbf{q}} \cdot \left( \int_{\mathfrak{b}} (\mathbf{x} - \mathbf{x}_0) \times \mathbf{b} + \int_{\partial \mathfrak{b}} ((\mathbf{x} - \mathbf{x}_0) \times \mathbf{T}\mathbf{n} + \mathcal{A}^T \mathcal{S}\mathbf{n}) \right) = 0 \end{aligned} \tag{51}$$

follows for any choice of $\mathbf{c}(t)$ and $\dot{\mathbf{q}}(t)$. The product $\mathcal{A}^T \mathcal{S}$ is a second-order tensor whose covariant components are given by $\mathcal{A}^T_{lij} \mathcal{S}_{ijk}$. Thanks to the arbitrariness of $\mathbf{c}(t)$ and $\dot{\mathbf{q}}(t)$, the validity of the integral balances of momentum and moment of momentum are, respectively, obtained as

$$\int_{\mathfrak{b}} \mathbf{b} + \int_{\partial \mathfrak{b}} \mathbf{T}\mathbf{n} = 0, \tag{52}$$

$$\int_{\mathfrak{b}} (\mathbf{x} - \mathbf{x}_0) \times \mathbf{b} + \int_{\partial \mathfrak{b}} ((\mathbf{x} - \mathbf{x}_0) \times \mathbf{T}\mathbf{n} + \mathcal{A}^T \mathcal{S}\mathbf{n}) = 0. \tag{53}$$

The arbitrariness of $\mathfrak{b}$ allows one to obtain the standard pointwise balance of momentum

$$\mathbf{b} + \operatorname{div} \mathbf{T} = 0, \tag{54}$$

which can be clarified further when the vector of the bulk forces is considered as the sum of an inertial and a noninertial contribution that (49) may be written in the standard way as

$$\mathbf{b}^{\mathrm{ni}} + \operatorname{div} \mathbf{T} = \rho \dot{\mathbf{u}}. \tag{55}$$

More delicate is the treatment of the integral balance of moment of momentum. Thanks to the arbitrariness of $\mathfrak{b}$ and the validity of (53), the pointwise balance reads

$$\mathsf{e}\mathbf{T} = \mathcal{A}^T \operatorname{div} \mathcal{S} + (\nabla \mathcal{A}^T) \mathcal{S}, \tag{56}$$

where, as in the previous sections, $\nabla$ is the gradient with respect to $x$. To assure the validity of (56), the existence of a second-order tensor such that

$$\mathcal{A}^T \mathbf{z}' = \mathsf{e}\mathbf{T} - (\nabla \mathcal{A}^T) \mathcal{S} \tag{57}$$

is necessary. This allows one to write (56) as

$$\operatorname{div} \mathcal{S} - \mathbf{z}' = \mathbf{z}'' \tag{58}$$

with $\mathbf{z}''$ an arbitrary second-order tensor such that

$$\mathcal{A}^T \mathbf{z}'' = 0. \tag{59}$$

The sum $\mathbf{z}' + \mathbf{z}''$ is indicated by $\mathbf{z}$ and is called the *self-force* in the current jargon of multifield theories [3, 4, 9]. As the microstress represents the interactions between neighboring material elements, the self-force accounts for the interactions between chains in the same material element. In other words, in the absence of bulk motion, to maintain at the (dynamical) equilibrium the population of polymeric chains in a given material element, it is necessary that the interactions exchanged between elements of the population at the given element are balanced by the interactions exchanged with the populations of the neighboring elements.

### 6.5.3. The dissipation inequality and the evolution equation for the descriptor

A mechanical dissipation inequality (an appropriate isothermal version of the second law) is here used to obtain constitutive restrictions relating the measures of interaction listed above to the free energy of the fluid. It is consistent with the developments in Section 6.3: Basically, it is a generalization of (26) written in isothermal conditions. The inequality states that *for any part* $\mathfrak{b}$ and *for any choice of the rates and their gradients*

$$\frac{d}{dt}\int_{\mathfrak{b}} A - \mathcal{P}^{\text{ext}}_{\mathfrak{b}}(\mathbf{u}, \dot{\mathcal{R}}) \leq 0, \tag{60}$$

where $A$ is the total *free energy density* per unit volume (i.e., $A$ accounts for $A^p$ and the bulk free energy of the fluid flow). The free energy is assumed to have a state dependence of the type

$$A = \hat{A}(\mathbf{F}, \mathcal{R}, \nabla\mathcal{R}). \tag{61}$$

Taking into account the arbitrariness of the part $\mathfrak{b}$ and the validity of the balance equations (55) and (58), and developing the time derivative of $A$ from (61) at thermodynamic *equilibrium*, equation (60) reduces to

$$\begin{aligned}&(\partial_{\mathbf{F}} A\mathbf{F}^T - \mathbf{T} - (\nabla\mathcal{R})^T\mathcal{S}) \cdot \mathbf{K} + (\partial_{\nabla\mathcal{R}} A - \mathcal{S}) \cdot \dot{\overline{\nabla\mathcal{R}}}\\ &\quad + (\partial_{\mathcal{R}} A - \mathbf{z}) \cdot \dot{\mathcal{R}} \leq 0,\end{aligned} \tag{62}$$

where $\partial_y$ represents the partial derivative with respect to the argument $y$ and, in developing calculations, the identities $\dot{\mathbf{F}} = \mathbf{FK}$ and $\dot{\overline{\nabla\mathcal{R}}} = \nabla\dot{\mathcal{R}} + (\nabla\mathcal{R})\mathbf{K}$ were used. Because the inequality (62) is linear in their rates and gradients and must hold for any choice of them, the following identities must hold:

$$\mathbf{T} = \partial_{\mathbf{F}} A\mathbf{F}^T - (\nabla\mathcal{R})^T\mathcal{S}, \quad \mathcal{S} = \partial_{\nabla\mathcal{R}} A, \quad \mathbf{z} = \partial_{\mathcal{R}} A. \tag{63}$$

However, the free energy must remain unchanged under the action of the group of material symmetries. In the case of a fluid this group coincides with the unimodular group of all second-order tensors $\mathbf{H}$ with $|\det\mathbf{H}| = 1$. Consequently, (61) must remain unchanged with respect to the transformation $\mathbf{F} \to \mathbf{FH}$. However, as the tensor $\mathbf{H} = |\det\mathbf{F}|\mathbf{F}^{-1}$ is obviously unimodular, (61) reduces to

$$\hat{A}(\mathbf{F}, \mathcal{R}, \nabla\mathcal{R}) = \hat{A}(\iota, \mathcal{R}, \nabla\mathcal{R}), \tag{64}$$

where $\iota = |\det\mathbf{F}|$ is the specific volume of the fluid. Then the constitutive restrictions (63) reduce to

$$\mathbf{T} = \iota\partial_\iota\hat{A}(\iota, \mathcal{R}, \nabla\mathcal{R})\mathbf{I} - (\nabla\mathcal{R})^T\partial_{\nabla\mathcal{R}}\hat{A}(\iota, \mathcal{R}, \nabla\mathcal{R}), \tag{65}$$

$$\mathcal{S} = \partial_{\nabla\mathcal{R}}\hat{A}(\iota, \mathcal{R}, \nabla\mathcal{R}), \tag{66}$$

$$\mathbf{z} = \partial_{\mathcal{R}}\hat{A}(\iota, \mathcal{R}, \nabla\mathcal{R}). \tag{67}$$

Note that the simplest expression of (61) is a decomposed free energy of the form

$$\hat{A}(\iota, \mathcal{R}, \nabla\mathcal{R}) = \hat{A}_1(\iota) + \hat{A}_2(\mathcal{R}) + \frac{1}{2}\alpha|\nabla\mathcal{R}|^2 \tag{68}$$

with $\alpha$ some constant. In this case, the balance of substructural interactions (61) reduces to a Ginzburg–Landau equation for $\mathcal{R}$:

$$\alpha\Delta\mathcal{R} - \partial_{\mathcal{R}}\hat{A}_2(\mathcal{R}) = 0. \tag{69}$$

When a partially coupled form like $\hat{A}(\iota, \mathcal{R}, \nabla\mathcal{R}) = \hat{A}_3(\iota, \mathcal{R}) + \frac{1}{2}\alpha|\nabla\mathcal{R}|^2$ is allowed, the expression

$$\alpha\Delta\mathcal{R} - \partial_{\mathcal{R}}\hat{A}_3(\iota, \mathcal{R}) = 0 \tag{70}$$

is obtained. Far from thermodynamic equilibrium, some measures of interaction are the sum of a rate dependent nonequilibrium part (ne) and a rate-independent equilibrium part (eq). In particular, the decompositions

$$\mathbf{T} = \mathbf{T}^{\text{ne}}(\iota, \mathcal{R}, \nabla\mathcal{R}; \mathbf{K}, \dot{\mathcal{R}}) + \mathbf{T}^{\text{eq}}(\iota, \mathcal{R}, \nabla\mathcal{R}), \tag{71}$$

$$\mathbf{z} = \mathbf{z}^{\text{eq}}(\iota, \mathcal{R}, \nabla\mathcal{R}) + \mathbf{z}^{\text{ne}}(\iota, \mathcal{R}, \nabla\mathcal{R}; \mathbf{K}, \dot{\mathcal{R}}) \tag{72}$$

are considered, where $\mathbf{T}^{\text{eq}}$ and $\mathbf{z}^{\text{eq}}$ are given by (65) and (67). By inserting these decompositions into (62), the reduced dissipation inequality

$$\mathbf{T}^{\text{ne}} \cdot \mathbf{K} + \mathbf{z}^{\text{ne}} \cdot \dot{\mathcal{R}} \leq 0 \tag{73}$$

is obtained.

#### 6.5.4. A revised form of the evolution equation for the conformation tensor

Physical instances suggest the transformation of the local dissipation inequality (73) into

$$\mathbf{T}^{\text{ne}} \cdot \mathbf{K} + \mathbf{z}^{\text{ne}} \cdot \overset{\triangledown}{\mathcal{R}} \leq 0, \tag{74}$$

where $\overset{\triangledown}{\mathcal{R}}$ is the upper-convected derivative of the tensor $\mathcal{R}$ following the velocity field $\mathbf{u}$. In other words, $\overset{\triangledown}{\mathcal{R}} = \dot{\mathcal{R}} - \mathbf{K}\mathcal{R} - \mathcal{R}\mathbf{K}^T$.

Since the self-force represents the self-interaction of each polymeric chain on itself and the interactions between the elements of the population of a given material element, $\mathbf{z}^{\text{ne}}$ does not develop power in the collective motion of the population due to the transport of the fluid. Rather, $\mathbf{z}^{\text{ne}}$ develops power in the relative motion of the chains with respect to the fluid and one another. This is the basic reason of the substitution of the term $\mathbf{z}^{\text{ne}} \cdot \dot{\mathcal{R}}$ with $\mathbf{z}^{\text{ne}} \cdot \overset{\triangledown}{\mathcal{R}}$ and the corresponding substitution of $\mathbf{z}^{\text{ne}}(\iota, \mathcal{R}, \nabla\mathcal{R}; \mathbf{K}, \dot{\mathcal{R}})$ with $\mathbf{z}^{\text{ne}}(\iota, \mathcal{R}, \nabla\mathcal{R}; \mathbf{K}, \overset{\triangledown}{\mathcal{R}})$. Reasoning of this kind is consistent with relation (1), which indicates the relative motion of the polymeric chain with respect to the fluid as a source of dissipative behavior. A solution of the inequality (74) is

$$\mathbf{T}^{ne} = \mathbf{B}_1\mathbf{K} + \mathbf{B}_2\overset{\nabla}{\mathcal{R}}, \tag{75}$$

$$\mathbf{z}^{ne} = \mathbf{B}_2\mathbf{K} + \mathbf{B}_3\overset{\nabla}{\mathcal{R}} \tag{76}$$

with $\mathbf{B}_1, \mathbf{B}_2$, and $\mathbf{B}_3$ appropriate definite negative constant fourth-order tensors. When $\mathbf{B}_2$ vanishes identically and $\mathbf{B}_1$ and $\mathbf{B}_3$ reduce to scalars, equation (70) may be changed into the time-dependent Ginzburg–Landau equation

$$B_3\overset{\nabla}{\mathcal{R}} = \alpha\Delta\mathcal{R} - \partial_{\mathcal{R}}\hat{A}_3(\iota, \mathcal{R}), \tag{77}$$

and the balance of momentum reduces to the standard Navier–Stokes equation. The evolution equation (14) is a special case of (77), which is also able to capture more complicated situations due to the coupling in $\hat{A}_3$ between $\iota$ and $\mathcal{R}$. The term $\alpha\Delta\mathcal{R}$ is a natural consequence of a rather detailed picture of substructural interactions and accounts for weakly nonlocal interactions among the polymeric chains. However, it should be emphasized that its origin is different and complementary to the diffusion of macromolecules as discussed in Section 6.1. An extention of the present multifield approach to account for diffusion can be developed following the same guidelines but including a continuity equation for the density number of the polymeric chains.

## 6.6. Concluding remarks and perspectives

The results obtained by the DNS for a dilute polymer solution using a FENE-P model have been investigated to study the interaction of the microstructure with turbulence. The instantaneous fields show that the viscoelastic reaction is organized on the same scale of the structures responsible for the velocity fluctuations. Large positive values of the stress power are observed in correspondence with intense instability events in the wall layer.

The role of diffusion in stabilizing the ultraviolet instability of polymer dynamics in a turbulent environment has been discussed at length. The capability of the multifield approach to model from a general viewpoint, complex microstructural dynamics has been exploited to explore weakly nonlocal interactions between polymer populations of neighboring elements even when diffusion effects are negligible. Though not conclusive, some indication is found that dissipative interactions different from pure diffusion may play a role in turbulent polymer solutions.

## References

[1] A. N. Beris and V. G. Mavrantzas, On the compatibility between various macroscopic formalisms for the concentration and flow of dilute polymer solutions, *J. Rheology*, **38** (1994), 1235–1250.

[2] R. B. Bird, C. F. Curtiss, R. C. Armstrong, and O. Hassager, *Dynamics of Polymeric Liquids*, Wyley–Interscience, New York, 1987.

[3] G. Capriz, Continua with latent microstructure, *Arch. Rational Mech. Anal.*, **90** (1985), 43–56.

[4] G. Capriz, *Continua with Microstructure*, Springer-Verlag, Berlin, 1989.

[5] E. De Angelis, C. M. Casciola, and R. Piva, DNS of wall turbulence: dilute polymers and self-sustaining mechanisms, *Computers and Fluids*, **31**-4-7 (2002), 495–507.

[6] A. W. El-Kareh and G. L. Leal, Existence of solutions for all Deborah numbers for a non-Newtonian model modified to include diffusion, *J. Non-Newtonian Fluid Mech.*, **33** (1989), 257.

[7] D. D. Joseph, *Fluid Dynamics of Viscoelastic Liquids*, Applied Mathematical Sciences 84, Springer-Verlag, New York, 1990.

[8] C. N. Likos, Effective interactions in soft condensed matter physics, *Phys. Reposts*, **348** (2001), 267–439.

[9] P. M. Mariano, Multifield theories in mechanics of solids, *Adv. Appl. Mech.*, **38** (2001), 1–93.

[10] H. C. Öttinger, Incorporation of polymer diffusivity and migration into constitutive equations, *Rheologica Acta*, **31** (1992), 14–21.

[11] S. K. Robinson, Coherent motion in the turbulent boundary layer, *Ann. Rev. Fluid Mech.*, **23** (1996), 601–639.

[12] G. C. Sarti and G. Marrucci, Thermomechanics of dilute polymer solutions: multiple beads-spring model, *Chem. Engrg. Sci.*, **28** (1995), 1053–1058.

[13] R. Sureskumar, A. N. Beris, and A. H. Handler, Direct numerical simulation of the turbulent channel flow of a polymer solution, *Phys. Fluids*, **9** (1997), 743–755.

[14] T. Min, J. Y. Yoo, and H. Choi, Effect of spatial discretization schemes on numerical solutions of viscoelastic fluid flows, *J. Non-Newtonian Fluid Mech.*, **100** (2001), 27.

[15] W. G. Tiederman, T. S. Luchik, and D. G. Bogard, Wall-layer structure and drag reduction., *J. Fluid Mech.*, **156** (1985), 419–437.

[16] F. Waleffe, On a self-sustaining process in shear flows, *Phys. Fluids*, **9** (1997), 883–900.

[17] L. E. Wedgewood and R. B. Bird, From molecular models to the solution of flow problems, *Ind. Engrg. Chem. Res.*, **27** (1988), 1313–1320.

[18] W. W. Willmarth, T. Wei, and C. O. Lee, Laser anemometer measurements of Reynolds stress in a turbulent channel flow with drag reducing polymer additives, *Phys. Fluids*, **30**-4 (1987), 933–935.

# 7. Fluxes and Flux-Conjugate Stresses

Reuven Segev*

## 7.1. Introduction

The work presented here is concerned with stress theory on differentiable manifolds. In addition to the theoretical interest as to the geometrical structure needed for the formulation of stress theory, such a general geometrical setting is used in theories of materials with microstructure to model the space where the "order parameters" are valued (see [2]). This sequel to our previous work on this subject (see [5, 6, 7, 8, 9]) follows [11] and focuses on two aspects of stress theory:

(a) the stresses are associated with bodies whose material structure is induced by an extensive property rather than assumed a priori;

(b) considering the electric charge as the extensive property, the general setting allows us to study the Maxwell stress tensor of electromagnetism for the case where a metric structure is not available on spacetime and without any reference to a constitutive relation for the electromagnetic fields (e.g., the aether relations).

Naturally, the theory is formulated on spacetime, taken to be an $m$-dimensional differentiable manifold $\mathscr{U}$. Spacetime is assumed to be orientable and a specific orientation is assumed to be chosen. It is noted that we do not use a connection in the analysis.

Before the presentation of these aspects in the main body of the paper, we present in Section 7.2 the motivation for the formulation of continuum mechanics on manifolds using differential forms even when microstructure is not present. The presentation in Section 7.2 is elementary and it is hoped that it will convey the main ideas to a wide readership.

Following the short introduction of notation in Section 7.3, Section 7.4 describes the way an extensive property induces a material structure (see [10]). Following [3], the point of view adopted is that the material points may be defined as integral manifolds of the flux vector field. Here, we do not assume the existence of a velocity vector field but obtain the flux field for the property under consideration from Cauchy's theorem for scalars. Furthermore, the general geometrical structure assumed implies that the flux field is an $(m-1)$-form on the $m$-dimensional manifold rather than a vector field. Nevertheless, we show that the relevant aspects of the material structure

*Department of Mechanical Engineering, Ben-Gurion University, Beer-Sheva, Israel, rsegev@bgumail.bgu.ac.il.

may be defined. It is noted that in a spacetime formulation the flux object contains information regarding the density of the property.

Section 7.5 gives a short description of stress theory on manifolds (see [7, 8, 9]) and Section 7.6 combines it with Section 7.3 by considering stress theory for bodies whose material structure is induced by an extensive property. The resulting stress object is a tensor that transforms the flux $(m-1)$-form of the property under consideration into the energy flux. Section 7.7 discusses the representation of such stresses and it is shown that there is a simple isomorphism between tensors on the space of $(m-1)$-forms and tensors on the tangent spaces. Thus, the stress-energy tensor may indeed be represented by a linear mapping of vectors as in the classical case.

Finally, we apply the foregoing framework to electromagnetism in Section 7.8. The property under consideration is the electric charge. While traditional formulations of electromagnetism rely on a metric structure, our formulation of the Maxwell stress tensor uses only a volume element. We conclude by giving a generalized equation for the Lorentz force.

## 7.2. Motivation

Consider an extensive property that we denote by $p$ such as the internal energy or the electric charge. One of the basic notions of continuum mechanics is that of the flux vector $\mathbf{j}$ associated with the property. Thus, for the internal energy property we have the heat flux vector field and for the electric charge we have the current density, etc. The flux vector field allows one to compute the rate at which the property $p$ leaves any region $\mathcal{R}$ through its boundary as

$$\int_{\partial\mathcal{R}} \mathbf{j}\cdot\mathbf{n}\,da,$$

where $\mathbf{n}$ is the unit normal to the boundary $\partial\mathcal{R}$ pointing out of the region. So far, we have not indicated whether the region $\mathcal{R}$ is the image of the current configuration of a body $\mathcal{B}$ in space $\mathscr{U}$ or a certain reference configuration. Thus, we will reserve the present notation for the case where values are measured with respect to current configuration $\kappa$ in space. For the variables relative to the reference configuration we will use the corresponding uppercase letters. So, for example, $\mathbf{N}$ is the unit vector normal to the boundary at the reference configuration. If $\kappa$ is the deformation of the body from its reference configuration then the corresponding integral over the reference configuration is

$$\int_{\kappa^{-1}(\partial\mathcal{R})} \mathbf{J}\cdot\mathbf{N}\,dA.$$

Traditionally, $\mathbf{J}$ is referred to as the Piola–Kirchhoff flux vector field and $\mathbf{j}$ is the Cauchy flux. The relation between the Cauchy flux and the Piola–Kirchhoff flux is obtained by the requirement that the two integrals should be equal on any subset of the boundary $\partial\mathcal{R}$ so that with some abuse of notation we may write

$$\mathbf{J}\cdot\mathbf{N}\,dA = \mathbf{j}\cdot\mathbf{n}\,da.$$

Consider an arbitrary infinitesimal vector $\mathbf{U}$ at the reference configuration, the image of which at the configuration $\kappa$ is $\mathbf{u} = F(\mathbf{U})$, where $F = D\kappa$ is the deformation gradient. Then the volumes

$$dV = \mathbf{U} \cdot \mathbf{N}\, dA$$

and

$$dv = \mathbf{u} \cdot \mathbf{n} da$$

of the infinitesimal parallelepipeds generated by $\mathbf{U}$, $dA$ and $\mathbf{u}$, $da$, respectively, are related by

$$dv = |F|\, dV,$$

where $|F|$ denotes the determinant of the deformation gradient. Hence,

$$\begin{aligned} \mathbf{U} \cdot \mathbf{N}\, |F|\; dA &= \mathbf{u} \cdot \mathbf{n}\, da \\ &= F(\mathbf{U}) \cdot \mathbf{n}\, da \\ &= \mathbf{U} \cdot F^T(\mathbf{n})\, da, \end{aligned}$$

and the fact the $\mathbf{U}$ is arbitrary implies that

$$|F|\, \mathbf{N}\, dA = F^T(\mathbf{n})\, da.$$

One then writes

$$\begin{aligned} \mathbf{j} \cdot \mathbf{n}\, da &= \mathbf{J} \cdot \mathbf{N}\, dA \\ &= \frac{\mathbf{J} \cdot F^T(\mathbf{n})\, da}{|F|} \\ &= \frac{F(\mathbf{J}) \cdot \mathbf{n}\, da}{|F|}. \end{aligned}$$

Since the last equality has to hold for any region so that $\mathbf{n}da$ is arbitrary, one finally obtains the traditional relation between the two forms of the flux, i.e.,

$$\mathbf{J} = |F|\, F^{-1}(\mathbf{j}).$$

The point of view we want to emphasize here is that both Piola–Kirchhoff's and Cauchy's flux vector fields actually represent the same physical quantity, the flux of the property $p$ inside the body. The current configuration, reference configuration, and any other configuration merely assign different coordinate systems on the body. Hence, a single mathematical object should represent the flux. That mathematical object may be represented by the Cauchy flux vector field with respect to the current configuration or the Piola–Kirchhoff flux with respect to the reference configuration just like a vector is represented by different sets of components when one considers two bases for a vector space. Thus, for the mathematical flux object, the relation between the Cauchy and Piola–Kirchhoff vector fields will be a transformation rule between the two representations. The relation between the Cauchy and Piola–Kirchhoff fluxes clearly indicates that the flux object is not a vector field (since the transformation does not involve only the application of $F^{-1}$).

### 7.2.1. The flux density form on the boundary of a body

The basic mathematical objects we can handle are scalars that remain invariant under a deformation, vectors that transform using the deformation gradient, and the various linear mappings whose transformation rules are induced by the transformation rule for vectors. We note that the area of a material subset of the boundary of a body is not an invariant scalar. In addition, the unit normals **n** and **N** are not related as vectors. Hence, we do not have an invariant unit normal associated with a body that is deformation independent. Consider a fixed body $\mathcal{R}$ and a material point $X$ on its boundary. The tangent plane to the boundary will be denoted by $T_X\partial\mathcal{R}$ and vectors in it are conceived as pointing from $X$ to infinitesimally close neighboring points. Note that if we do not want to represent the vectors by components, the particular configuration of the body is immaterial. For this reason, we may use $u$, for example, to denote such a tangent vector and its representations in the current and reference configurations are the previous $\mathbf{u}$ and $\mathbf{U}$, respectively. Consider an infinitesimal triangle on the boundary whose edges are two tangent vectors $u$ and $v$. We will denote by $\tau(u, v)$ the infinitesimal flux of the property through the parallelogram formed by the two vectors, i.e., the flux through the triangle multiplied by 2. If we study $\tau(u, v)$ as the vectors $u$ and $v$ vary, it is natural to assume that for such an infinitesimal triangle the value depends linearly on both arguments. In other words $\tau(\cdot, \cdot)$ is bilinear and we have

$$\begin{aligned}&\tau(a_1u_1 + a_2u_2, b_1v_1 + b_2v_2)\\ &\qquad = a_1b_1\tau(u_1, v_1) + a_1b_2\tau(u_1, v_2) + a_2b_1\tau(u_2, v_1) + a_2b_2\tau(u_2, v_2).\end{aligned}$$

In addition to being bilinear we expect $\tau(u, v)$ to vanish if the edges of the triangle are situated along one line. In other words, $\tau(u, v) = 0$ if the two vectors are not linearly independent, and in particular, $\tau(u, u) = 0$ for any vector $u$. These two requirements imply that for any two vectors

$$\begin{aligned}\tau(u, v) + \tau(v, u) &= \tau(u, u) + \tau(u, v) + \tau(v, u) + \tau(v, v)\\ &= \tau(u + v, u + v)\\ &= 0.\end{aligned}$$

Hence, $\tau(u, v) = -\tau(v, u)$ and we conclude that as a tensor, $\tau$ is antisymmetric. Thus, we expect that the flux density on the boundary is represented by an antisymmetric 2-tensor, a 2-differential form on the boundary $\partial\mathcal{R}$. Note that since the tangent space is two dimensional, the dimension of the space of antisymmetric 2-tensors is one. It follows that it is enough to know the value of the flux for the triangle constructed by any two vectors to determine the value of $\tau$ at $X$.

### 7.2.2. Orientation

The antisymmetry of $\tau$ causes a complication. We have to decide which of the arguments of $\tau$ comes first. On the other hand, this was expected, since we did not indicate what direction of flux across the surface element we consider to be positive. Pick any three linearly independent vectors $v_1$, $v_2$, $v_3$ and declare every other triplet of linearly

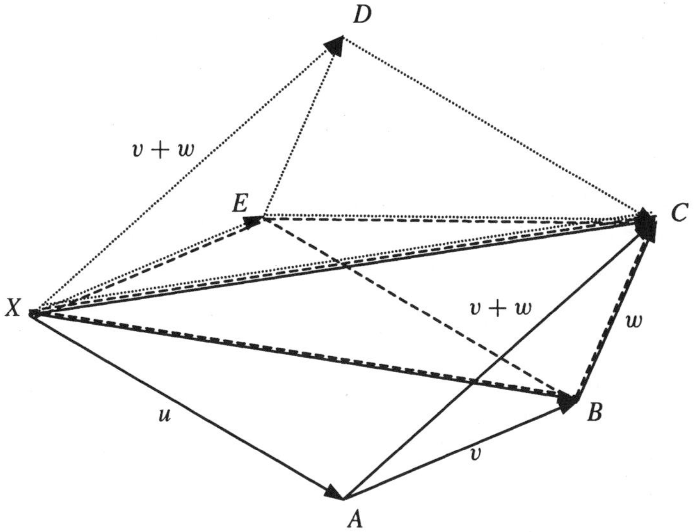


FIGURE 1. Linearity, motivation.

independent vectors $u_1, u_2, u_3$ to be positively oriented if the determinant of the matrix representing them as linear combinations of $v_1, v_2, v_3$ has a positive determinant. Since the determinant of a matrix is an invariant quantity, the orientation of a triplet of vectors is well defined. The orientation of the body induces an orientation on the boundary, i.e., a way to decide in which order two tangent vectors should appear as arguments of $\tau$. If $u$ at $X$ on the boundary is any vector pointing outwards, we say that the pair of tangent vectors $v, w$ is positively oriented on the boundary if the triplet $u, v, w$ is positively oriented in the body. Hence, only positively oriented (or ordered) pairs of vectors may be used in $\tau$ if we want to evaluate the flux out of the body.

### 7.2.3. The flux form

Let $X$ be a point in the body. The flux object $J$ at $X$ should assign the flux rate $J(u, v)$ through an arbitrary infinitesimal triangle defined in terms of some two vectors $u, v$ at $X$. That is, unlike the flux density form, we do not know a priori the plane where the vectors are situated. We expect that in case the vectors $u$ and $v$ are on a tangent plane for which the flux density form is $\tau$ then $J(u, v) = \tau(u, v)$ for consistency. Hence in particular $J(u, v) = -J(v, u)$, $J(au, v) = aJ(u, v)$ and $J(v, v) = 0$. Consider the infinitesimal tetrahedron $X, A, B, C$ having a vertex at $X$ and generated by the three vectors $u, v, w$ originating at $X$) (see the tetrahedron with solid lines in Figure 1). Assume we choose a right-hand orientation in space so that, for example, the pair of vectors $u, v$ on the $XAB$ face is negatively oriented. Since the tetrahedron's volume tends to zero faster than the total area of its boundary in any configuration of the body, the balance of the property $p$ implies that the total flux should vanish. Hence, taking into account the orientations of the various faces, we have

$$J(v, u) + J(v, w) + J(u, v + w) - J(u + v, w) = 0.$$

If we carry out the analogous calculations for the tetrahedrons $X, B, C, E$ (dashed lines) and $X, C, D, E$ (dotted lines) we obtain the additional equations

$$J(u, w) + J(u + v, w) + J(v, u) - J(v, w + u) = 0,$$
$$J(w, u) - J(v + w, u) - J(v, w) + J(v, w + u) = 0.$$

When we add the three equations we obtain the balance of flux for the prism $ABCDEX$ and using the antisymmetry of $J$ we arrive at

$$J(u, v + w) = J(u, v) + J(u, w).$$

Hence, it follows that the flux object $J$ is an antisymmetric bilinear tensor on the three-dimensional vector space of vectors at $X$. Since the dimension of the space of antisymmetric bilinear tensors is three, it is clear that $J$ is uniquely determined by the flux densities pertaining to three independent planes.

### 7.2.4. Stresses

We now want to use the flux object in order to define the material stress object. Our starting point is the expression for the virtual power

$$\int_{\mathcal{R}} \mathbf{b} \cdot \mathbf{u}\, dv + \int_{\partial\mathcal{R}} \mathbf{t}_{\mathcal{R}} \cdot \mathbf{u}\, da = \int_{\mathcal{R}} \mathbf{b} \cdot \mathbf{u}\, dv + \int_{\partial\mathcal{R}} \mathbf{T}(\mathbf{n}) \cdot \mathbf{u}\, da$$
$$= \int_{\mathcal{R}} \mathbf{b} \cdot \mathbf{u}\, dv + \int_{\partial\mathcal{R}} \mathbf{T}^T(\mathbf{u}) \cdot \mathbf{n}\, da,$$

where $\mathbf{u}$ denotes the virtual velocity field, $\mathbf{b}$ is the body force field, and $\mathbf{t}_{\mathcal{R}}$ is the surface force field that depends on the region $\mathcal{R}$ under consideration through the Cauchy formula $\mathbf{t}_{\mathcal{R}} = \mathbf{T}(\mathbf{n})$ using the Cauchy stress $\mathbf{T}$. We regard the last integrand as the flux of power through the boundary $\partial\mathcal{R}$ so that the foregoing discussion holds for the energy extensive property. Hence, the corresponding Cauchy flux vector field is $\mathbf{j}^{(e)} = \mathbf{T}^T(\mathbf{u})$, which depends linearly on $\mathbf{u}$. We denote by $J^{(e)}$ the corresponding flux density so that

$$J^{(e)}(v, w) = \mathbf{j}^{(e)} \cdot \mathbf{n}\, da = \mathbf{T}^T(\mathbf{u}) \cdot \mathbf{n}\, da$$

is the flux of power through the infinitesimal surface element $da$ associated with the vectors $v$ and $w$. Because $\mathbf{j}^{(e)}$ depends on the spatial velocity field $\mathbf{u}$ linearly, this property is inherited by the flux form $J^{(e)}$. In other words, one has to show that if $u$ is the invariant velocity vector represented by $\mathbf{u}$, then the dependence of $J^{(e)}$ on $u$ is linear. If we write $J^{(e)}(u)(v, w)$ to emphasize the dependence on the particular velocity $u$ so $J^{(e)}(u)(v, w) = \mathbf{T}^T(\mathbf{u}) \cdot \mathbf{n}\, da$, then the linear relation between $u$ and $\mathbf{u}$ implies that

$$J^{(e)}(b_1 u_1 + b_2 u_2)(v, w) = \mathbf{T}^T(b_1 \mathbf{u}_1 + b_2 \mathbf{u}_2) \cdot \mathbf{n}\, da$$
$$= b_1 \mathbf{T}^T(\mathbf{u}_1) \cdot \mathbf{n}\, da + b_2 \mathbf{T}^T(\mathbf{u}_2) \cdot \mathbf{n}\, da$$
$$= b_1 J^{(e)}(u_1)(v, w) + b_2 J^{(e)}(u_2)(v, w),$$

which yields the required linearity, since the vectors $v$ and $w$ are arbitrary. In other words, there is a linear mapping $\sigma$ at $X$ such that $J^{(e)} = \sigma(u)$. This mapping is a generalization of the stress tensor and it is a mapping that takes the virtual velocity at $X$ and gives the power flux.

### 7.2.5. Conclusions

We conclude from the above introductory remarks that multilinear mappings, particularly the completely antisymmetric or alternating ones, play a crucial role in the analysis of fluxes and stresses.

## 7.3. Notation

The language used for the formulation is that of differential forms on manifolds and their integration (see [1]). We recall that spacetime $\mathscr{U}$ is assumed to be an $m$-dimensional orientable manifold with a definite orientation chosen. An $m$-dimensional compact submanifold with boundary $\mathcal{R}$ of $\mathscr{U}$ will be referred to as a *control region*. A *frame* is a diffeomorphism of spacetime onto the Cartesian product $\mathbb{R} \times \mathscr{S}$, where $\mathbb{R}$ represents the time axis and $\mathscr{S}$ represents the space manifold—a "slice" of spacetime. Although our treatment does not use a frame, we will sometimes describe the physical meaning of variables in the setting of a frame. Thus, if a frame is given, a control region in spacetime may be conceived as a time-varying control region in space. The tangent space to $\mathscr{U}$ at $x$ will be denoted by $T_x\mathscr{U}$ and the tangent bundle to $\mathscr{U}$ will be denoted by $T\mathscr{U}$. A typical coordinate system in a neighborhood of $x \in \mathscr{U}$ will be denoted by $\{x^1, \dots, x^m\}$. The induced local basis for the tangent spaces will be denoted by $\{\partial/\partial x^i\}$, and the basis for the dual space $T_x^*\mathscr{U}$ will be denoted by $\{dx^i\}$.

Denote by $\bigwedge^r(T_x^*\mathscr{U})$ the vector space of $r$-alternating multilinear forms on the tangent space $T_x\mathscr{U}$. The bundle of $r$-forms is denoted by $\bigwedge^r(T^*\mathscr{U})$ so that an $r$-differential form $\alpha$ is a section $\alpha\colon \mathscr{U} \to \bigwedge^r(T^*\mathscr{U})$. An $m$-form $\rho$ may be interpreted intuitively as a density of an extensive property such that for $x \in \mathscr{U}$, $\rho_x(u_1, \dots, u_m)$ is interpreted as the amount of the property in the infinitesimal element generated by the positively oriented collection of tangent vectors $\{u_1, \dots, u_m\}$ at $x$. Such a form may be integrated over a control region to give $\int_{\mathcal{R}} \rho$. We recall that the dimension of $\bigwedge^m(T_x^*\mathscr{U})$ is one. An $m$-form that does not vanish anywhere on $\mathscr{U}$ is a *volume element*. The existence of a volume element is equivalent to orientability of $\mathscr{U}$ and a choice of a volume element determines an orientation on $\mathscr{U}$.

An $(m-1)$-form $\omega$ may be integrated over $(m-1)$-dimensional oriented submanifolds of $\mathscr{U}$, in particular, over boundaries of control regions. The orientation of the boundary of a region is induced by that of $\mathscr{U}$ and the outward-pointing vectors. We recall that the dimension of $\bigwedge^{m-1}(T_x^*\mathscr{U})$ is $m$ and $\omega_x$ may be evaluated on the various collections of $m-1$ tangent vectors $\{u_1, \dots, u_{m-1}\}$ situated on the various $(m-1)$-dimensional subspaces of $T_x\mathscr{U}$, or *hyperplanes*, at $x$. Hence, the evaluation of $\omega$ for $m$ collections of $m-1$ vectors such that the various collections are situated on linearly independent hyperplanes determines $\omega$ uniquely.

We denote the exterior derivative of an $(m-1)$-form $\omega$ as $d\omega$. The $m$-form $d\omega$ satisfies Stokes's theorem, i.e., for each control region $\mathcal{R}$, we have

$$\int_{\partial\mathcal{R}} \iota^*\omega = \int_{\mathcal{R}} d\omega,$$

where $\iota\colon \partial\mathcal{R} \to \mathscr{U}$ is the natural inclusion and $\iota^*$ is the pull-back of forms it induces (i.e., its adjoint) so that

$$(\iota^*\omega)_x(u_1, \ldots, u_{m-1}) = \omega_x(u_1, \ldots, u_{m-1})$$

for $u_1, \ldots, u_{m-1} \in T_x\partial\mathcal{R}$.

## 7.4. Material structures and extensive properties

Consider the conservation of an extensive property $p$ in spacetime $\mathscr{U}$. The balance of the property $p$ is assumed to be given in terms of boundary integrals on the various control regions. Specifically, it is assumed that for each control region $\mathcal{R}$ there is an $(m-1)$-form $\tau_{\mathcal{R}}$ on $\partial\mathcal{R}$, the *flux density*. The integral $\int_{\partial\mathcal{R}} \tau_{\mathcal{R}}$ is interpreted as the flux of the property out of the control region in spacetime relative to the positive orientation induced on $\partial\mathcal{R}$ by the orientation on $\mathscr{U}$ and the outward-pointing vectors.[1]

Regarding $\tau_{\mathcal{R}}$ as the value of a set function defined on the collection of control regions, Cauchy's postulates of continuum mechanics can be generalized to the present setting (see [6], [8] for the details) to yield an analogous generalization of Cauchy's theorem. The traditional assumption of dependence on the normal to the boundary is replaced by dependence on the tangent space to the boundary. The theorem states that there is a unique $(m-1)$-form $J$ on $\mathscr{U}$, to which we will refer as the *flux form* associated with the property $p$, that induces the flux densities for the various subbodies. Thus, in the absence of a metric tensor, the value of the flux at a point is an $(m-1)$-alternating tensor rather then a vector. It is noted that both objects are elements of $m$-dimensional vector spaces. The flux form $J$ induces flux densities $\tau_{\mathcal{R}}$ on the boundaries of the various control regions by the analog of the Cauchy formula

$$\tau_{\mathcal{R}}(u_1, \ldots, u_{m-1}) = J(u_1, \ldots, u_{m-1}).$$

Equivalently, we may write

$$\tau_{\mathcal{R}} = \iota^*(J).$$

In other words, the flux form is restricted to a hyperplane through $x$ to give the flux density associated with it. As expected, one can determine the flux form uniquely by knowing the flux densities corresponding to $m$ independent hyperplanes.

[1] In case a frame is given, the flux density through a space-like slice is interpreted as the density of the property $p$ in space. The flux through a hyperplane containing $\partial/\partial t$, the natural tangent vector to the time axis, is interpreted as the classical flux density of $p$ into the corresponding region in space $\mathscr{S}$.

Usually, it is assumed that there is a source density term $s$ for the property, an $m$-form on $\mathscr{U}$, so that the conservation equation of the property is

$$\int_{\partial\mathcal{R}} \tau_{\mathcal{R}} = \int_{\mathcal{R}} s$$

for every control region $\mathcal{R}$. In this case, Stokes's theorem and the generalized version of Cauchy's formula imply that

$$\int_{\mathcal{R}} dJ = \int_{\mathcal{R}} s$$

for every $\mathcal{R}$. Thus, the conservation equation may be written in a differential form as $dJ = s$.[2]

We now consider the particular case where a volume element $\theta$ is given on $\mathscr{U}$. In physical terms such a volume element may be interpreted as a positive extensive property that fills spacetime, e.g., mass density. We recall that, given the $m$-form $\theta_x$ and a vector field $u$, the $(m-1)$-form $u \lrcorner \theta$, the *contraction* of $\theta$ with $u$, is defined by

$$u \lrcorner \theta(u_1, \ldots, u_{m-1}) = \theta(u, u_1, \ldots, u_{m-1})$$

for every collection of $m-1$ vectors. For a given volume element, at each point this relation is an isomorphism $T_x\mathscr{U} \to \bigwedge^{m-1} T_x^*\mathscr{U}$ and we use

$$i_\theta \colon \bigwedge^{m-1} (T_x^*\mathscr{U}) \to T_x\mathscr{U}$$

to denote its inverse. Thus, if $u = i_\theta(J)$, then $u \lrcorner \theta = J$. Let $\theta$ be represented locally by

$$r(x^i)\, dx^1 \wedge \cdots \wedge dx^m,$$

where a "wedge" ($\wedge$) denotes the exterior product so $dx^1 \wedge \ldots \wedge dx^m$ is a single basis element of $\bigwedge^m (T^*\mathscr{U})$. Then $i_\theta(J)$ is represented by

$$u^i = \frac{(-1)^{i+1} J_i}{r}.$$

If $J$ is a flux form of an extensive property $p$ and a volume element is given, we will refer to $u = i_\theta(J)$ as the *kinematic flux* associated with $p$. The kinematic flux is the analog of the 4-velocity field. In other words, in order that the flux is represented by a vector field, one needs the additional structure of a volume element. If $\mathscr{L}$ denotes

[2] Again, if a frame is given on spacetime, then the time component of $J$ (that involving $dx^2 \wedge \cdots \wedge dx^m$, where $x^1$ is the time coordinate) is interpreted as the density in space of the property $p$ and the term in $dJ$ containing it is the time derivative of that density. The space-like components of $J$ (those associated with the basis vectors $dx^1 \wedge \cdots$) describe the three-dimensional flux and the terms in $dJ$ involving the space-like components make its $(m-1)$-dimensional divergence. In a particular frame, for every time $t$, the classical conservation law has the integral form

$$\int_{\mathcal{R}} \beta_{\mathcal{R}} + \int_{\partial\mathcal{R}} \tau_{\mathcal{R}} = \int_{\mathcal{R}} s,$$

where here $\mathcal{R}$ is interpreted now as a region in space $\mathscr{S}$ (a slice of spacetime) and $\beta_{\mathcal{R}}$ is the rate of change of the density of the property—a 3-form. In order that the previous Cauchy assumptions apply, it is usually assumed that $\beta_{\mathcal{R}}$ is actually independent of $\mathcal{R}$.

the Lie derivative, then the differential balance equation can now be written in the form $\mathcal{L}_u\theta = s$.

A flux form $J$ induces at each point $x \in \mathcal{U}$ a one-dimensional subspace, the *flux space*, $E_x$ as follows. The flux space is the intersection of all hyperplanes for which the induced flux densities vanish. A tangent vector $u$ belongs to the flux space if and only if $u \lrcorner J = 0$. This is analogous to the velocity vector at a point. The flux density will vanish through any hyperplane containing it. Alternatively, the flux space is the one-dimensional vector space obtained by the relation $u = i_\theta(J)$ when the flux form $J$ is kept fixed and the volume element $\theta$ is allowed to vary. The collection $E$ of the subspaces $E_x$ for the various elements of spacetime will be referred to as the *flux bundle*.

Since the flux bundle is one dimensional, for each $x \in \mathcal{U}$ there is a unique one-dimensional submanifold, the integral submanifold $Y$ of $\mathcal{U}$ so that $E_x = T_xY$ (see [1]). We will refer to $Y$ as a *worldline*. Thus, even in the general case where the flux cannot be represented by a vector field, its integral submanifolds and worldlines may be defined. Following the ideas of Capriz & Trebeschi [3] we can now identify a *body point* with a worldline, a *material body* with a collection of worldlines, and the *material universe* with the collection of all worldlines (see [10] for further details).

It is noted that within this general framework, one can easily define the flow potential. If the source $s$ vanishes, then the flux satisfies the equation $dJ = 0$. Hence, at least locally, there is an $(m-2)$-form $\psi$, the *flow potential*, such that $J = d\psi$.

## 7.5. Stress theory for manifolds

We consider a $\mu$-dimensional vector bundle $\pi : W \to \mathcal{U}$ over $\mathcal{U}$. The vector bundle is interpreted as the bundle of generalized velocities over $\mathcal{U}$.[3] For the case of a general differentiable manifold one cannot model forces by vectors and integrate them. Forces should be defined in terms of the power they produce for virtual velocities. Hence, the stress theory for manifolds, presented in [7, 8], considers for each compact $m$-dimensional submanifold with boundary $\mathcal{R}$ of $\mathcal{U}$ a linear functional of the generalized velocity fields containing a volume term and a boundary term of the form

$$F_{\mathcal{R}}(w) = \int_{\mathcal{R}} \mathbf{b}_{\mathcal{R}}(w) + \int_{\partial\mathcal{R}} \mathbf{t}_{\mathcal{R}}(w).$$

Here, $w$ is a section of $W$ interpreted as a virtual velocity field, $\mathbf{b}_{\mathcal{R}}$, the *body force*, is a section of $L\big(W, \bigwedge^m(T^*\mathcal{R})\big)$, and $\mathbf{t}_{\mathcal{R}}$, the *boundary force*, is a section of $L\big(W, \bigwedge^{m-1}(T^*\partial\mathcal{R})\big)$ so the integrals make sense. The functional $F_{\mathcal{R}}$ is interpreted as the force, or power, functional and the value $F_{\mathcal{R}}(w)$ is classically interpreted as

[3] In classical continuum mechanics, if $\mathcal{U}$ is interpreted as the physical space (a slice of spacetime), then in many cases $W$ is the tangent bundle $T\mathcal{U}$. If $\mathcal{U}$ is interpreted as the material body, then $W$ is usually the pull-back of the tangent bundle of the space manifold under the configuration mapping that embeds the material universe in space. This is the interpretation used in previous works (e.g., [5, 8]). In either case, a section of the bundle $\pi$ is interpreted as a generalized velocity field from the respective Eulerian or Lagrangian point of view.

the power of the force for the generalized velocity field $w$. Thus, for example, the boundary force is applied to the virtual velocity to produce the power flux density pertaining to the region under consideration.

The Cauchy postulates for forces are analogous to those pertaining to the scalar-valued properties (see [8]) and will not be described here. Again, the Cauchy theorem asserts the existence and uniqueness of the stress object that induces the various boundary forces. The resulting generalized stress $\sigma$ is a section of the vector bundle $L\big(W, \bigwedge^{m-1}(T^*\mathscr{U})\big)$; i.e., the value of a stress at a point is a linear mapping that assigns an $(m-1)$-multilinear mapping on the tangent space to any virtual velocity at that point. The resulting $(m-1)$-multilinear mapping $\sigma(w)$ is interpreted naturally as the flux form for the power associated with the given generalized velocity. The *Cauchy formula* assumes now the form

$$\mathbf{t}_{\mathcal{R}}(w) = \iota^*(\sigma \circ w),$$

and we will also write $\sigma(w)$ for $\sigma \circ w$ and $\iota^*(\sigma)$ for $\iota^* \circ \sigma$. Thus, the Cauchy formula may be written as $\mathbf{t}_{\mathcal{R}} = \iota^*(\sigma)$ in analogy with the scalar case. In other words, the power flux density associated with the generalized velocity $w$ at $x \in \partial\mathcal{R}$ is the restriction of the flux obtained from the stress.

Using Stokes's theorem, the action of $F_{\mathcal{R}}$ may now be rewritten using an integral over $\mathcal{R}$ as

$$F_{\mathcal{R}}(w) = \int_{\mathcal{R}} \big(d\sigma(w) + \mathbf{b}(w)\big).$$

It follows from this representation that the power depends on both the values of $w$ and its derivatives.

Assume that $(x^i, w^\alpha)$ are local vector bundle coordinates in a neighborhood $\pi^{-1}(U) \subset W$, $U \subset \mathscr{U}$ with local basis elements $\{e_\alpha\}$ so that a section of $W$ is represented locally by $w^\alpha e_\alpha$. Then denoting the dual basis vectors by $\{e^\alpha\}$, a stress $\sigma$ is represented locally by

$$\sigma_{\alpha 1 \ldots \hat{k} \ldots m} e^\alpha \otimes dx^1 \wedge \cdots \wedge \widehat{dx^k} \wedge \cdots \wedge dx^m,$$

where a "hat" ($\widehat{\ }$) indicates the omission of an item (an index or a factor) so that

$$\{dx^1 \wedge \cdots \wedge \widehat{dx^i} \wedge \cdots \wedge dx^{m-1} \mid i = 1, \ldots, m\}$$

is a basis for $\bigwedge^{m-1}(T^*\mathscr{U})$. The value of $\sigma(w)$ is represented locally by

$$\sigma_{\alpha 1 \ldots \hat{k} \ldots m} w^\alpha dx^1 \wedge \cdots \wedge \widehat{dx^k} \wedge \cdots \wedge dx^m.$$

## 7.6. Stress-energy tensors

We now combine the points of view of the previous two sections. That is, we consider stresses for bodies whose material structure is induced by an extensive property $p$. Although we do not have a velocity field over a manifold for the case of such bodies in general, we can use the the flux form of the property as a generalization. In particular, this is motivated by the fact that a volume element induces an isomorphism between the space of flux forms and the space of tangent vectors at each point.

Thus, we consider stress theory on spacetime $\mathscr{U}$ where we set

$$W = \bigwedge^{m-1}(T^*\mathscr{U}).$$

In addition, adopting the point of view of Section 7.3, we formulate the theory in spacetime so there is no need for the term containing $\mathbf{b}_{\mathcal{R}}$ and it is omitted. Therefore, we write

$$F_{\mathcal{R}}(J) = \int_{\partial\mathcal{R}} \mathbf{t}_{\mathcal{R}}(J),$$

where the boundary term $\mathbf{t}_{\mathcal{R}}$ is a section of $L\big(\bigwedge^{m-1}(T^*\mathscr{U}), \bigwedge^{m-1}(T^*\partial\mathcal{R})\big)$. Thus, we conclude from the results of Section 7.4 that the stress $\sigma$ is a section of

$$L\big(\bigwedge^{m-1}(T^*\mathscr{U}), \bigwedge^{m-1}(T^*\mathscr{U})\big).$$

The situation may be described generally as follows. We started with an extensive property $p$ given in terms of the flux densities $\tau_{\mathcal{R}}$ for the various control regions $\mathcal{R}$ in spacetime. The source term for property $p$ is $s$ and assuming the Cauchy postulates are satisfied the property $p$ has a flux form $J$. We now consider a second property, the $q$ property, whose flux densities $\tau_{\mathcal{R}}^{(q)}$ for the various control regions and source term $s^{(q)}$ satisfy the balance equation

$$\int_{\partial\mathcal{R}} \tau_{\mathcal{R}}^{(q)} = \int_{\mathcal{R}} s^{(q)}.$$

Again, assuming the Cauchy postulates hold for the property $q$, we have the corresponding flux form $J^{(q)}$ satisfying $\tau_{\mathcal{R}}^{(q)} = \iota^*(J^{(q)})$ and the balance equation has the differential representation $dJ^{(q)} = s^{(q)}$.

We will say that the property $q$ is a *resource* for the property $p$ if the flux density $\tau_{\mathcal{R}}^{(q)}$ depends pointwise linearly on the flux form $J$ of the property $p$. Thus, there is a section $\mathbf{t}_{\mathcal{R}}$ as above such that $\tau_{\mathcal{R}}^{(q)} = \mathbf{t}_{\mathcal{R}}(J)$.

In this framework, the Cauchy theorem implies that

$$\begin{aligned} \iota^*(J^{(q)}) &= \tau_{\mathcal{R}}^{(q)} \\ &= \mathbf{t}_{\mathcal{R}}(J) \\ &= \iota^*\big(\sigma(J)\big), \end{aligned}$$

for the inclusion $\iota$ of an arbitrary region, so $J^{(q)} = \sigma(J)$. In other words, the value $\sigma_x$ of the stress at a point is the linear transformation that assigns to the flux of the property $p$ the value of the flux of the resource $q$ used by $p$. The source term for the property $q$ is now given by $s^{(q)} = d\sigma(J)$.

Naturally, in what follows we will be concerned primarily with the energy resource.

## 7.7. Representation of stress-energy tensors

Since the stress-energy tensor is now a section of

$$L\Big(\bigwedge^{m-1}(T^*\mathscr{U}),\ \bigwedge^{m-1}(T^*\mathscr{U})\Big),$$

it is locally represented with respect to a local basis of $\bigwedge^{m-1}(T^*\mathscr{U})$ by a matrix. We will make below some further observations regarding the representations of stresses.

Assume that a volume element $\theta$ is given on $\mathscr{U}$. Then we may use the vector bundle isomorphism

$$i_\theta\colon\ \bigwedge^{m-1}(T^*\mathscr{U}) \to T\mathscr{U}$$

to represent the section $\sigma$ of $L\big(\bigwedge^{m-1}(T^*\mathscr{U}), \bigwedge^{m-1}(T^*\mathscr{U})\big)$ by a section $\tilde{\sigma}$ of $L\big(T\mathscr{U}, T\mathscr{U}\big)$ satisfying $\tilde{\sigma} \circ i_\theta = i_\theta \circ \sigma$.

Let us consider the relation between the local representation of $\sigma$ and the local representation $\tilde{\sigma}_i^j\, dx^i \otimes \frac{\partial}{\partial x^j}$ of $\tilde{\sigma}$. To represent $\sigma$ locally, we will use the notation $\hat{e}^i$ for the basis element $dx^1 \wedge \cdots \wedge \widehat{dx^i} \wedge \cdots \wedge dx^m$ of $\bigwedge^{m-1}(T^*\mathscr{U})$. Thus, the flux form $J$ is represented locally by $\hat{J}_i \hat{e}^i$, and the stress is represented locally in the form $\hat{\sigma}_i{}^j\, \hat{e}_j \otimes \hat{e}^i$, where $\{\hat{e}_j\}$ is the dual basis to $\{\hat{e}^i\}$.

If the volume element $\theta$ is represented locally by $r\, dx^1 \wedge \cdots \wedge dx^m$, the action of $i_\theta$ is given locally by

$$\hat{J}_i \hat{e}^i \mapsto \sum_i (-1)^{i+1} \frac{1}{r}\, \hat{J}_i \frac{\partial}{\partial x^i}.$$

(We use the summation symbol since the summation convention cannot be used on the right.) Thus, $i_\theta(\sigma(J))$ is represented by

$$\sum_j (-1)^{j+1} \frac{1}{r}\, \hat{\sigma}_j{}^i\, \hat{J}_i \frac{\partial}{\partial x^j},$$

and $\tilde{\sigma}(i_\theta(J))$ is represented by

$$\sum_i (-1)^{i+1} \frac{1}{r}\, \tilde{\sigma}_i^j\, \hat{J}_i \frac{\partial}{\partial x^j}.$$

Hence, the relation between $\sigma$ and $\tilde{\sigma}$ is represented locally as

$$\tilde{\sigma}_k^j = (-1)^{j+k} \hat{\sigma}_j{}^k.$$

It is interesting to note that the volume element does not enter the last relation and in fact we can write a natural isomorphism between the bundles $L\big(\bigwedge^{m-1}(T^*\mathscr{U}),$ $\bigwedge^{m-1}(T^*\mathscr{U})\big)$ and $L\big(T\mathscr{U}, T\mathscr{U}\big)$ as follows. Consider the tensor product $T^*\mathscr{U} \otimes_{\mathscr{U}}$

$T\mathscr{U}$. This tensor product is naturally isomorphic to $L(T\mathscr{U}, T\mathscr{U})$. For an element $\tilde{\sigma} = \tilde{\sigma}_i^j\, \phi^i \otimes v_j$ in the tensor product $T^*\mathscr{U} \otimes_{\mathscr{U}} T\mathscr{U}$, $v_j \in T_x\mathscr{U}$, $\phi^i \in T_x^*\mathscr{U}$, set

$$\sigma : \bigwedge^{m-1}(T^*\mathscr{U}) \to \bigwedge^{m-1}(T^*\mathscr{U})$$

by

$$\begin{aligned}\sigma(J) &= \tilde{\sigma}_i^j\, v_j \lrcorner (\phi^i \wedge J) \qquad (*)\\ &= \tilde{\sigma}_i^j \big(\phi^i(v_j)J - \phi^i \wedge (v_j \lrcorner J)\big).\end{aligned}$$

We note that $\sigma$ is indeed linear in $J$. Since $\sigma$ depends linearly on the $v^i$ and on the $\phi^j$, it depends linearly on the elements of the tensor product.

For the local coordinates $\{x^i\}$, let us determine the stress $\sigma$ induced by the linear mapping $\tilde{\sigma} \in L(T\mathscr{U}, T\mathscr{U})$ represented locally by the tensor $\tilde{\sigma}_i^j dx^i \otimes \frac{\partial}{\partial x^j}$. By definition, $\sigma(J)$ is represented by (the sum on $i$ is explicitly written)

$$\begin{aligned}\sum_i \tilde{\sigma}_i^j \frac{\partial}{\partial x^j} \lrcorner (dx^i \wedge J) &= \sum_i \tilde{\sigma}_i^j \frac{\partial}{\partial x^j} \lrcorner \big(dx^i \wedge (\hat{J}_k\, dx^1 \wedge \cdots \wedge \widehat{dx^k} \wedge \cdots \wedge dx^m)\big)\\ &= \sum_i \tilde{\sigma}_i^j \frac{\partial}{\partial x^j} \lrcorner \big((-1)^{i+1} \hat{J}_i\, dx^1 \wedge \cdots \wedge dx^m\big)\\ &= \sum_i \tilde{\sigma}_i^j (-1)^{i+j} \hat{J}_i\, dx^1 \wedge \cdots \wedge \widehat{dx^j} \wedge \cdots \wedge dx^m\\ &= \sum_i \tilde{\sigma}_i^j (-1)^{i+j} \hat{J}_i\, \hat{e}^j\\ &= \sum_i \hat{\sigma}_j^i\, \hat{J}_i \hat{e}^j.\end{aligned}$$

Hence, the matrix representing $\sigma$ is $\hat{\sigma}_j^i = (-1)^{i+j} \tilde{\sigma}_i^j$. We conclude that equation $(*)$ is indeed the natural, invariant representation of the isomorphism

$$L\Big(\bigwedge^{m-1}(T^*\mathscr{U}), \bigwedge^{m-1}(T^*\mathscr{U})\Big) \longleftrightarrow L(T\mathscr{U}, T\mathscr{U}).$$

This motivates even further the interpretation of the Cauchy stress as a transformation operating on the flux or velocity field of the property $p$ to give the flux form for the energy or velocity of the generalized energy points.

## 7.8. Electromagnetic stresses

The theory of the Maxwell stress-energy tensor of electromagnetism does not follow usually the lines of stress theory in continuum mechanics. As an example for the use of the foregoing analysis, we can now present a definition of the Maxwell stress-energy tensor. Furthermore, while electromagnetism utilizes the metric of relativity theory, we will assume here only the existence of a volume element on the four-dimensional $\mathscr{U}$. The following setting is also independent of any relation between the Maxwell 2-form and the Faraday 2-form, such as the aether relations. The extensive property under consideration is of course the electric charge and $J$ is the charge-current density—a 3-form. The conservation of charge implies that $dJ = 0$ and the Maxwell 2-form $\mathfrak{g}$ is a flow potential for the flux form so $J = d\mathfrak{g}$. For a 1-form $A$, the vector potential, the energy source density is $A \wedge J$. It follows that the Faraday 2-form $\mathfrak{f} = dA$ satisfies $d\mathfrak{f} = 0$.

Thus, assuming that a volume element $\theta$ is given on $\mathscr{U}$, we set $u = i_\theta(J)$ and define the stress-energy tensor as the section $\sigma$ of

$$L\Big(\bigwedge^{m-1}(T^*\mathscr{U}), \bigwedge^{m-1}(T^*\mathscr{U})\Big)$$

by (cf. [12, p. 36])

$$\sigma(J) = \big(i_\theta(J) \lrcorner \mathfrak{g}\big) \wedge \mathfrak{f} - \big(i_\theta(J) \lrcorner \mathfrak{f}\big) \wedge \mathfrak{g}.$$

Alternatively, using

$$u \lrcorner (\mathfrak{g} \wedge \mathfrak{f}) = (u \lrcorner \mathfrak{g}) \wedge \mathfrak{f} + \mathfrak{g} \wedge (u \lrcorner \mathfrak{f}),$$

the definition of the electromagnetic stress-energy tensor may also be written as

$$\sigma(J) = i_\theta(J) \lrcorner (\mathfrak{g} \wedge \mathfrak{f}) - 2\big(i_\theta(J) \lrcorner \mathfrak{f}\big) \wedge \mathfrak{g}.$$

Note that the matrix of the stress with respect to the natural basis of the space of $(m-1)$-forms is related to the usual matrix of the stress-energy-momentum tensor as discussed in the previous section.

Following the traditional formulations, we write the local representation $\hat{\mathfrak{f}}_{ij}\, dx^i \wedge dx^j$ of the Faraday 2-form $\mathfrak{f}$ in the form

$$\{\hat{\mathfrak{f}}_{ij}\} = \begin{pmatrix} 0 & -E_1 & -E_2 & -E_3 \\ E_1 & 0 & B_3 & -B_2 \\ E_2 & -B_3 & 0 & B_1 \\ E_3 & B_2 & -B_1 & 0 \end{pmatrix},$$

and the corresponding representation $\hat{\mathfrak{g}}_{ij}\, dx^i \wedge dx^j$ for the Maxwell 2-form as

$$\{\hat{\mathfrak{g}}_{ij}\} = \begin{pmatrix} 0 & H_1 & H_2 & H_3 \\ -H_1 & 0 & D_3 & -D_2 \\ -H_2 & -D_3 & 0 & D_1 \\ -H_3 & D_2 & -D_1 & 0 \end{pmatrix}.$$

For simplicity of notation, we assume that locally the volume element $\theta$ is of the form $dx^1 \wedge \cdots \wedge dx^4$. Then $u = i_\theta(J)$ is represented by $u^i = (-1)^{i+1}\hat{J}_i$. With this notation, the matrix $\{\hat{\sigma}_i{}^j\}$ representing the stress-energy tensor is

$$\{\hat{\sigma}_i{}^j\} = \big(\ \{\hat{\sigma}_i{}^1\}\ \ \{\hat{\sigma}_i{}^2\}\ \ \{\hat{\sigma}_i{}^3\}\ \ \{\hat{\sigma}_i{}^4\}\ \big),$$

where

$$\{\hat{\sigma}_i{}^1\} = \left\{\begin{array}{c} H_1B_1 + H_2B_2 + H_3B_3 + D_1E_1 + D_2E_2 + E_3D_3 \\ 2(E_3H_2 - E_2H_3) \\ 2(E_3H_1 - E_1H_3) \\ 2(E_2H_1 - E_1H_2) \end{array}\right\},$$

$$\{\hat{\sigma}_i{}^2\} = \left\{\begin{array}{c} 2(B_3D_2 - B_2D_3) \\ H_1B_1 - H_2B_2 - H_3B_3 + E_1D_1 - E_2D_2 - E_3D_3 \\ 2(E_1D_2 - B_2H_1) \\ 2(E_1D_3 + B_3H_1) \end{array}\right\},$$

$$\{\hat{\sigma}_i{}^3\} = \left\{\begin{array}{c} 2(B_3D_1 - B_1D_3) \\ 2(B_1H_2 - E_2D_1) \\ -H_1B_1 + H_2B_2 - H_3B_3 - E_1D_1 + E_2D_2 - E_3D_3 \\ 2(-E_2D_3 - B_3H_2) \end{array}\right\},$$

$$\{\hat{\sigma}_i{}^4\} = \left\{\begin{array}{c} 2(B_2D_1 - B_1D_2) \\ 2(B_1H_3 - E_3D_1) \\ 2(-E_3D_2 - B_2H_3) \\ -H_1B_1 - H_2B_2 + H_3B_3 - E_1D_1 - E_2D_2 + E_3D_3 \end{array}\right\}.$$

We now consider the energy source term $d\sigma(J)$. Using $u = i_\theta(J)$ one obtains

$$\begin{aligned} d\sigma(J) &= d\Big(\big(i_\theta(J) \lrcorner \mathfrak{g}\big) \wedge \mathfrak{f} - \big(i_\theta(J) \lrcorner \mathfrak{f}\big) \wedge \mathfrak{g}\Big) \\ &= d(u \lrcorner \mathfrak{g}) \wedge \mathfrak{f} - (u \lrcorner \mathfrak{g}) \wedge d\mathfrak{f} + (u \lrcorner \mathfrak{f}) \wedge d\mathfrak{g} - d(u \lrcorner \mathfrak{f}) \wedge \mathfrak{g} \\ &= d(u \lrcorner \mathfrak{g}) \wedge \mathfrak{f} + (u \lrcorner \mathfrak{f}) \wedge J - d(u \lrcorner \mathfrak{f}) \wedge \mathfrak{g}, \end{aligned}$$

where Maxwell's equations were used to arrive at the last line. Using the identity $d(u \lrcorner \alpha) = \mathscr{L}_u\alpha - u \lrcorner d\alpha$, for any differential form $\alpha$, we have

$$d\sigma(J) = (u \lrcorner \mathfrak{f}) \wedge J + (\mathscr{L}_u\mathfrak{g} - u \lrcorner d\mathfrak{g}) \wedge \mathfrak{f} - (\mathscr{L}_u\mathfrak{f} - u \lrcorner d\mathfrak{f}) \wedge \mathfrak{g}.$$

Finally, since $u \lrcorner J = 0$, Maxwell's equations give for the energy source term

$$d\sigma(J) = (u \lrcorner \mathfrak{f}) \wedge J + (\mathscr{L}_u\mathfrak{g}) \wedge \mathfrak{f} - (\mathscr{L}_u\mathfrak{f}) \wedge \mathfrak{g}.$$

It is noted that the term $(u \lrcorner \mathfrak{f}) \wedge J$ represents the power of the Lorentz force. In addition, in the classical formulation where a metric is available and $\mathfrak{g} = *\mathfrak{f}$ ($*$ denotes the Hodge operator), the terms $(\mathscr{L}_u\mathfrak{g}) \wedge \mathfrak{f}$ and $(\mathscr{L}_u\mathfrak{f}) \wedge \mathfrak{g}$ are equal and the energy source density contains the power of the Lorentz force (and energy balance) only. For an analogous expression where the constitutive relation between $\mathfrak{g}$ and $\mathfrak{f}$ is not specified but a metric is used, see [4, p. 91].

## Acknowledgment

This research was partially supported by the Paul Ivanier Center for Robotics Research and Production Management at Ben-Gurion University.

## References


[1] R. Abraham, J. E. Marsden, and R. Ratiu, *Manifolds, Tensor Analysis, and Applications*, Springer-Verlag, New York, 1988.

[2] G. Capriz, *Continua with Microstructure*, Springer-Verlag, New York, 1989.

[3] G. Capriz and P. Trebeschi, Reflections upon the axioms of continuum mechanics prompted by the study of complex materials, *Bull. Tech. Univ. Istanbul*, **47** (1994), 1–12.

[4] A. Lichnerowicz, *Relativistic Hydrodynamics and Magnetohydrodynamics*, Benjamin, New York, 1967,

[5] R. Segev, Forces and the existence of stresses in invariant continuum mechanics, *J. Math. Phys.*, **27** (1986), 163–170.

[6] R. Segev, The geometry of Cauchy's fluxes, *Arch. Rational Mech. Anal.*, **154** (2000), 183–198.

[7] R. Segev and G. Rodnay, Interactions on manifolds and the construction of material structure, *Internat. J. Solids Structures*, **38** (2000), 997–1018.

[8] R. Segev and G. Rodnay, Cauchy's theorem on manifolds, *J. Elasticity*, **56** (2000), 129–144.

[9] R. Segev and G. Rodnay, Divergences of stresses and the principle of virtual work on manifolds, *Tech. Mech.*, **20** (2000), 129–136.

[10] R. Segev and G. Rodnay, Worldlines and body points associated with an extensive property, *Internat. J. Non-Linear Mech.*, **38** (2003), 1–9.

[11] R. Segev, Metric independent properties of the stress-energy tensor, *J. Math. Phys.*, **43** (2002), 3220–3231.

[12] W. Thirring, *A Course in Mathematical Physics* 2: *Classical Field Theory*, Springer-Verlag, New York, 1979.

# 8. Algebra, Geometry, and Computations of Exact Relations for Effective Moduli of Composites

Yury Grabovsky*

**Abstract.** In this paper we will review and extend the results of [21], which covered the case of 3D thermopiezoelectric polycrystals. In that context the settings of conductivity, elasticity, pyroelectricity, piezoelectricity, thermoelectricity, and thermoelasticity can be viewed as particular cases. We will consider a class of composites more general than polycrystals, where the set of allowable materials is not constrained in any way. In addition, the tensors of material properties are not assumed to be symmetric—an assumption we made in [21]. For example, the Hall effect for conduction in a weak magnetic field is described by a nonsymmetric conductivity tensor. We explain the step-by-step process of finding all exact relations for the simple example of the 2D Hall effect. The paper concludes with a discussion of new algebraic and geometric questions posed by the theory of exact relations.

## 8.1. Introduction

Composite materials play an increasingly important role in our everyday life and technology. They are used everywhere from skis and golf clubs to sensors and actuators in high-tech components. By "composite" we mean a perfectly bonded mixture of two or more materials, where the mixing occurs on length scales much smaller than human size but much larger than interatomic distances. The physical properties of a composite (thermal, electric, elastic, etc.) are described by a tensor—the *effective tensor* of a composite. In order to create a composite with desired properties two basic problems become important: prediction of the effective properties of composite materials and determination of the effective tensor of a given composite by as few measurements as possible.

The problem of analytical prediction of effective properties of composite materials is important for both practical and theoretical points of view. Understanding the effective behavior mathematically may help save many costly and time-consuming measurements or may prevent spending time and money on many trial-and-error approaches to material design. Recent years have seen a lot of success in that direction.

*Department of Mathematics, Temple University, Philadelphia, PA 19122.

I would like to describe in this chapter a theory of exact relations for effective moduli of composites. The theory combines algebra, geometry, analysis, and mechanics in a beautiful symbiosis. It also achieves a high level of generality encompassing conductivity, elasticity, piezoelectricity, and many other coupled problem contexts. The theory and many of the physical results of this paper are also discussed in the new book by Milton [39].

The principal difficulty in the prediction of effective properties is the universally recognized fact that the effective tensors of composites depend on the microstructure (spatial arrangement of component materials) in general. Therefore, the object of importance is the set of all possible effective properties of a composite made with given materials taken in prescribed volume fractions (the so-called G-closure set). Unfortunately, aside from a few particular cases the G-closures are extremely difficult to compute analytically.

Exact relations are dependencies between various material properties that "survive" homogenization. For example, if we mix two isotropic elastic materials the resulting composite does not have to be isotropic. So, the relations defining isotropy are not preserved under homogenization. Yet isotropy survives if both isotropic materials have the same shear modulus. In fact, the composite will also have that same shear modulus, *no matter how the materials are mixed*. Statements about effective tensors that do not depend on the geometric arrangement of the constituent materials are called exact relations. The example described above is the exact relation due to Hill [22, 23] in 1963.

Exact relations and, more generally, G-closures provide information about all composites regardless of their origin, details, and complexity of microstructure. If we want to move beyond the G-closure and obtain a more detailed information about the effective tensors of composites we need to distinguish composites according to the type of microstructure they have.

One class of composites is where the (possibly infinitely many) scales of inhomogeneity are well separated. For example, imagine a material that looks homogeneous to the naked eye. Yet when we examine this material under a microscope, we may observe that it is, in fact, composed of several seemingly homogeneous materials, each of which, upon examination under a more powerful microscope, is observed to be composed of other seemingly homogeneous materials and so on for several or possibly an infinite number of steps. We also assume that on each step of our microscopic examination of an apparently homogeneous component, we find a rather simple geometric arrangement of phases, each occupying regions of approximately the same size. This broad class of composites is adequately described by the homogenization theory (see, e.g., [5, 24]), which we briefly review in this paper.

On the other end of the spectrum are disordered media where the microstructure at one point is almost uncorrelated to the microstructure at a somewhat distant point. In this situation it is usually assumed that the composite is random, and the stochastic approach [18, 30, 42] works well here.

The third and very important type of composite has a power law distribution of sizes and properties of the constituents. The power law is the telltale sign of self-organized criticality—a theory proposed by Bak, Tang, and Weisenfeld [3, 4]. According to this theory many open systems with constant influx of energy self-organize into poised states on the border between order and chaos. Examples include earthquakes and sandpiles, extinction of species and traffic jams [2]. I can speculate that the structure of real geological media is critically self-organized. The upper crust of our home planet can be considered an open system with constant energy flow from inner layers of Earth in the forms of seismic and volcanic activity. As a result the Earth's crust is a complicated multiscale structure with inhomogeneities on a vast range of scales. Under the microscope we can see tiny particles making up clay. With a naked eye we can see small individual grains of sand. We can also see boulders—the heterogeneities are on the scale of meters. The scale staircase goes on and on. The Earth's crust is a heterogeneous mixture of materials on a continuous spectrum of length scales from microscopic to global (continental plates, oceans). I would like to call such composites *critical*. The appropriate mathematical tools for a rigorous discussion of effective properties of such composites are only beginning to emerge. On the one hand, there is a nonrigorous renormalization group approach [27, 44]. On the other, there is a more rigorous micromechanics-based approach leading to nonlocal constitutive laws [10, 11, 41]. Yet, hydrologists, for example, have to deal with the effective hydrolic permittivity of such media as a rule [8].

As we mentioned earlier, we will focus on the microstructure-independent aspect of the theory of composite materials, which applies to all composites equally well. The discussion of the different types of composites was needed in order to place the subject of this paper in a broader context.

## 8.2. G-closures

We start this discussion with the example of conductivity before going on to the abstract framework encompassing a variety of physical contexts.

The conduction of the electric current is described by two fields, the electric field $\boldsymbol{e}$ and the current field $\boldsymbol{j}$. The two fields satisfy differential constraints: The electric field is curl-free, the current field is divergence-free. The two fields are related via a tensorial Ohm's law. Thus, we have the following equations:

$$\nabla \times \boldsymbol{e} = 0, \quad \nabla \cdot \boldsymbol{j} = 0, \quad \boldsymbol{j} = \boldsymbol{\sigma} \boldsymbol{e}, \tag{1}$$

where the $3 \times 3$ matrix $\boldsymbol{\sigma}$ is symmetric and positive definite.

An $n$-phase composite made with $n$ materials $\boldsymbol{\sigma}_1, \ldots, \boldsymbol{\sigma}_n$ is a Borel measurable matrix field $\boldsymbol{\sigma}(\boldsymbol{x})$, such that $\boldsymbol{\sigma}(\boldsymbol{x}) \in D = \{\boldsymbol{\sigma}_1, \ldots, \boldsymbol{\sigma}_n\}$. An $n$-phase polycrystal is a Borel measurable matrix field $\boldsymbol{\sigma}(\boldsymbol{x})$, such that $\boldsymbol{\sigma}(\boldsymbol{x}) \in D = \bigcup SO(3) \cdot \boldsymbol{\sigma}_j$. In general, we fix a set $D$ (finite or infinite) of materials and consider the set

$$\mathcal{D} = \{\boldsymbol{\sigma}(\boldsymbol{x}) \in L^\infty(\Omega) \mid \boldsymbol{\sigma}(\boldsymbol{x}) \in D \text{ a.e. } \boldsymbol{x} \in \Omega\}. \tag{2}$$

From the applied point of view one needs to solve the elliptic boundary value problem in a domain $\Omega$ occupied by the composite

$$\begin{cases} \nabla \cdot (\sigma(x)\nabla\phi) = f, & x \in \Omega, \\ \phi = 0, & x \in \partial\Omega. \end{cases} \tag{3}$$

However, if $\sigma(x)$ has a very complicated geometry, the numerical solution of (3) is not feasible. The fruitful idea is to represent the local conductivity tensor $\sigma(x)$ as a member in a sequence $\sigma^\epsilon(x)$ for small $\epsilon$. As $\epsilon \to 0$ the length scales in the microstructure become more and more separated. In the limit as $\epsilon \to 0$, the material properties are described by a *homogenized* tensor $\sigma^*(x)$, which may be much simpler than the local conductivity tensor. This is especially true in the case of composites with inhomogeneities on well-separated length scales. If $\sigma^*(x)$ is particularly simple then (3) may become solvable numerically.

Putting the practical questions aside and turning to mathematical justification of the homogenization procedure, we need to answer the following question. Given a sequence of bounded measurable functions $\sigma^\epsilon(x)$, is there a sense in which we can say that $\sigma^\epsilon \to \sigma^*$, so that the solutions $\phi^\epsilon$ of (3) with $\sigma = \sigma^\epsilon$ converge to the solution $\phi^*$ of the homogenized equation? Such a notion was proposed by Spagnolo [45, 46] and further developed by De Giorgi and Spagnolo [7]. Murat and Tartar [40] (the English translation of the French original) extended G-convergence to the case of nonsymmetric tensors $\sigma$ and proved that G-convergence of symmetric elliptic operators in (3) implies the convergence of fluxes

$$\sigma^\epsilon \nabla\phi^\epsilon \rightharpoonup \sigma^* \nabla\phi^* \tag{4}$$

weakly in $L^2(\Omega)$.

The nonsymmetric tensors of material properties may arise in a variety of contexts. For example, the Hall effect in the electric current conduction in a very weak magnetic field is governed by the same basic equations of conductivity (1), except the tensor $\sigma$ is no longer symmetric

$$\sigma = \sigma_s + \pi(r),$$

where $\sigma_s$ is the symmetric and positive definite conductivity tensor, $r$ is the Hall vector, and $\pi$ is the "cross-product" mapping between vectors in $\mathbb{R}^3$ and $3 \times 3$ skew symmetric matrices such that for any $\{u, x\} \subset \mathbb{R}^3$ we have $\pi(u)x = u \times x$.

Following Murat and Tartar, the name H-convergence is attached to the kind of G-convergence that is appropriate for nonsymmetric material tensors.

**Definition 1.** A sequence of not necessarily symmetric tensors $\sigma^\epsilon$ H-converges to $\sigma^*$ if for any $f$ in (3),

(i) $\phi^\epsilon \rightharpoonup \phi^*$ weakly in $H_0^1(\Omega)$,
(ii) $\sigma^\epsilon \nabla\phi^\epsilon \rightharpoonup \sigma^* \nabla\phi^*$ weakly in $L^2(\Omega)$.

Murat and Tartar [40] noticed that for symmetric matrices $\sigma$ the second condition is redundant. It follows from the first.

The key result here is the compactness theorem proved by De Giorgi and Spagnolo for the case of symmetric $\sigma$ and by Murat and Tartar in general. The theorem states that any sequence of uniformly positive definite and bounded matrix fields $\sigma^\epsilon(x)$ contains an H-convergent subsequence. In this connection the problem of the closure of the set $\mathcal{D}$, defined in (2) with respect to the H-convergence, becomes important. In order to formulate a fundamental result, we must first give the basic nontrivial example of an $H$-converging sequence $\boldsymbol{\sigma}^{\epsilon}$. Suppose that the matrix-valued function $\boldsymbol{\sigma}(\boldsymbol{y})$ is periodic with a parallelepiped of periods $Q = [0, 1]^3$. We further assume that $\boldsymbol{\sigma}(\boldsymbol{y})$ is uniformly positive definite and uniformly bounded. Then $\boldsymbol{\sigma}^{\epsilon}(\boldsymbol{x}) = \boldsymbol{\sigma}(\boldsymbol{x}/\epsilon)$ H-converges to a constant positive definite matrix $\boldsymbol{\sigma}^*$, defined via a solution of the so-called periodic cell problem. We will write the cell problem in a form that will be useful later:

$$\nabla \cdot \boldsymbol{j} = 0, \quad \nabla \times \boldsymbol{e} = 0, \quad \boldsymbol{j}(\boldsymbol{y}) = \boldsymbol{\sigma}(\boldsymbol{y})\boldsymbol{e}(\boldsymbol{y}), \tag{5}$$

where all functions and differential operators are $Q$-periodic. Equations (5) have a unique solution if we fix the mean value of the electric field $\boldsymbol{e}$ over the period cell. Suppose $\langle \boldsymbol{e} \rangle = \boldsymbol{e}^*$, where $\langle \cdot \rangle$ denotes the average over the period cell $Q$. Then the unique field $\boldsymbol{j}$ satisfying (5) depends linearly on $\boldsymbol{e}^*$, and therefore there exists a $3 \times 3$ matrix $\boldsymbol{\sigma}^*$ such that

$$\boldsymbol{\sigma}^* \boldsymbol{e}^* = \langle \boldsymbol{j} \rangle. \tag{6}$$

**Definition 2.** Let $D$ be the set of matrices representing conductivities of the materials constituting the composite. Then the *G-closure* [31, 49] $G(D)$ of the set of materials $D$ is the set of all effective tensors $\boldsymbol{\sigma}^*$ of *periodic* composites made with materials from the set $D$.

Kohn and Dal Maso realized that the set $G(D)$ is sufficient to describe the H-closure (closure with respect to the H-convergence topology) of the set $\mathcal{D}$ defined in (2). The result has been recently rigorously proved in a very general context by Raitums [43]. The theorem states that the matrix field $\boldsymbol{\sigma}^*(\boldsymbol{x})$ is in the H-closure of $\mathcal{D}$ if and only if $\boldsymbol{\sigma}^*(\boldsymbol{x}) \in G(D)$ for a.e. $\boldsymbol{x} \in \Omega$.

Our primary interest is the G-closure sets of materials in a variety of physical contexts. We must mention that the explicit form of the G-closure is known only in a very few cases [15, 19, 20, 29, 32, 33, 34, 35, 37, 49]. Only the paper [15] characterizes G-closure sets for *arbitrary* sets of materials $D$ for 2D conducting polycrystals.

At this point, we would like to make several remarks. Observe that the G-closure problem is the problem of computing a subset in a *finite dimensional* space, given another subset of the same space. Yet, it involves solving a periodic PDE problem (5). It would be nice to have a completely geometric description of how to construct $G(D)$ knowing $D$, as was done in [15] for 2D conducting polycrystals. The first step towards such a geometric description in the general case was made in [16] for sets closed under lamination (the cornerstone of this paper), but a major new geometric breakthrough is still needed for the method to be truly useful.

Another remark is that G-closure lives in a rather high dimensional space (6D for conductivity, 9D for the Hall effect). Therefore, even if we have exact formulas for the boundary of $G(D)$, they are not immediately useful in applications without further (often very complicated) algebra. Instead, a more useful goal would be an efficient numerical algorithm for computing quantities like

$$f_D = \min_{\sigma \in G(D)} f(\sigma)$$

for a class of functions $f$. For each $f$ the answer is a *number* $f_D$—something an engineer can relate to.

Getting closer to the subject of the paper, there is a fundamental dichotomy in the shapes of G-closure sets. Most G-closure sets (aside from single points) have a nonempty interior. In exceptional cases the G-closure sets lie on surfaces (of various codimensions). It is these exceptional cases that we call *exact relations* and it is these cases that we are after. The trick here is that we are no longer in the losing game of computing the boundaries of the G-closure sets. Instead, we just want to identify the surfaces that contain G-closure sets. To give a trivial example, consider the Hall effect and assume that we are mixing materials whose Hall tensors are all zero. From the physical point of view, it is obvious that the Hall effect will not arise in a composite if every component does not have it. Geometrically, a 6D surface (all $3 \times 3$ symmetric matrices) in a 9D space (all $3 \times 3$ matrices) is an exact relation. In identifying this exact relation we do not need to compute any G-closure sets; yet the result is not entirely devoid of useful information.

## 8.3. Hilbert space formalism

Milton [38] has observed that the periodic cell problems in various physical contexts follow the same abstract pattern, which is best described using the language of Hilbert spaces (see also [9, 18, 25, 28, 42, 50] for similar formal approaches). Milton observed that regardless of the particular physics of the problem there are two fields: the intensity field $\boldsymbol{E}$ and the flux field $\boldsymbol{J}$. They take their values in a certain finite dimensional tensor space $\mathcal{T}$. For the Hall effect example, the electric field $\boldsymbol{e}$ is the intensity field and the current field $\boldsymbol{j}$ is the flux field (in fact, the names are purely formal, since the theory is symmetric with respect to the swapping of the fields). The fields are vectors so that $\mathcal{T} = \mathbb{R}^3$ for the Hall effect. All intensity fields and fluxes will be assumed to belong to the ambient Hilbert space $\mathcal{H} = L^2(Q) \otimes \mathcal{T}$. The most important observation of Milton concerns the general structure of the differential constraints satisfied by the fields $\boldsymbol{E}$ and $\boldsymbol{J}$. If we denote by $\mathcal{U} = \mathbb{R} \otimes \mathcal{T}$ the subspace of constant fields in $\mathcal{H}$, then the Hilbert space $\mathcal{H}$ is split into the orthogonal sum $\mathcal{H} = \mathcal{E} \oplus \mathcal{J} \oplus \mathcal{U}$, where $\mathcal{E}$ and $\mathcal{J}$ are the subspaces of mean zero intensity fields and fluxes, respectively. The subspaces $\mathcal{E}$ and $\mathcal{J}$ correspond to the differential constraints on $\boldsymbol{E}$ and $\boldsymbol{J}$, respectively. For that reason Milton required that the orthogonal projection $\Gamma$ onto $\mathcal{E}$ be local in Fourier space. In other words, there is a degree zero function $\Gamma(\boldsymbol{k})$ such that for each $\boldsymbol{k} \in \mathbb{S}^2$ the matrix $\Gamma(\boldsymbol{k})$ is an orthogonal projection onto a subspace $\mathcal{E}_{\boldsymbol{k}}$ of $\mathcal{T}$ and such that

$$\widehat{\Gamma f}(\boldsymbol{k}) = \begin{cases} \Gamma(\boldsymbol{k})\widehat{f}(\boldsymbol{k}), & \text{if } \boldsymbol{k} \neq \boldsymbol{0}, \\ \boldsymbol{0}, & \text{if } \boldsymbol{k} = \boldsymbol{0}. \end{cases}$$

The subspaces $\mathcal{E}_{\boldsymbol{k}}$ cannot be arbitrary. For example, they have to be of the same dimension. The basic $O(3)$ symmetry of our space implies that the subspaces $\mathcal{E}_{\boldsymbol{k}}$ are permuted by rotations. In other words, the vector space $\mathcal{T}$ is not just a linear space, it is a representation space for the group $O(3)$. For $\boldsymbol{\tau} \in \mathcal{T}$ we denote the action of $\boldsymbol{R} \in O(3)$ by $\boldsymbol{R} \cdot \boldsymbol{\tau}$. With this notation we can write the basic property of the subspaces $\mathcal{E}_{\boldsymbol{k}}$: $\boldsymbol{R} \cdot \mathcal{E}_{\boldsymbol{k}} = \mathcal{E}_{\boldsymbol{Rk}}$. We use the same notation for the action of rotations on $\mathrm{End}(\mathcal{T})$, the space of linear operators on $\mathcal{T}$. For any $\boldsymbol{A} \in \mathrm{End}(\mathcal{T})$ and any $\boldsymbol{\tau} \in \mathcal{T}$ we define $\boldsymbol{R} \cdot \boldsymbol{A}$ by $(\boldsymbol{R} \cdot \boldsymbol{A})\boldsymbol{\tau} = \boldsymbol{R} \cdot (\boldsymbol{A}(\boldsymbol{R}^{-1} \cdot \boldsymbol{\tau}))$. With this notation the basic property of the subspaces $\mathcal{E}_{\boldsymbol{k}}$ can also be written as $\boldsymbol{R} \cdot \Gamma(\boldsymbol{k}) = \Gamma(\boldsymbol{Rk})$. This last formula tells us that the function $\Gamma(\boldsymbol{k})$ is uniquely determined by a single matrix $\Gamma_0 = \Gamma((1, 0, 0))$, which in turn is uniquely determined by a subspace $\mathcal{E}_0 = \mathcal{E}_{(1,0,0)}$. The only restriction on this subspace is that the subgroup $O(2)$ of $SO(3)$ that leaves the $\boldsymbol{e}_1 = (1, 0, 0)$ direction invariant should leave the subspace $\mathcal{E}_0$ invariant. Thus, a standard representation theory of $SO(3)$ can give us a list of all possible subspaces $\mathcal{E}_0$ satisfying all of our constraints. For example, for $\mathcal{T} = \mathbb{R}^3$, there are just two proper subspaces $\mathcal{E}_0$ satisfying our constraints. One is $\mathbb{R}\boldsymbol{e}_1$, and the other is its orthogonal complement. Thus, for $\boldsymbol{k} \in \mathbb{S}^2$ there are two choices for the function $\Gamma(\boldsymbol{k})$. Either $\Gamma(\boldsymbol{k}) = \boldsymbol{k} \otimes \boldsymbol{k}$ or $\Gamma(\boldsymbol{k}) = \boldsymbol{I} - \boldsymbol{k} \otimes \boldsymbol{k}$, where $\boldsymbol{I}$ is a $3 \times 3$ identity matrix. The first choice corresponds to the equations (5), and so does the second, with $\boldsymbol{e}$ and $\boldsymbol{j}$ interchanged. Thus, in the context of conductivity, equations (5) are the only possibility consistent with a general Hilbert space framework described here.

The composite microstructure is given by an $L^\infty$ mapping $\boldsymbol{L}(\boldsymbol{x})$ of $Q$ into $\mathrm{End}(\mathcal{T})$. The function $\boldsymbol{L}(\boldsymbol{x})$ can also be viewed as an operator $\mathsf{L}$ mapping $\mathcal{H} = L^2(Q) \otimes \mathcal{T}$ into $\mathcal{H}$: for any $\boldsymbol{f} \in \mathcal{H}$ we have $(\mathsf{L}\boldsymbol{f})(\boldsymbol{x}) = \boldsymbol{L}(\boldsymbol{x})\boldsymbol{f}(\boldsymbol{x})$.

The effective tensor $\boldsymbol{L}^*$ is defined by analogy with (6):

$$\boldsymbol{J}^* = \boldsymbol{L}^* \boldsymbol{E}^*, \tag{7}$$

where $\boldsymbol{E}^* = \langle \boldsymbol{E} \rangle$ and $\boldsymbol{J}^* = \langle \boldsymbol{J} \rangle$.

## 8.4. Lamination formula

In this section, we derive a lamination formula that has the form of convex combination. This formula was first derived by Milton [38] and independently by Zhikov [51]. Other linear lamination formulas were derived by Backus [1] and Tartar [48] for elasticity. Their idea was to rewrite the constitutive relation so that the continuous and discontinuous components of the elastic fields are separated. The laminate of two materials $\boldsymbol{L}_1$ and $\boldsymbol{L}_2$ with layer normal $\boldsymbol{n}$ and volume fractions $\theta_1, \theta_2$ $(\theta_1 + \theta_2 = 1)$ is a periodic stricture with the period cell shown in Figure 1.

More generally, a laminate is a function $\boldsymbol{L}(\boldsymbol{y})$ that depends only on $\boldsymbol{y} \cdot \boldsymbol{n}$. In order to formulate the theorem we introduce the W-transformation of Milton [38]. Let $\boldsymbol{L}_0$ be a reference medium that is assumed to be positive definite but which otherwise is completely arbitrary. Let $\Gamma'(\boldsymbol{n})$ denote the nonorthogonal projection onto the subspace $\boldsymbol{L}_0\mathcal{E}_{\boldsymbol{n}}$ along the subspace $\mathcal{J}_{\boldsymbol{n}}$. In fact, the projection $\Gamma'(\boldsymbol{n})$ is well defined as long as $\boldsymbol{L}_0$ is positive definite on $\mathcal{E}_{\boldsymbol{n}}$ (to ensure that $\boldsymbol{L}_0\mathcal{E}_{\boldsymbol{n}}$ and $\mathcal{J}_{\boldsymbol{n}}$ have trivial intersection). One can check that

$$\Gamma'(\boldsymbol{n}) = \boldsymbol{L}_0(\boldsymbol{I} - \Gamma(\boldsymbol{n}) + \Gamma(\boldsymbol{n})\boldsymbol{L}_0)^{-1}\Gamma(\boldsymbol{n}).$$

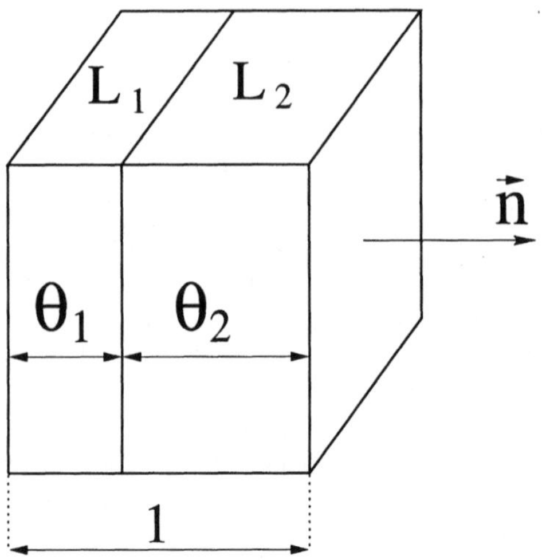


FIGURE 1. The period cell of the laminate.

First, define the S-transformation

$$S(L) = (I - LL_0^{-1})^{-1}. \tag{8}$$

Now define

$$W_n(L) = \left[S(L) - \Gamma'(n)\right]^{-1}. \tag{9}$$

**Theorem 1.** *Let $L(y)$ be the laminate with the normal $n$ and let $L_0$ be an arbitrary positive definite reference medium (positive definite on $\mathcal{E}_n$ is enough). Then*

$$\langle W_n(L(y))\rangle = W_n(L^*).$$

The proof is analogous to the proof of the corresponding theorem in [21, Theorem 3.1] for symmetric matrices.

*Proof.* Taking the average of the polarization field $P = (L - L_0)E$ we obtain $P^* = (L^* - L_0)E^*$. These relations can also be written using the $S$-transformation defined in (8):

$$L_0E = -S(L)P, \quad L_0E^* = -S(L^*)P^*. \tag{10}$$

Applying the projection operator $\Gamma'$ to $P$ we obtain

$$\Gamma'(n)(P - P^*) = -L_0(E - E^*).$$

Substituting the values for $L_0E$ and $L_0E^*$ from (10), we obtain

$$\Gamma'(n)(P - P^*) = S(L)P - S(L^*)P^*.$$

Solving for $P$ and using the definition of the W-transformation, we obtain

$$P(y) = W_n(L(y))(W_n(L^*))^{-1}P^*.$$

Taking averages, we get

$$P^* = \langle W_n(L(y))\rangle (W_n(L^*))^{-1} P^*.$$

Theorem 1 now follows because the constant field $P^*$ can be arbitrary. □

For the Hall effect $\mathcal{E}_n = \mathbb{R}n$ and the requirement that $L_0$ is positive definite on $\mathcal{E}_n = \mathbb{R}n$ for all $n$ is equivalent to $L_0$ being positive definite. (For elasticity the corresponding condition is equivalent to the Legendre–Hadamard condition for $L_0$.) In this case,

$$\Gamma'(n) = \frac{L_0 n \otimes n}{L_0 n \cdot n}. \tag{11}$$

The W-transformation maps the classical lamination formula [17, 49] into a convex combination. Namely, if $L^*$ is an effective tensor of a laminate made with materials $L_1$ and $L_2$ taken in volume fractions $\theta_1$ and $\theta_2$ with lamination normal $n$, then

$$W_n(L^*) = \theta_1 W_n(L_1) + \theta_2 W_n(L_2). \tag{12}$$

A corollary is that for any direction $n$ a $W_n$-image of any set stable under lamination must be a convex set. The idea to use this property to study the geometry of sets stable under lamination is due to Francfort and Milton [16].

## 8.5. The main ideas

Theorem 1 implies that if a set $G$ is G-closed then $W_n(G)$ is a convex set for all $n \in \mathbb{S}^2$. If, on the other hand, $W_n(G)$ is a convex set for all $n \in \mathbb{S}^2$ then the set $G$ is L-closed (or stable under lamination). In almost all cases, G-closure and L-closure coincide; however, there is an example by Milton of a set that is L-closed but not G-closed [39]. In any case, from the practical point of view L-closure would provide a very good approximation to G-closure. Thus, the *geometric* problem of finding the smallest set $G(D)$ containing $D$ and such that $W_n(G(D))$ is convex replaces the problem of solving the cell problem. In the important polycrystalline case, the geometric problem admits a very attractive formulation. (Is there an attractive answer?)

Assume that if a material $L$ is in $D$ then all of its rotations $\{R \cdot L \mid R \in SO(3)\}$ are also in $D$. If we choose an isotropic reference medium then we can easily verify that

$$R \cdot W_n(L) = W_{Rn}(R \cdot L). \tag{13}$$

It follows, then, that if a set $G \subset \mathrm{End}(\mathcal{T})$ is rotationally invariant and $W_n(G)$ is convex for *a single choice of* $n$ then $W_n(G)$ is a convex set for all $n \in \mathbb{S}^2$. Thus, we can fix one unit vector $n$, for example, $n = (1, 0, 0)$, and consider just one map $W(L) = W_n(L)$. The question is this:

> Find the smallest rotationally invariant set $G(D)$ containing a given set $D$ such that $W(G(D))$ is convex.

We can ask a different question:

> Can we characterize convex functions $g(\boldsymbol{K})$ such that the functions $f(\boldsymbol{L}) = g(W(\boldsymbol{L}))$ are rotationally invariant, $f(\boldsymbol{R} \cdot \boldsymbol{L}) = f(\boldsymbol{L})$ for all $\boldsymbol{R} \in SO(3)$?

This question sounds very similar to the one answered by Chandler Davis [6]. Davis proved that a rotationally invariant function defined on hermitean matrices is convex if and only if its restriction to the diagonal matrices is convex. The method of Davis does not apply to our problem because the action of $SO(3)$ on the $W$ variables is no longer linear. Yet, a similar result is hoped for. We conjecture that there is a subspace in $\mathrm{End}(\mathcal{T})$ transversal to the action of the rotation group $SO(3)$, such that the convexity of the restriction of $g$ to this transversal is equivalent to the convexity of $g$. (Of course, the restriction of $g$ to the transversal has to be invariant under the action of the finite Weil group $S_3$ of $SO(3)$.) It is our guess that the geometric results of [15] can be reformulated in terms of convexity in the transversal.

## 8.6. Exact relations

We now turn to another implication of convexity discussed in Section 8.5 above. If we are searching for submanifolds $\mathbb{M}$ of $\mathrm{End}(\mathcal{T})$ corresponding to exact relations, then convexity of the submanifolds $\Pi_{\boldsymbol{n}} = W_{\boldsymbol{n}}(\mathbb{M})$ is equivalent to saying that the $\Pi_{\boldsymbol{n}}$ are affine subspaces of $\mathrm{End}(\mathcal{T})$. Observe now that $W_{\boldsymbol{n}}(\boldsymbol{L}_0) = \boldsymbol{0}$. Thus, choosing $\boldsymbol{L}_0$ to lie on $\mathbb{M}$ we make sure that $\Pi_{\boldsymbol{n}}$ is a subspace for each $\boldsymbol{n}$. The following theorem gives necessary conditions for $\mathbb{M}$ to be an exact relation.

**Theorem 2.** *Suppose the submanifold $\mathbb{M}$ is an exact relation. Then the subspaces $\Pi_{\boldsymbol{n}} = W_{\boldsymbol{n}}(\mathbb{M})$ do not depend on $\boldsymbol{n}$. Moreover, this single subspace $\Pi$ is closed with respect to the family of Jordan multiplications defined by*

$$\boldsymbol{K}_1 *_A \boldsymbol{K}_2 = \frac{1}{2}(\boldsymbol{K}_1 \boldsymbol{A} \boldsymbol{K}_2 + \boldsymbol{K}_2 \boldsymbol{A} \boldsymbol{K}_1), \tag{14}$$

*where $\boldsymbol{A}$ can be any matrix from the subspace*

$$\mathcal{A} = \mathrm{Span}\{\Gamma'(\boldsymbol{n}) - \Gamma'(\boldsymbol{e}_1) \mid \boldsymbol{n} \in \mathbb{S}^2\}. \tag{15}$$

The proof of this theorem follows word for word the proof of the corresponding theorem in [21, Theorem 3.5] for the symmetric case. From now on we will say that a theorem is proved in [21] if the proof for the general case is the same as for the symmetric case with obvious modifications.

Another important question is about stability under homogenization. To this end, in [21], we have derived a formula for the effective tensor $\boldsymbol{L}^*$.

**Theorem 3.** *Let $\boldsymbol{L}^*$ be the effective tensor for the composite with the local tensor $\boldsymbol{L}(\boldsymbol{x})$. Then for any $\boldsymbol{n} \in \mathbb{S}^2$*

$$W_{\boldsymbol{n}}(\boldsymbol{L}^*) = \langle (\mathsf{I} - \mathsf{W}_{\boldsymbol{n}} \Lambda_{\boldsymbol{n}})^{-1} W_{\boldsymbol{n}}(\boldsymbol{L}(\boldsymbol{x})) \rangle, \tag{16}$$

*where $\mathsf{I}$ denotes the identity operator on the Hilbert space $\mathbb{H} = L^2(Q) \otimes \mathrm{End}(\mathcal{T})$, $\mathsf{W}_{\boldsymbol{n}}$ denotes the multiplication operator on $\mathbb{H}$: $(\mathsf{W}_{\boldsymbol{n}} \boldsymbol{H})(\boldsymbol{x}) = W_{\boldsymbol{n}}(\boldsymbol{L}(\boldsymbol{x}))\boldsymbol{H}(\boldsymbol{x})$, and*

$\Lambda_{\boldsymbol{n}}$ *is defined by* $\widehat{\Lambda_{\boldsymbol{n}}\boldsymbol{H}}(\boldsymbol{k}) = \boldsymbol{A}_{\boldsymbol{n}}(\boldsymbol{k})\widehat{\boldsymbol{H}}(\boldsymbol{k})$, *where*

$$\boldsymbol{A}_{\boldsymbol{n}}(\boldsymbol{m}) = \begin{cases} \Gamma'(\frac{\boldsymbol{m}}{|\boldsymbol{m}|}) - \Gamma'(\boldsymbol{n}), & \text{if } \boldsymbol{m} \neq 0, \\ 0, & \text{if } \boldsymbol{m} = 0. \end{cases} \tag{17}$$

Using this formula, we can prove the following rather messy necessary and sufficient condition for a subspace $\Pi$ to correspond to an exact relation.

**Theorem 4.** *For* $\boldsymbol{v} = (\boldsymbol{l}_1, \ldots, \boldsymbol{l}_k) \in (\mathbb{Z}^3)^k$ *and for* $\sigma \in S_k$*, the permutation group of* $k$ *elements, we define*

$$\sigma(\boldsymbol{v}) = (\boldsymbol{l}_{\sigma(1)}, \ldots, \boldsymbol{l}_{\sigma(k)}) \in (\mathbb{Z}^3)^k. \tag{18}$$

*Let* $\mathcal{O}(\boldsymbol{v}) = \{\sigma(\boldsymbol{v}) \in (\mathbb{Z}^3)^k \mid \sigma \in S_k\}$ *be the orbit of* $\boldsymbol{v}$ *under the action of the permutation group* $S_k$*. The subspace* $\Pi \subset \mathrm{End}(\mathcal{T})$ *corresponds to an exact relation if and only if for any* $k \in \mathbb{N}$*, any* $\boldsymbol{n} \in \mathbb{S}^2$*, any* $\boldsymbol{v} = (\boldsymbol{l}_1, \ldots, \boldsymbol{l}_k) \in (\mathbb{Z}^3)^k$ *such that*

$$\sum_{i=1}^{k} \boldsymbol{l}_i = \boldsymbol{0}, \tag{19}$$

*and for any function* $\boldsymbol{K} : \mathbb{Z}^3 \to \Pi$*, we have*

$$\sum_{(\boldsymbol{p}_1,\ldots \boldsymbol{p}_k) \in \mathcal{O}(\boldsymbol{v})} \left( \prod_{s=1}^{k-1} \boldsymbol{K}(\boldsymbol{p}_s)\boldsymbol{A}_{\boldsymbol{n}} \left( \sum_{j=1}^{s} \boldsymbol{p}_j \right) \right) \boldsymbol{K}(\boldsymbol{p}_k) \in \Pi, \tag{20}$$

*where* $\boldsymbol{A}_{\boldsymbol{n}}$ *is defined in* (17) *above.*

The analogous theorem was formulated in [21, Theorem 3.9], where some steps of the proof were indicated. Unfortunately, the formulation of the theorem was not entirely correct. The sum in the formula [21, (3.38)] corresponding to (20) extended over the set $S_k$ of all permutations instead of the elements of the orbit $\mathcal{O}(\boldsymbol{v})$. In [21] the theorem was not used anywhere else and was included only for the purposes of a discussion. Here we give the correct formulation and a complete proof of the theorem.

*Proof.* The proof is based on the formula (16) and the analyticity properties enjoyed by the effective tensor. In [21, Appendix] we have shown that the map $W_{\boldsymbol{n}}$ is an analytic diffeomorphism, defined everywhere on the set of positive definite matrices. Therefore, all exact relation manifolds $\mathbb{M}$, being analytic images of subspaces $\Pi$, are analytic. Also, if we take an analytic family $\boldsymbol{L}_\lambda(\boldsymbol{x})$ of local tensors, it will generate an analytic family of effective tensors $\boldsymbol{L}_\lambda^*$ [18]. The principle of analytic continuation implies that if $\boldsymbol{L}_\lambda^* \in \mathbb{M}$ for some small interval of $\lambda$ then $\boldsymbol{L}_\lambda^* \in \mathbb{M}$ for all $\lambda$ in the interval of analyticity of $\boldsymbol{L}_\lambda^*$. This argument shows that if we want to prove that $\mathbb{M}$ is an exact relation, then it is enough to show that $\mathbb{M}$ satisfies necessary conditions of Theorem 2 and that for one choice of $\boldsymbol{n} \in \mathbb{S}^2$ and for any $\boldsymbol{W}(\boldsymbol{x}) = W_{\boldsymbol{n}}(\boldsymbol{L}(\boldsymbol{x})) \in \Pi$ sufficiently small, we have $\boldsymbol{W}^* = W_{\boldsymbol{n}}(\boldsymbol{L}^*) \in \Pi$. We can ensure that $\boldsymbol{W}(\boldsymbol{x})$ is sufficiently small if we choose the values of $\boldsymbol{L}(\boldsymbol{x})$ sufficiently close to the reference medium $\boldsymbol{L}_0 \in \mathbb{M}$.

Let $\Pi_{\mathbb{C}} = \{\boldsymbol{K}_1 + i\boldsymbol{K}_2 \mid \{\boldsymbol{K}_1, \boldsymbol{K}_2\} \subset \Pi\}$ be the complexification of $\Pi$. The analytic continuation principle also implies that if $\Pi$ corresponds to an exact relation then so does $\Pi_{\mathbb{C}}$ in the sense that if $\boldsymbol{W}(\boldsymbol{x}) \in \Pi_{\mathbb{C}}$ then also $\boldsymbol{W}^* \in \Pi_{\mathbb{C}}$, where $\boldsymbol{W}^*$ is computed from $\boldsymbol{W}(\boldsymbol{x})$ by the formula (16). The argument is due to Milton (private communication). Let $\boldsymbol{W}_\lambda(\boldsymbol{x}) = \boldsymbol{W}_1(\boldsymbol{x}) + \lambda \boldsymbol{W}_2(\boldsymbol{x})$, where $\boldsymbol{W}_1(\boldsymbol{x}) \in \Pi$ and $\boldsymbol{W}_2(\boldsymbol{x}) \in \Pi$ are assumed to be small enough. Then, since $\Pi$ corresponds to an exact relation we conclude that for real values of $\lambda$ in some finite interval, we have $\boldsymbol{W}_\lambda^* \in \Pi$. Let $\boldsymbol{P} \in \mathrm{End}(\mathcal{T})$ be such that $\mathrm{Tr}(\boldsymbol{P}\boldsymbol{K}) = 0$ for all $\boldsymbol{K} \in \Pi$. Then the function $f(\lambda) = \mathrm{Tr}(\boldsymbol{P}\boldsymbol{W}_\lambda^*)$ is analytic in $\lambda$ and is zero on an interval on the real axis. Since $\boldsymbol{W}$ is assumed to be small, the region of analyticity of $f(\lambda)$ in the complex plane includes $\lambda = i$. Thus, we conclude that $\mathrm{Tr}(\boldsymbol{P}\boldsymbol{W}_i^*) = 0$ for all real matrices $\boldsymbol{P}$ such that $\mathrm{Tr}(\boldsymbol{P}\boldsymbol{K}) = 0$ for all $\boldsymbol{K} \in \Pi$. But

$$\Pi_{\mathbb{C}} = \{\boldsymbol{K} \in \mathrm{End}_{\mathbb{C}}(\mathcal{T}) \mid \mathrm{Tr}(\boldsymbol{P}\boldsymbol{K}) = 0 \,\forall \boldsymbol{P} : \mathrm{Tr}(\boldsymbol{P}\boldsymbol{W}) = 0 \,\forall \boldsymbol{W} \in \Pi\}.$$

Thus, $\boldsymbol{W}_i^* \in \Pi_{\mathbb{C}}$.

If $\boldsymbol{W}(\boldsymbol{x})$ is complex valued and sufficiently close to zero, then we can expand (16) in a convergent power series:

$$\boldsymbol{W}^* = \langle \boldsymbol{W}(\boldsymbol{x}) \rangle + \langle \mathsf{W}\Lambda_{\boldsymbol{n}} \boldsymbol{W}(\boldsymbol{x}) \rangle + \cdots + \langle (\mathsf{W}\Lambda_{\boldsymbol{n}})^k \boldsymbol{W}(\boldsymbol{x}) \rangle + \cdots, \tag{21}$$

where $\mathsf{W}$ is the operator of multiplication by $\boldsymbol{W}(\boldsymbol{x})$. It follows from the argument above that $\mathbb{M}$ is an exact relation if and only if each term of the expansion (21) belongs to $\Pi_{\mathbb{C}}$, when $\boldsymbol{W}(\boldsymbol{x}) \in \Pi_{\mathbb{C}}$ is small enough.

Let $\boldsymbol{T}_k(\boldsymbol{x}) = (\mathsf{W}\Lambda_{\boldsymbol{n}})^k \boldsymbol{W}(\boldsymbol{x})$. Writing $\boldsymbol{T}_k(\boldsymbol{x}) = \boldsymbol{W}(\boldsymbol{x})(\Lambda_{\boldsymbol{n}}\boldsymbol{T}_{k-1})(\boldsymbol{x})$ recursively, then taking the Fourier transform and using induction in $k$, we can prove that

$$\widehat{\boldsymbol{T}}_k(\boldsymbol{m}) = \sum_{\boldsymbol{l}_1 + \cdots + \boldsymbol{l}_k = \boldsymbol{m}} \left( \prod_{s=1}^{k-1} \widehat{\boldsymbol{W}}(\boldsymbol{l}_s) \boldsymbol{A}_{\boldsymbol{n}} \left( \sum_{j=s+1}^{k} \boldsymbol{l}_j \right) \right) \widehat{\boldsymbol{W}}(\boldsymbol{l}_k),$$

where $\boldsymbol{m}, \boldsymbol{l}_j \in \mathbb{Z}^3$. Thus, we get

$$\langle (\mathsf{W}\Lambda_{\boldsymbol{n}})^k \boldsymbol{W}(\boldsymbol{x}) \rangle = \widehat{\boldsymbol{T}}_k(\boldsymbol{0}) = \sum_{\boldsymbol{l}_1 + \cdots + \boldsymbol{l}_k = \boldsymbol{0}} \left( \prod_{s=1}^{k-1} \widehat{\boldsymbol{W}}(\boldsymbol{l}_s) \boldsymbol{A}_{\boldsymbol{n}} \left( \sum_{j=1}^{s} \boldsymbol{l}_j \right) \right) \widehat{\boldsymbol{W}}(\boldsymbol{l}_k). \tag{22}$$

Observe that the sum in (22) can be split into parts. The summation in each part goes over all the distinct permutations of the same set of vectors $\boldsymbol{l}_j$. In order to say this more rigorously we define the action of the permutation group $S_k$ on $(\mathbb{Z}^3)^k$ by (18). Thus, $(\mathbb{Z}^3)^k$ splits into the disjoint union of orbits of the group action. Let $\mathcal{Z}$ denote the set of orbits whose elements satisfy the constraint (19) (invariant under the group action). Then we can write

$$\langle \boldsymbol{T}_k \rangle = \sum_{\mathcal{O} \in \mathcal{Z}} \sum_{(\boldsymbol{l}_1, \ldots, \boldsymbol{l}_k) \in \mathcal{O}} \left( \prod_{s=1}^{k-1} \widehat{\boldsymbol{W}}(\boldsymbol{l}_s) \boldsymbol{A}_{\boldsymbol{n}} \left( \sum_{j=1}^{s} \boldsymbol{l}_j \right) \right) \widehat{\boldsymbol{W}}(\boldsymbol{l}_k). \tag{23}$$

If $\mathbb{M}$ is an exact relation then we can choose an arbitrary $k$-tuple $(\boldsymbol{l}_1, \ldots, \boldsymbol{l}_k) \in (\mathbb{Z}^3)^k$ satisfying (19) and let

$$W(\boldsymbol{x}) = \sum_{j=1}^{k} \boldsymbol{K}_j e^{i\boldsymbol{l}_j \cdot \boldsymbol{x}},$$

where $\{\boldsymbol{K}_1, \ldots, \boldsymbol{K}_k\} \subset \Pi$ are arbitrary. Then we will obtain (20). Conversely, if (20) is satisfied for all $\{\boldsymbol{K}_1, \ldots, \boldsymbol{K}_k\} \subset \Pi$ then, by the analytic continuation principle, it is satisfied for all $\{\boldsymbol{K}_1, \ldots, \boldsymbol{K}_k\} \subset \Pi_{\mathbb{C}}$. Thus, each term under the exterior sum in (23) is in $\Pi_{\mathbb{C}}$. Therefore, $\boldsymbol{W}^* \in \Pi_{\mathbb{C}}$ and is real, so $\boldsymbol{W}^* \in \Pi$. The theorem is proved. □

The above theorem does provide algebraic conditions for $\Pi$ that guarantee that $\Pi$ corresponds to an exact relation. The practical utility of such a theorem is minimal, since it is virtually impossible to check infinitely many conditions that are as complicated as (20). Therefore, in [21] we proved a convenient sufficient condition for $\Pi$ to correspond to an exact relation. In order to formulate it we introduce the following terminology. The expression $\boldsymbol{K}_1\boldsymbol{A}_1\boldsymbol{K}_2\boldsymbol{A}_2\boldsymbol{K}_3\boldsymbol{A}_3 \ldots \boldsymbol{K}_{j-1}\boldsymbol{A}_{j-1}\boldsymbol{K}_j$ will be called a $j$-chain. We say that a subspace $\Pi$ satisfies a $j$-chain property if for every $\{\boldsymbol{K}_1, \ldots, \boldsymbol{K}_j\} \subset \Pi$ and every $\{\boldsymbol{A}_1, \ldots, \boldsymbol{A}_{j-1}\} \subset \mathcal{A}$ we have

$$\boldsymbol{K}_1\boldsymbol{A}_1\boldsymbol{K}_2\boldsymbol{A}_2 \ldots \boldsymbol{K}_{j-1}\boldsymbol{A}_{j-1}\boldsymbol{K}_j + \boldsymbol{K}_j\boldsymbol{A}_{j-1}\boldsymbol{K}_{j-1} \ldots \boldsymbol{A}_2\boldsymbol{K}_2\boldsymbol{A}_1\boldsymbol{K}_1 \in \Pi. \tag{24}$$

We remark that if $\Pi$ is a subspace corresponding to an exact relation then Theorem 2 says that $\Pi$ satisfies a 2-chain property. Now the analogue of the sufficient condition established in [21] can be formulated as follows.

**Theorem 5.** *If $\Pi$ satisfies the $j$-chain property for $j = 2, 3, 4$, then $\Pi$ corresponds to an exact relation.*

*Proof.* In [21, proof of Lemma 3.7] we proved that if $\Pi$ satisfies the $j$-chain property for $j = 2, 3, 4$, then $\Pi$ satisfies the $j$-chain property for all $j \geq 2$. It remains to show that if $\Pi$ satisfies the $j$-chain property for every $j \geq 2$ then (20) will be satisfied.

Let $\eta$ be the permutation defined by $\eta(j) = k + 1 - j$. The permutation $\eta$ is an element of order two in the group $S_k$. It acts on $\mathcal{O}(v)$ and splits $\mathcal{O}(v)$ in a disjoint union of orbits. If an orbit contains two elements then their sum has the form (24). If the orbit contains a single element then this element is $1/2$ of the sum of two copies of itself, which is again of the form (24). □

In two space dimensions, however, we can say a little bit more.

**Theorem 6.** *In two space dimensions the* 3*-chain property is necessary for stability under homogenization.*

*Proof.* We will show that in 2D the necessary and sufficient condition (20) from Theorem 4 for $k = 3$ implies the 3-chain property.

Let $\boldsymbol{n} = (1, 0)$. Fix $N \in \mathbb{Z} \setminus \{0\}$. Choose $\boldsymbol{l}_1 \in \mathbb{Z}^2$ to be linearly independent with $\boldsymbol{n}$ and define $\boldsymbol{l}_2 = -N\boldsymbol{n} - \boldsymbol{l}_1$, so that $\boldsymbol{l}_3 = N\boldsymbol{n}$. Observe that $\boldsymbol{l}_1$, $\boldsymbol{l}_2$, and $\boldsymbol{l}_3$ are distinct and that according to (17), $\boldsymbol{A}(\boldsymbol{l}_3) = \boldsymbol{A}(\boldsymbol{l}_1 + \boldsymbol{l}_2) = \boldsymbol{0}$. For simplicity we use

$A(\cdot)$ notation instead of $A_n(\cdot)$, since $n = (1, 0)$ is fixed. Condition (20) for $k = 3$ then becomes

$$K_2A(l_1 + Nn)K_3A(l_1)K_1 + K_1A(l_1)K_3A(l_1 + Nn)K_2 \in \Pi$$

for all $\{K_1, K_2, K_3\} \subset \Pi$. Now, if we vary $N$ then we observe that the vectors $\{\pm(l_1 + Nn)/|l_1 + Nn| : N \in \mathbb{Z}\}$ form a dense subset of the unit circle. (The same construction does not yield a dense subset of the unit sphere in 3D.) Therefore, by continuity, we conclude that for any $\{q_1, q_2\} \subset \mathbb{S}^1$ we have

$$K_2A(q_2)K_3A(q_1)K_1 + K_1A(q_1)K_3A(q_2)K_2 \in \Pi$$

for all $\{K_1, K_2, K_3\} \subset \Pi$, which easily implies the 3-chain condition. □

We remark that in all the examples that we have worked out so far our sufficient condition was satisfied whenever the necessary conditions in Theorem 2 were.

We would like to conclude this section by stating the results and problems obtained in [21] for the symmetric case. For a subspace $\mathcal{X}$ of $\mathrm{End}(\mathcal{T})$ let $\mathcal{X}_{\mathrm{sym}}$ denote the set of symmetric parts of matrices in $\mathcal{X}$. If $\Pi \subset \mathrm{Sym}(\mathcal{T})$ then Theorem 2 can be restated as follows.

**Theorem 7.** *If the subspace $\Pi \subset \mathrm{Sym}(\mathcal{T})$ corresponds to an exact relation, then it satisfies*

$$(\Pi\mathcal{A}\Pi)_{\mathrm{sym}} \subset \Pi. \tag{25}$$

The sufficient condition for $\Pi$ to correspond to an exact relation stated in Theorem 5 can be cast into a more attractive algebraic form. Let $\Pi' \subset \mathrm{End}(\mathcal{T})$ be the smallest associative algebra containing the Jordan algebra $\Pi$. In other words $\Pi'$ is the smallest subspace in $\mathrm{End}(\mathcal{T})$ containing $\Pi$ that satisfies

$$\Pi'\mathcal{A}\Pi' \subset \Pi'. \tag{26}$$

Theorem 5 can then be restated as follows.

**Theorem 8.** *If the subspace $\Pi \subset \mathrm{Sym}(\mathcal{T})$ solves (25) and has the additional property that $\Pi = \Pi'_{\mathrm{sym}}$, then $\Pi$ corresponds to the exact relation.*

The important open question is whether every Jordan algebra $\Pi$ (understood in the sense of (25)) is the set of all symmetric matrices of the smallest associative algebra $\Pi'$ (understood in the sense of (26)) containing $\Pi$. The question, in other words, is "Are sufficient conditions necessary?"

## 8.7. Finding exact relations

The preceding section established simple algebraic conditions that the subspace $\Pi$ has to satisfy in order to correspond to an exact relation. Now the question is: Can we characterize all solutions $\Pi$ of

$$K_1AK_2 + K_2AK_1 \in \Pi, \tag{27}$$

for all $\{K_1, K_2\} \subset \Pi$ and all $A \in \mathcal{A}$? Unfortunately, at present we do not have an efficient way to solve equation (27). In this section, we will discuss an inefficient way of solving (27) for the 2D Hall effect. The method simply consists of picking a matrix and computing by brute force the smallest Jordan algebra $\Pi$ containing that matrix. Then we compute all other Jordan algebras extending the smallest one, again, by brute force. This method is applicable to rather small dimensional cases (or small block-dimensional as in [21]).

For the 2D Hall effect, the space $\mathcal{T} = \mathbb{R}^2$. Let

$$L_0 = \sigma_0 + \alpha S \tag{28}$$

be the positive definite reference medium, where $\sigma_0$ is the symmetric part of $L_0$ and

$$S = \begin{bmatrix} 0 & 1 \\ -1 & 0 \end{bmatrix}. \tag{29}$$

We recall that for the Hall effect the matrix $\Gamma'(n)$ is given by (11). Therefore, the subspace $\mathcal{A}$ defined in (15) is

$$\mathcal{A} = \{A \in \operatorname{End}(\mathbb{R}^2) \mid \operatorname{Tr} A = 0,\ AL_0^T = L_0 A^T\}. \tag{30}$$

### 8.7.1. Solving (27)

Observe that equation (27) behaves nicely with respect to the following "change of variables." Let $X$ and $Y$ be invertible matrices. Let $\overline{\Pi} = X\Pi Y$. Then $\Pi$ solves (27) if and only if $\overline{\Pi}$ solves (27) with $\mathcal{A}$ replaced by $\overline{\mathcal{A}} = Y^{-1}\mathcal{A}X^{-1}$. Applying this observation to

$$X = \sigma_0^{-1/2}, \quad Y = L_0 X, \tag{31}$$

we obtain that $\overline{\mathcal{A}}$ is the space of trace-free symmetric $2 \times 2$ matrices. From now on we will work only with subspaces $\overline{\mathcal{A}}$ and $\overline{\Pi}$, and therefore we will rename them $\mathcal{A}$ and $\Pi$, respectively, for notational convenience. We will return to the original notation when we have solved equation (27). Thus, $\mathcal{A}$ now denotes the subspace of $2 \times 2$ symmetric trace-free matrices.

Observe that for $X = R$ and $Y = R^T$, where $R$ is a $2 \times 2$ rotation matrix, the equation $K\mathcal{A}K \subset \Pi$, $K \in \Pi$ remains invariant. Now fix $K \in \Pi$. We can find a rotation $R \in O(2)$ such that

$$R \cdot K = K_0 = \rho \begin{bmatrix} 1 & \beta \\ -\beta & \alpha \end{bmatrix}. \tag{32}$$

Thus, without loss of generality, $\Pi$ contains a matrix $K_0$ of the form (32). We have

$$\begin{bmatrix} 1 & \beta \\ -\beta & \alpha \end{bmatrix}\begin{bmatrix} s & t \\ t & -s \end{bmatrix}\begin{bmatrix} 1 & \beta \\ -\beta & \alpha \end{bmatrix} = \begin{bmatrix} (1+\beta^2)s & (\alpha+\beta^2)t + \beta(1-\alpha)s \\ (\alpha+\beta^2)t - \beta(1-\alpha)s & -(\alpha^2+\beta^2)s \end{bmatrix}. \tag{33}$$

*Case* I: *There exists* $\boldsymbol{K}_0 \in \Pi$ *with* $\alpha + \beta^2 \neq 0$. In this case, we easily see from (33) (setting $s = 0, t = 1$) that

$$\begin{bmatrix} 0 & 1 \\ 1 & 0 \end{bmatrix} \in \Pi.$$

Consequently,

$$\begin{bmatrix} 0 & 1 \\ 1 & 0 \end{bmatrix}\begin{bmatrix} s & t \\ t & -s \end{bmatrix}\begin{bmatrix} 0 & 1 \\ 1 & 0 \end{bmatrix} = \begin{bmatrix} -s & t \\ t & s \end{bmatrix}$$

implies that $\mathcal{A} \subset \Pi$. Simple manipulations show that the right-hand side of (33) is equal to $(1 - \alpha)s\boldsymbol{K}_0 \bmod \mathcal{A}$. Thus, we get that $\Pi$ contains the subspace $\mathcal{A}$ and the matrix

$$\boldsymbol{K}_0' = \begin{bmatrix} (1+\alpha)/2 & \beta \\ -\beta & (1+\alpha)/2 \end{bmatrix}. \tag{34}$$

*Case* I(a): *There exists* $\boldsymbol{K}_0 \in \Pi$ *such that* $\alpha \neq -1$, *in addition to* $\alpha + \beta^2 \neq 0$. Then we infer that

$$\begin{bmatrix} 1 & \beta' \\ -\beta' & 1 \end{bmatrix} \in \Pi,$$

where $\beta' = 2\beta/(1+\alpha)$. Again an easy calculation shows that

$$\begin{bmatrix} 1 & \beta' \\ -\beta' & 1 \end{bmatrix}\begin{bmatrix} s & t \\ t & -s \end{bmatrix}\begin{bmatrix} x & y \\ y & -x \end{bmatrix} + \begin{bmatrix} x & y \\ y & -x \end{bmatrix}\begin{bmatrix} s & t \\ t & -s \end{bmatrix}\begin{bmatrix} 1 & \beta' \\ -\beta' & 1 \end{bmatrix}$$

is equal to

$$2(sx + ty)\begin{bmatrix} 1 & \beta' \\ -\beta' & 1 \end{bmatrix} (\bmod \mathcal{A}).$$

Thus, the minimal subspace $\Pi$ satisfying (27) in Case I(a) is a 3D subspace

$$\Pi = \mathcal{A} \oplus \begin{bmatrix} 1 & \beta' \\ -\beta' & 1 \end{bmatrix}\mathbb{R}. \tag{35}$$

Since this subspace is of codimension 1, there are no larger proper subspaces containing $\Pi$. We note that (35) gives a one-parameter family of solutions $\Pi$, labeled by $\beta' \in \mathbb{R}$.

*Case* I(b): $\alpha = -1$ *for all* $\boldsymbol{K}_0 \in \Pi$. Then, returning to (34) we obtain that

$$\begin{bmatrix} 0 & \beta \\ -\beta & 0 \end{bmatrix} \in \Pi.$$

Now if all $\boldsymbol{K}_0 \in \Pi$ are symmetric, then $\beta = 0$ for all $\boldsymbol{K}_0 \in \Pi$ and $\Pi = \mathcal{A}$ or $\Pi = \mathrm{Sym}(\mathbb{R}^2)$—the space of symmetric $2 \times 2$ matrices.

If there is $\boldsymbol{K}_0 \in \Pi$ which is nonsymmetric, then

$$\Pi = \mathcal{A} \oplus \begin{bmatrix} 0 & 1 \\ -1 & 0 \end{bmatrix}\mathbb{R}. \tag{36}$$

This subspace is of codimension 1 and therefore is not contained in any other proper subspace.

*Case* II: $\alpha = -\beta^2$ *for all* $\boldsymbol{K}_0 \in \Pi$. In this case, we easily see that

$$\Pi = \begin{bmatrix} 1 & \beta \\ -\beta & -\beta^2 \end{bmatrix} \mathbb{R} \tag{37}$$

is a one-parameter family of 1D solutions of (27). Now we would like to see if there are other solutions $\Pi$ containing (37). Let $\boldsymbol{K}_1$, not a multiple of $\boldsymbol{K}_0$, be in $\Pi$. Then for all $\lambda \in \mathbb{R}$ the matrices $\boldsymbol{K}(\lambda) = \boldsymbol{K}_1 + \lambda \boldsymbol{K}_0$ belong to $\Pi$ and to Case II (we have already analyzed Case I). Thus, $\boldsymbol{K}(\lambda)$ can be reduced to the form (37) with a different $\beta$ by a rotation and scaling:

$$\boldsymbol{K}(\lambda) = \mu(\lambda)\boldsymbol{R}_\lambda \begin{bmatrix} 1 & \beta(\lambda) \\ -\beta(\lambda) & -\beta(\lambda)^2 \end{bmatrix} \boldsymbol{R}_\lambda^T.$$

It will be convenient to represent $\boldsymbol{K}_1$ in the form $\boldsymbol{K}_1 = \boldsymbol{P} + \gamma \boldsymbol{S}$, where $\boldsymbol{P}$ is a symmetric matrix and $\boldsymbol{S}$ is given by (29). Taking the antisymmetric part in the formula for $\boldsymbol{K}(\lambda)$ we obtain $\mu(\lambda)\beta(\lambda) = \gamma + \lambda\beta$, while taking the determinant of the symmetric part of $\boldsymbol{K}(\lambda)$ we get

$$\det \boldsymbol{P} + \lambda(p_{22} - \beta^2 p_{11}) - \lambda^2\beta^2 = -\mu(\lambda)^2\beta(\lambda)^2,$$

for all $\lambda \in \mathbb{R}$. Thus, we obtain the following equations for $\gamma$ and for the components $p_{ij}$ of $\boldsymbol{P}$:

$$p_{11}p_{22} - p_{12}^2 = -\gamma^2, \quad p_{22} - \beta^2 p_{11} = -2\gamma\beta.$$

Eliminating $p_{22}$ we get $(p_{11}\beta - \gamma)^2 = p_{12}^2$. So either

$$p_{12} = p_{11}\beta - \gamma \tag{38}$$

or

$$p_{12} = \gamma - p_{11}\beta. \tag{39}$$

Assume first that $\boldsymbol{K}_1$ satisfies (38). Then

$$\boldsymbol{K}_1 = \begin{bmatrix} p_{11} & p_{11}\beta \\ p_{11}\gamma - 2\gamma & p_{11}\beta^2 - 2\gamma\beta \end{bmatrix}.$$

Now observe that

$$\boldsymbol{K}_1 - p_{11}\boldsymbol{K}_0 = 2(p_{11}\beta - \gamma)\begin{bmatrix} 0 & 0 \\ 1 & \beta \end{bmatrix}.$$

Notice that if $\gamma = p_{11}\beta$ then $\boldsymbol{K}_1 = p_{11}\boldsymbol{K}_0$ in contradiction to our assumption that $\boldsymbol{K}_1$ is not a multiple of $\boldsymbol{K}_0$. Thus, we conclude that

$$\Pi_1 = \text{Span}\left\{\begin{bmatrix} 0 & 0 \\ 1 & \beta \end{bmatrix}, \begin{bmatrix} 1 & \beta \\ 0 & 0 \end{bmatrix}\right\} \subset \Pi. \tag{40}$$

One may easily check that $\Pi_1$ is a one-parameter family of 2D solutions.

Making a similar analysis for case (39), we obtain a solution $\Pi_2 \subset \Pi$, where

$$\Pi_2 = \text{Span}\left\{\begin{bmatrix} 1 & 0 \\ -\beta & 0 \end{bmatrix}, \begin{bmatrix} 0 & 1 \\ 0 & -\beta \end{bmatrix}\right\}. \tag{41}$$

If $\Pi$ contains $\Pi_1$ but is larger than $\Pi_1$ then our previous analysis shows that $\boldsymbol{K} \in \Pi$ must be either in $\Pi_1$ or in $\Pi_2$. But $\Pi_1 \bigcup \Pi_2$ is not a subspace.

In summary, a proper subspace $\Pi$ is a solution of (27) if and only if it is a rotated image of either $\mathcal{A}$, $\mathrm{Sym}(\mathbb{R}^2)$, (35), (36), (37), (40), or (41). We can now state the result.

**Theorem 9.** *If a proper subspace $\Pi$ in the space of all $2 \times 2$ matrices satisfies* (27) *with $\mathcal{A}$ being the space of symmetric, trace-free $2 \times 2$ matrices, then $\Pi$ is a subspace from the following list.*

1. $\mathrm{Sym}(\mathbb{R}^2)$—*the space of symmetric matrices;*
2. $\Pi_0$—*the space of trace-free matrices;*
3. $\mathcal{A}$—*the space of symmetric, trace-free matrices;*
4. $$\Pi_\beta = \left\{ \begin{bmatrix} a & b \\ c & d \end{bmatrix} : a + d = \beta(b - c) \right\}, \quad \beta \neq 0;$$
5. $\Pi_{\boldsymbol{a}} = \{\boldsymbol{v} \otimes \boldsymbol{a} \mid \boldsymbol{v} \in \mathbb{R}^2\}$;
6. $\Pi_{\boldsymbol{a}}^T$;
7. $\Pi_{\boldsymbol{a},\boldsymbol{b}} = \mathbb{R}(\boldsymbol{a} \otimes \boldsymbol{b})$.

Observe that in our list items 1 and 2 are limiting cases of item 4: $\Pi_\beta \to \Pi_0$, when $\beta \to 0$ and $\Pi_\beta \to \mathrm{Sym}(\mathbb{R}^2)$, when $\beta \to \infty$. We also observe that $\mathcal{A}$ is the intersection of $\Pi_{\beta_1}$ and $\Pi_{\beta_2}$ for any $\beta_1 \neq \beta_2$. Also $\Pi_{\boldsymbol{a},\boldsymbol{b}} = \Pi_{\boldsymbol{b}} \bigcap \Pi_{\boldsymbol{a}}^T$. Thus, we need to focus only on three subspaces from items 4, 5, and 6.

### 8.7.2. Checking sufficient conditions

According to Theorem 5 and our discussion above we need to check the 3- and 4-chain property of subspaces $\Pi$ from items 4, 5, and 6 in the list above. Indeed, the intersection of subspaces satisfying these properties must also satisfy them. The subspaces $\Pi_{\boldsymbol{a}}$ and $\Pi_{\boldsymbol{a}}^T$ are closed with respect to associative multiplication: $\boldsymbol{K}_1\boldsymbol{A}\boldsymbol{K}_2 \in \Pi$ whenever $\{\boldsymbol{K}_1, \boldsymbol{K}_2\} \subset \Pi$. Thus the $j$-chain property is clearly satisfied for those subspaces. The actual checking needs to be done only for the subspaces $\Pi_\beta$. The checking can be easily done with Maple—the symbolic algebra package. And indeed, we find that $\Pi_\beta$ does satisfy 3- and 4-chain properties.

### 8.7.3. Returning to $\boldsymbol{L}$ variables

Now that we have a list of subspaces $\Pi$ corresponding to exact relations, we need to return to $\boldsymbol{L}$ variables. At the first glance the task before us is simply to compute

$$\mathbb{M} = \{\boldsymbol{L}_0 - [\boldsymbol{I} + \boldsymbol{K}\Gamma'(\boldsymbol{n})]^{-1}\boldsymbol{K}\boldsymbol{L}_0 \mid \boldsymbol{K} \in \Pi\}, \tag{42}$$

which is the inverse of W-transformation. However, we quickly realize that in order to get explicit results, the computation in (42) is not so easy. Fortunately, we can often simplify our job and sometimes avoid it altogether. One obvious observation that we have already made is that if we have computed two exact relations then there is no need to compute their intersection.

In [21] we identified an especially simple class of exact relations: uniform field relations (UFR). In general these are defined as

$$\mathbb{M} = \{L \in \mathrm{End}(\mathcal{T}) \mid La_1 = b_1, \ldots, La_s = b_s\}, \tag{43}$$

for fixed uniform fields $\{a_1, \ldots, a_s, b_1, \ldots, b_s\} \subset \mathcal{T}$. These are easily recognizable at the level of subspaces $\Pi$.

**Theorem 10.** *The uniform field relations* (43) *are in one-to-one correspondence with subspaces $V \subset \mathcal{T}$. The corresponding subspace $\Pi$ is the annihilator of $V$:*

$$\Pi = \{K \in \mathrm{End}(\mathcal{T}) \mid Kv = 0 \,\forall v \in V\}.$$

The subspaces $\Pi_a$ are annihilators of subspaces $V_a = \mathbb{R}a^{\perp}$, where $a^{\perp} = Sa$. Thus, the exact relations corresponding to subspaces $\Pi_a$ are the sets of positive definite matrices $L$ such that $Lu = v$ for fixed vectors $\{u, v\} \subset \mathbb{R}^2$. Obviously $\Pi_a^T$ corresponds to the same class of exact relations, where $L$ is replaced with $L^T$.

The only exact relation here where we do need to compute something is $\Pi_\beta$. The following theorems were proved in [21] to facilitate our task.

**Theorem 11.** *Fix $n \in \mathbb{S}^2$. Let $M$ be such that $K(\Gamma'(n) - M)K \in \Pi$ for all $K \in \Pi$. Then the invertible transformation $W_M = [S(L) - M]^{-1}$ maps $\mathbb{M}$ into $\Pi$.*

Recall that in our example of the 2D Hall effect, we found it easier not to work with subspaces $\Pi$ and $\mathcal{A}$ directly but rather with subspaces $\overline{\Pi} = X\Pi Y$ and $\overline{\mathcal{A}} = Y^{-1}\mathcal{A}X^{-1}$. Accordingly, we fix $n$ and define $\overline{\Gamma} = Y^{-1}\Gamma'(n)X^{-1}$.

**Theorem 12.** *Let $M$ be such that*

$$K(\overline{\Gamma} - M)K \in \overline{\Pi} \tag{44}$$

*for all $K \in \overline{\Pi}$. Then the invertible transformation $W_M = [Y^{-1}S(L)X^{-1} - M]^{-1}$ maps $\mathbb{M}$ into $\overline{\Pi}$.*

Applying this theorem to the case at hand with $X$ and $Y$ defined by (31), we obtain the inversion formula that we will use:

$$L = \sigma_0 + \alpha S - \sigma_0^{1/2}[I + KM]^{-1}K\sigma_0^{1/2}, \tag{45}$$

where $\sigma_0$ and $\alpha$ are related to the reference medium $L_0$ via (28). The utility of (45) is in the fact that it allows us to compute the manifold

$$\overline{\mathbb{M}} = \{[I + KM]^{-1}K \mid K \in \overline{\Pi}\}$$

using simplified objects $M$ and $\overline{\Pi}$. The actual exact relation is just an affine image of $\overline{\mathbb{M}}$.

In our example, $\overline{\Gamma} = u \otimes u$, where $u = \sigma_0^{1/2}n/|\sigma_0^{1/2}n|$. For $\overline{\Pi} = \mathrm{Sym}(\mathbb{R}^2)$ we easily see that $M = 0$ satisfies (44). So the exact relation $\mathbb{M}$ says that if we mix materials with the same Hall coefficient $r_0$ then the mixture will have the same Hall coefficient $r_0$.

Now let us compute the exact relation corresponding to $\Pi_\beta$. Here $M = 0$ does not work, so we choose $M = \overline{\Gamma} = u \otimes u$. Observe that the choice $M = \overline{\Gamma}$ always satisfies the conditions of Theorem 12. In our simple example of the 2D Hall effect, the choice $M = \overline{\Gamma}$ is still simple enough for practical purposes. In other contexts such as elasticity, thermoelectricity, or other coupled problems the matrix $\overline{\Gamma}$ is not so simple, and other choices for $M$ work (see [21] for such formidable examples as 3D thermopiezoelectricity). Observe that the subspace $\Pi_\beta$ is a hyperplane defined by the equation $\mathrm{Tr}(\boldsymbol{K}\boldsymbol{Q}) = 0$, where

$$\boldsymbol{Q} = \begin{bmatrix} 1 & -\beta \\ \beta & 1 \end{bmatrix}.$$

Solving (45) for $\boldsymbol{K}$, we obtain

$$\boldsymbol{K} = \left[(\boldsymbol{P} - \boldsymbol{L}')^{-1} - \boldsymbol{u} \otimes \boldsymbol{u}\right]^{-1}, \tag{46}$$

where $\boldsymbol{L}' = \boldsymbol{\sigma}_0^{-1/2} \boldsymbol{L} \boldsymbol{\sigma}_0^{-1/2}$, $\boldsymbol{P} = \boldsymbol{I} + \alpha' \boldsymbol{S}$ and $\alpha' = \alpha \det \boldsymbol{\sigma}_0^{-1/2}$. Since the answer does not depend on the choice of the vector $\boldsymbol{u}$, we can set $\boldsymbol{u} = (1, 0)$ and simplify the equation $\mathrm{Tr}(\boldsymbol{K}\boldsymbol{Q}) = 0$ with Maple. The result is written most conveniently in terms of the conductivity tensor $\boldsymbol{\sigma}$ and Hall coefficient $r$, so that $\boldsymbol{L} = \boldsymbol{\sigma} + r\boldsymbol{S}$:

$$\det \boldsymbol{\sigma} + (r - \alpha - \beta\sqrt{\det \boldsymbol{\sigma}_0})^2 = (1 + \beta^2) \det \boldsymbol{\sigma}_0. \tag{47}$$

The exact relation we obtain can be written most concisely as

$$\mathbb{M} = \{(\boldsymbol{\sigma}, r) \mid \det \boldsymbol{\sigma} + (r - r_0)^2 = \mathrm{const}\}. \tag{48}$$

This relation was first derived by Milton [37] (see also [14]). When $\beta \to 0$ the exact relation (47) still retains the form (48). If $\beta \to \infty$ the exact relation (47) becomes $r = r_0$, first derived by Stroud and Bergman [47]. We note that the intersection of exact relations (48) and $r = 0$ results in the well-known Keller–Dykhne–Mendelson exact relation [13, 26, 36].

### 8.7.4. Exact relations with volume fractions

Very often exact relations are supplemented by other relations involving volume averages. For example, Hill's exact relation for elasticity mentioned in the Introduction has an extra part:

$$(3\kappa^* + 4\mu)^{-1} = \langle (3\kappa(\boldsymbol{x}) + 4\mu)^{-1} \rangle, \tag{49}$$

where $\mu$ is the common shear modulus, $\kappa(\boldsymbol{x})$ and $\kappa^*$ are the local and effective bulk moduli, respectively. In order to obtain these additional relations, we need to compute the *derived Jordan ideals* for each subspace $\Pi$ satisfying (27).

**Definition 3.** The derived Jordan ideal of the solution $\Pi$ of (27) is the subspace

$$\Pi^2 = \mathrm{Span}\{\boldsymbol{K}_1 *_{\boldsymbol{A}} \boldsymbol{K}_2 \mid \{\boldsymbol{K}_1, \boldsymbol{K}_2\} \subset \Pi,\ \boldsymbol{A} \in \mathcal{A}\},$$

where the Jordan multiplication $*_{\boldsymbol{A}}$ is defined in (14).

**Theorem 13.** *Let $\Pi^2$ be the derived Jordan ideal of $\Pi$ and suppose $\Pi^2 \neq \Pi$. Let $\mathcal{N}$ be the orthogonal complement of $\Pi^2$ in $\Pi$. Then we have*

$$\mathcal{P}_{\mathcal{N}} W_{\boldsymbol{M}}(\boldsymbol{L}^*) = \mathcal{P}_{\mathcal{N}} \langle W_{\boldsymbol{M}}(\boldsymbol{L}(\boldsymbol{x})) \rangle, \tag{50}$$

*where $\mathcal{P}_{\mathcal{N}}$ denotes the orthogonal projection onto $\mathcal{N}$ and $\boldsymbol{M}$ satisfies the conditions of Theorem* 11.

We actually computed the derived Jordan ideals for each subspace $\Pi$ when we were computing all the solutions of equation (27). The reader may go back and verify that for each solution $\Pi$ for the 2D Hall effect we have $\Pi^2 = \Pi$. The simplest context where nontrivial derived Jordan ideals appear is 2D elasticity, which is a bit more involved and therefore less suitable for the purposes of the present review than the 2D Hall effect. We refer the reader to [21], where there are plenty of exact relations with nontrivial derived Jordan ideals and corresponding volume average relations.

### 8.7.5. Links between uncoupled problems

Now we can ask the following question: Is there a link between conducting and thermal properties of a composite? More generally, is there a link between $\boldsymbol{L}^*$ and $\boldsymbol{F}^*$—the effective tensor for $\boldsymbol{f}(\boldsymbol{L}(\boldsymbol{x}))$. This question has also been investigated in [21], where $f$ is some nonlinear map between $\mathrm{End}(\mathcal{T}_1)$ and $\mathrm{End}(\mathcal{T}_2)$. The answer uses the concept of the Jordan ideal.

**Definition 4.** A subspace $\mathcal{K}$ of a solution $\Pi$ of (27) is called a Jordan ideal in $\Pi$ if $\boldsymbol{K}_1 *_A \boldsymbol{K}_2 \in \mathcal{K}$ for all $\boldsymbol{A} \in \mathcal{A}$, all $\boldsymbol{K}_1 \in \mathcal{K}$, and all $\boldsymbol{K}_2 \in \Pi$.

For example, the derived Jordan ideal is a Jordan ideal. As is customary in algebra, an ideal is good for factoring over it. Let $\mathcal{F} = \Pi/\mathcal{K}$ be the factor space in the sense of vector spaces. Then $\mathcal{F}$ has a set of well-defined Jordan multiplications:

$$\overline{\boldsymbol{K}_1} *_A \overline{\boldsymbol{K}_2} \stackrel{\text{def}}{=} \overline{\boldsymbol{K}_1 *_A \boldsymbol{K}_2},$$

for all $\boldsymbol{A} \in \mathcal{A}$ and all $\{\boldsymbol{K}_1, \boldsymbol{K}_2\} \subset \Pi$, where $\overline{\boldsymbol{K}} \in \mathcal{F}$ denotes the equivalence class of $\boldsymbol{K}$.

In order to treat the links between uncoupled problems we need to discuss coupled problems, at least in passing. As we have argued in Section 8.3, each physical context gives rise to a vector space $\mathcal{T}$ where the pair of relevant physical fields takes its values. If we have two problems with spaces $\mathcal{T}_1$ and $\mathcal{T}_2$ then the vector space $\mathcal{T}$ corresponding to a coupled problem is $\mathcal{T} = \mathcal{T}_1 \oplus \mathcal{T}_2$. Exact relations for the coupled problem would correspond to subspaces of $\mathrm{End}(\mathcal{T}_1 \oplus \mathcal{T}_2)$.

**Definition 5.** The links between uncoupled problems are those exact relations for the coupled problem whose subspaces $\Pi$ lie in the "block-diagonal" part $\mathrm{End}(\mathcal{T}_1) \oplus \mathrm{End}(\mathcal{T}_2)$ of $\mathrm{End}(\mathcal{T}_1 \oplus \mathcal{T}_2)$.

The technical difficulty is that in general $\mathcal{A} \neq \mathcal{A}_1 \oplus \mathcal{A}_2$. In [21] we were able to avoid this difficulty by focusing on polycrystals, where we were able to replace the subspace $\mathcal{A}$ by a single matrix $\tilde{\Gamma}$. Nevertheless, we can still state the general theorem.

In what follows, we use the notation $[\boldsymbol{A}, \boldsymbol{B}]$ for

$$\begin{bmatrix} \boldsymbol{A} & \boldsymbol{0} \\ \boldsymbol{0} & \boldsymbol{B} \end{bmatrix}$$

in order to save space.

**Theorem 14.** *Let $\Pi \subset \mathrm{End}(\mathcal{T}_1) \oplus \mathrm{End}(\mathcal{T}_2)$ be an exact relation. Define*

$$\begin{aligned} \Pi_1 &= \{\boldsymbol{K}_1 \in \mathrm{End}(\mathcal{T}_1) : [\boldsymbol{K}_1, \boldsymbol{K}_2] \in \Pi \text{ for some } \boldsymbol{K}_2 \in \mathrm{End}(\mathcal{T}_2)\}, \\ \Pi_2 &= \{\boldsymbol{K}_2 \in \mathrm{End}(\mathcal{T}_2) : [\boldsymbol{K}_1, \boldsymbol{K}_2] \in \Pi \text{ for some } \boldsymbol{K}_1 \in \mathrm{End}(\mathcal{T}_1)\}, \\ \mathcal{K}_1 &= \{\boldsymbol{K}_1 \in \mathrm{End}(\mathcal{T}_1) : [\boldsymbol{K}_1, \boldsymbol{0}] \in \Pi\}, \\ \mathcal{K}_2 &= \{\boldsymbol{K}_2 \in \mathrm{End}(\mathcal{T}_2) : [\boldsymbol{0}, \boldsymbol{K}_2] \in \Pi\}. \end{aligned} \tag{51}$$

*Then the subspaces $\Pi_j$ of $\mathrm{End}(\mathcal{T}_j)$, $j = 1, 2$ are Jordan algebras in the sense of equations (27). The subspaces $\mathcal{K}_j \subset \Pi_j$, $j = 1, 2$ are Jordan ideals. There is a natural linear isomorphism $\Phi : \mathcal{F}_1 = \Pi_1/\mathcal{K}_1 \to \Pi_2/\mathcal{K}_2 = \mathcal{F}_2$ defined by the rule $\overline{\boldsymbol{K}_2} = \Phi(\overline{\boldsymbol{K}_1})$, whenever $[\boldsymbol{K}_1, \boldsymbol{K}_2] \in \Pi$. The map $\Phi$ is well defined and satisfies an important additional condition:*

$$\Phi(\overline{\boldsymbol{K}_1} *_{A_1} \overline{\boldsymbol{K}'_1}) = \Phi(\overline{\boldsymbol{K}_1}) *_{A_2} \Phi(\overline{\boldsymbol{K}'_1}) \tag{52}$$

*for every $\{\boldsymbol{K}_1, \boldsymbol{K}'_1\} \subset \Pi_1$ and every $[\boldsymbol{A}_1, \boldsymbol{A}_2] \in \mathcal{A}$.*

Please note that if $[\boldsymbol{A}_1, \boldsymbol{A}_2] \in \mathcal{A}$ then, in general, $\boldsymbol{A}_2$ is not determined by $\boldsymbol{A}_1$ uniquely nor are $\boldsymbol{A}_1$ and $\boldsymbol{A}_2$ independent. Therefore, the linear isomorphism $\Phi$ with the property (52) is a Jordan algebra homomorphism.

For the specific example of the 2D Hall effect, we simply used the brute-force Maple computation to figure out the subspace $\mathcal{A}$. Let $\boldsymbol{L}_0 = [\boldsymbol{L}_1, \boldsymbol{L}_2]$ be the reference medium through which the link $\mathbb{M}$ passes. As before it will be convenient to work not with the subspace $\mathcal{A}$ directly but with the subspace $\overline{\mathcal{A}} = \boldsymbol{Y}^{-1}\mathcal{A}\boldsymbol{X}^{-1}$, where $\boldsymbol{X} = [\boldsymbol{\sigma}_1^{-1/2}, \boldsymbol{\sigma}_2^{-1/2}]$ and $\boldsymbol{Y} = [\boldsymbol{L}_1, \boldsymbol{L}_2]\boldsymbol{X}$. Then $\overline{\mathcal{A}}$ is described in terms of $\mathcal{A}_0 = \{\boldsymbol{A} \in \mathrm{Sym}(\mathbb{R}^2) \mid \mathrm{Tr}(\boldsymbol{A}) = 0\}$, depending on what $\boldsymbol{L}_1$ and $\boldsymbol{L}_2$ are.

**Theorem 15.** *Let $\boldsymbol{\sigma}_j$, $j = 1, 2$ be the symmetric parts of $\boldsymbol{L}_j$. If there is a scalar $s > 0$ such that $\boldsymbol{\sigma}_2 = s\boldsymbol{\sigma}_1$, then $\overline{\mathcal{A}} = \{[\boldsymbol{A}, \boldsymbol{A}] : \boldsymbol{A} \in \mathcal{A}_0\}$. Otherwise, $\overline{\mathcal{A}} = \mathcal{A}_0 \oplus \mathcal{A}_0$.*

Now that we know the subspace $\overline{\mathcal{A}}$, we can apply Theorem 14. First, consider the case when $\boldsymbol{\sigma}_1$ and $\boldsymbol{\sigma}_2$ are not multiples of one another. Then we can set $\boldsymbol{A}_2 = \boldsymbol{0}$ in (52) and, recalling that $\Phi$ is a bijection, conclude that for every $\{\boldsymbol{K}_1, \boldsymbol{K}'_1\} \subset \Pi_1$ and every $\boldsymbol{A}_1 \in \mathcal{A}_0$ we have $\overline{\boldsymbol{K}_1} *_{A_1} \overline{\boldsymbol{K}'_1} = \overline{\boldsymbol{0}}$. The meaning of that last relation is that the ideal $\mathcal{K}_1$ contains the derived ideal $\Pi_1^2$. But we have already verified that there are no nontrivial derived ideals in the context of the 2D Hall effect. In this case, we have to conclude that $\mathcal{K}_1 = \Pi_1$ resulting in the trivial link: $[\boldsymbol{L}_1(\boldsymbol{x}), \boldsymbol{L}_2(\boldsymbol{x})]^* = [\boldsymbol{L}_1^*, \boldsymbol{L}_2^*]$ with no relation between $\boldsymbol{L}_1^*$ and $\boldsymbol{L}_2^*$.

Now let us assume that $\boldsymbol{\sigma}_2 = s\boldsymbol{\sigma}_1$. Here we will get some interesting results. We begin by finding all pairs of nested exact relations from the list in Theorem 9 (to which the space $\mathrm{End}(\mathbb{R}^2)$ is added) and checking if the smaller subspace in each pair is a proper Jordan ideal of the larger. The checking is routine and may be automated with Maple even for problems of large size. We find that none of the subspaces from Theorem 9 is a Jordan ideal in any of the other subspaces, including $\mathrm{End}(\mathbb{R}^2)$. Thus, we must necessarily have that $\mathcal{K}_1$ and $\mathcal{K}_2$ are zero subspaces and that $\Phi$ is a Jordan isomorphism $\Phi : \Pi_1 \to \Pi_2$

$$\Phi(\boldsymbol{K}_1 *_A \boldsymbol{K}_2) = \Phi(\boldsymbol{K}_1) *_A \Phi(\boldsymbol{K}_2) \tag{53}$$

for all $\boldsymbol{A} \in \mathcal{A}_0$.

We will now describe how to find all the Jordan isomorphisms $\Phi$ between the subspaces $\Pi_1$ and $\Pi_2$ satisfying (27). The method here is based on a nice relation between $2\times 2$ matrix algebra and complex arithmetic. To a complex number $z = a+ib$ we associate a vector $\pi(z) = (a, b) \in \mathbb{R}^2$ and two $2 \times 2$ matrices:

$$\phi(z) = \begin{bmatrix} a & -b \\ b & a \end{bmatrix} \text{ and } \psi(z) = \begin{bmatrix} a & b \\ b & -a \end{bmatrix}.$$

In the remaining part of this section, the bold lowercase letter will denote a 2D vector corresponding to a complex number denoted by the same nonbold letter. In situations where such simplified notation is inadequate we will use notation $\pi(\cdot)$ defined above.

The functions $\phi$ and $\psi$ enjoy many special properties. For example,

$$\phi(z)\boldsymbol{u} = \pi(zu), \quad \psi(z)\boldsymbol{u} = \pi(z\overline{u}).$$

As a corollary we have the following multiplicative identities:

$$\phi(z_1)\phi(z_2) = \phi(z_1z_2), \quad \phi(z_1)\psi(z_2) = \psi(z_1z_2),$$

$$\psi(z_1)\phi(z_2) = \psi(z_1\overline{z_2}), \quad \psi(z_1)\psi(z_2) = \phi(z_1\overline{z_2}).$$

Observe that if $\Phi$ is the Jordan isomorphism between $\Pi_1$ and $\Pi_2$ then $\dim \Pi_1 = \dim \Pi_2$. Therefore, our strategy is to go through every dimension 1 through 4 and determine all Jordan isomorphisms between subspaces $\Pi$ from Theorem 9 of that dimension.

*Dimension* 1. All 1D solutions of (27) have the form $\Pi_{\boldsymbol{a},\boldsymbol{b}}$.

**Proposition 1.** *The linear mapping $\Phi : \Pi_{\boldsymbol{a},\boldsymbol{b}} \to \Pi_{\boldsymbol{c},\boldsymbol{d}}$, defined by $\Phi(\boldsymbol{a}\otimes\boldsymbol{b}) = \boldsymbol{c}\otimes\boldsymbol{d}$, is a Jordan isomorphism between $\Pi_{\boldsymbol{a},\boldsymbol{b}}$ and $\Pi_{\boldsymbol{c},\boldsymbol{d}}$ if and only if $ab = cd$.*

*Proof.* For every $\boldsymbol{A} \in \mathcal{A}_0$ we have

$$(\boldsymbol{a} \otimes \boldsymbol{b})\boldsymbol{A}(\boldsymbol{a} \otimes \boldsymbol{b}) = (\boldsymbol{A}\boldsymbol{a} \cdot \boldsymbol{b})\boldsymbol{a} \otimes \boldsymbol{b}.$$

Applying the mapping $\Phi$ to this relation and using the property (53), we obtain

$$(\boldsymbol{c} \otimes \boldsymbol{d})\boldsymbol{A}(\boldsymbol{c} \otimes \boldsymbol{d}) = (\boldsymbol{A}\boldsymbol{a} \cdot \boldsymbol{b})\boldsymbol{c} \otimes \boldsymbol{d}.$$

Thus, we must have

$$\boldsymbol{A}\boldsymbol{a} \cdot \boldsymbol{b} = \boldsymbol{A}\boldsymbol{c} \cdot \boldsymbol{d} \tag{54}$$

for every $\boldsymbol{A} \in \mathcal{A}_0$. Now observe that $\mathcal{A}_0 = \{\psi(z) \mid z \in \mathbb{C}\}$. Therefore, equation (54) becomes $\Re(\overline{z}(ab - cd)) = 0$ for every $z \in \mathbb{C}$. Thus, $\Phi$ is a Jordan isomorphism if and only if $ab = cd$. □

*Dimension* 2. There are three classes of 2D solutions of (27). We denoted them $\mathcal{A}_0$, $\Pi_{\boldsymbol{a}}$, and $\Pi_{\boldsymbol{a}}^T$.

**Proposition 2.** *There are no Jordan isomorphisms between $\mathcal{A}_0$ and $\Pi_{\boldsymbol{a}}$ and between $\mathcal{A}_0$ and $\Pi_{\boldsymbol{a}}^T$. The only Jordan isomorphism between $\mathcal{A}_0$ and $\mathcal{A}_0$ is the identity mapping.*

*Proof.* Suppose $\Phi$ is the Jordan isomorphism between $\mathcal{A}_0$ and $\Pi_{\boldsymbol{a}}$. Then there exists a nonsingular matrix $\boldsymbol{V}$ such that $\Phi(\psi(u)) = \boldsymbol{V}\boldsymbol{u} \otimes \boldsymbol{a}$ for all $u \in \mathbb{C}$. Applying the Jordan isomorphism $\Phi$ to

$$\psi(u)\psi(z)\psi(u) = \psi(u^2\overline{z}), \tag{55}$$

we obtain

$$(\boldsymbol{V}\boldsymbol{u} \otimes \boldsymbol{a})\psi(z)(\boldsymbol{V}\boldsymbol{u} \otimes \boldsymbol{a}) = \boldsymbol{V}\pi(u^2\overline{z}) \otimes \boldsymbol{a}.$$

Therefore, $(\psi(z)\boldsymbol{a}\cdot\boldsymbol{V}\boldsymbol{u})u = u^2\overline{z}$, which implies that $u\overline{z}$ must be real for all $\{u, z\} \subset \mathbb{C}$. This is impossible, and therefore the Jordan isomorphism $\Phi$ between $\mathcal{A}_0$ and $\Pi_{\boldsymbol{a}}$ does not exist. The Jordan isomorphism between $\mathcal{A}_0$ and $\Pi_{\boldsymbol{a}}^T$ is also impossible. Indeed the mapping $\boldsymbol{K} \to \boldsymbol{K}^T$ is the Jordan isomorphism between $\Pi_{\boldsymbol{a}}$ and $\Pi_{\boldsymbol{a}}^T$. If there was a Jordan isomorphism between $\mathcal{A}_0$ and $\Pi_{\boldsymbol{a}}^T$ then there would be a Jordan isomorphism between $\mathcal{A}_0$ and $\Pi_{\boldsymbol{a}}$, which we proved does not exist.

Now let $\Phi$ be the Jordan isomorphism from $\mathcal{A}_0$ to itself. Then there exists a real linear invertible mapping $\vartheta : \mathbb{C} \to \mathbb{C}$ such that $\Phi(\psi(u)) = \psi(\vartheta(u))$. Applying $\Phi$ to (55) we get $(\vartheta(u))^2\overline{z} = \vartheta(u^2\overline{z})$. The linear map $\vartheta$ has the form $\vartheta(u) = \theta_1 u + \theta_2\overline{u}$ for some complex numbers $\theta_1$ and $\theta_2$. Equating the coefficients at $z$ and $\overline{z}$ in

$$(\theta_1 u + \theta_2\overline{u})^2\overline{z} = \theta_1 u^2\overline{z} + \theta_2\overline{u}^2 z,$$

we obtain that $\theta_2 = 0$ and $\theta_1 = 1$. Thus, $\Phi(\psi(u)) = \psi(u)$ for all $u \in \mathbb{C}$. □

**Proposition 3.** *The unique Jordan isomorphism $\Phi$ between $\Pi_{\boldsymbol{a}}$ and $\Pi_{\boldsymbol{b}}$ is given by*

$$\Phi(\boldsymbol{v} \otimes \boldsymbol{a}) = \phi(a/b)\boldsymbol{v} \otimes \boldsymbol{b}. \tag{56}$$

*The unique Jordan isomorphism $\Phi$ between $\Pi_{\boldsymbol{a}}$ and $\Pi_{\boldsymbol{b}}^T$ is given by*

$$\Phi(\boldsymbol{v} \otimes \boldsymbol{a}) = \boldsymbol{b} \otimes \phi(a/b)\boldsymbol{v}. \tag{57}$$

*The unique Jordan isomorphism $\Phi$ between $\Pi_{\boldsymbol{a}}^T$ and $\Pi_{\boldsymbol{b}}^T$ is given by*

$$\Phi(\boldsymbol{a} \otimes \boldsymbol{v}) = \boldsymbol{b} \otimes \phi(a/b)\boldsymbol{v}. \tag{58}$$

*Proof.* For a linear isomorphism $\Phi$ between $\Pi_{\boldsymbol{a}}$ and $\Pi_{\boldsymbol{b}}$ there exists a nonsingular matrix $\boldsymbol{V}$ such that $\Phi(\boldsymbol{v} \otimes \boldsymbol{a}) = \boldsymbol{V}\boldsymbol{v} \otimes \boldsymbol{b}$. Applying $\Phi$ to

$$(\boldsymbol{v} \otimes \boldsymbol{a})\psi(z)(\boldsymbol{v} \otimes \boldsymbol{a}) = (\psi(z)\boldsymbol{a} \cdot \boldsymbol{v})\boldsymbol{v} \otimes \boldsymbol{a},$$

we obtain

$$(\boldsymbol{V}\boldsymbol{v} \otimes \boldsymbol{b})\psi(z)(\boldsymbol{V}\boldsymbol{v} \otimes \boldsymbol{b}) = (\psi(z)\boldsymbol{a} \cdot \boldsymbol{v})\boldsymbol{V}\boldsymbol{v} \otimes \boldsymbol{b}.$$

Therefore, for every $z \in \mathbb{C}$, we have $\boldsymbol{V}^T\psi(z)\boldsymbol{b} = \psi(z)\boldsymbol{a}$. The map $\boldsymbol{V}$ as a map on $\mathbb{C}$ can be written as

$$\pi^{-1}(\boldsymbol{V}\boldsymbol{u}) = v_1 u + v_2\overline{u}$$

for some complex numbers $v_1$ and $v_2$. Therefore, we have

$$\overline{v_1}z\overline{b} + v_2\overline{z}b = z\overline{a}.$$

Equating coefficients at $z$ and $\overline{z}$ we obtain that $v_2 = 0$ and $v_1 = a/b$. Thus, (56) gives the unique Jordan isomorphism between $\Pi_a$ and $\Pi_b$.

Now, if $\Phi$ is a Jordan isomorphism between $\Pi_a$ and $\Pi_b^T$ then $\tau \circ \Phi$ is the Jordan isomorphism between $\Pi_a$ and $\Pi_b$, where $\tau(\boldsymbol{K}) = \boldsymbol{K}^T$. Hence there is a unique Jordan isomorphism between $\Pi_a$ and $\Pi_b^T$ given by (57). Similarly, if $\Phi$ is a Jordan isomorphism between $\Pi_a^T$ and $\Pi_b^T$, then $\tau \circ \Phi \circ \tau$ is a Jordan isomorphism between $\Pi_a$ and $\Pi_b$. □

*Dimension* 3. The 3D solutions of (27) are the subspaces $\Pi_\beta$ from Theorem 9.

**Proposition 4.** *For any pair* $\{\beta_1, \beta_2\} \subset \mathbb{R} \cup \{\infty\}$ *there are two Jordan isomorphisms between* $\Pi_{\beta_1}$ *and* $\Pi_{\beta_2}$ *given by*

$$\Phi(\boldsymbol{K}) = \phi(e^{i\alpha/2})\boldsymbol{K}\phi(e^{i\alpha/2}), \quad \Phi(\boldsymbol{K}) = \phi(ie^{i\alpha/2})\boldsymbol{K}\phi(ie^{i\alpha/2}), \tag{59}$$

*where* $\alpha$ *is the unique solution in* $[0, \pi)$ *of*

$$\tan(\alpha) = \frac{\beta_2 - \beta_1}{1 + \beta_1\beta_2}.$$

*Proof.* Suppose $\Phi$ is the Jordan isomorphism between $\Pi_{\beta_1}$ and $\Pi_{\beta_2}$. Observe that $\mathcal{A}_0 \subset \Pi_\beta$ for any $\beta$. We conclude that $\Phi(\mathcal{A}_0) = \mathcal{A}_0$, since $\mathcal{A}_0$ is not isomorphic to any other Jordan algebra but itself. Moreover, the restriction of $\Phi$ to $\mathcal{A}_0$ must be the identity map. Observe that any matrix $\boldsymbol{K} \in \Pi_{\beta_1}$ can be written as $\boldsymbol{K} = x\phi(e^{i\gamma_1}) + \psi(u)$, where $\beta_1 = -\cot\gamma_1$. Any linear map $\Phi$ from $\Pi_{\beta_1}$ into $\Pi_{\beta_2}$ that is the identity on $\mathcal{A}_0$ has the form

$$\Phi(x\phi(e^{i\gamma_1}) + \psi(u)) = ax\phi(e^{i\gamma_2}) + \psi(u + xp)$$

for some $a \in \mathbb{R}$ and $p \in \mathbb{C}$. Here $\beta_2 = -\cot\gamma_2$. Applying the Jordan isomorphism $\Phi$ to

$$(x\phi(e^{i\gamma_1}) + \psi(u))\psi(z)(x\phi(e^{i\gamma_1}) + \psi(u)) = 2x\Re(z\overline{u})\phi(e^{i\gamma_1}) + \psi(x^2z + u^2\overline{z}),$$

we obtain

$$\begin{aligned}(ax\phi(e^{i\gamma_2}) &+ \psi(u + xp))\psi(z)(ax\phi(e^{i\gamma_2}) + \psi(u + xp)) \\ &= 2xa\Re(z\overline{u})\phi(e^{i\gamma_2}) + \psi(x^2z + u^2\overline{z} + 2x\Re(z\overline{u})p).\end{aligned}$$

Since any $2 \times 2$ matrix can be uniquely written as $\phi(z_1) + \psi(z_2)$ we obtain, by equating arguments of $\phi$,

$$\Re(z\overline{u}) = \Re(z\overline{u} + xz\overline{p}).$$

It follows, therefore, that $p = 0$. Equating arguments of $\psi$, we get

$$x^2z + u^2\overline{z} = a^2x^2z + u^2\overline{z}.$$

It follows that either $a = 1$ or $a = -1$, corresponding to the two Jordan isomorphisms given by (59). □

*Dimension* 4.

**Theorem 16.** *If $\Phi$ is a Jordan isomorphism of* $\mathrm{End}(\mathbb{R}^2)$*, then either* $\Phi(\boldsymbol{K}) = \phi(e^{i\alpha})\boldsymbol{K}\phi(e^{i\alpha})$ *or* $\Phi(\boldsymbol{K}) = \phi(e^{i\alpha})\boldsymbol{K}^T\phi(e^{i\alpha})$*. Moreover, all Jordan isomorphisms between $\Pi_1$ and $\Pi_2$—solutions of* (27)*—are restrictions of the Jordan isomorphisms of* $\mathrm{End}(\mathbb{R}^2)$ *to* $\Pi_1$.

Before proving the theorem we remark that all Jordan isomorphisms $\Phi$ of $\mathrm{End}(\mathbb{R}^2)$ have the property

$$\Phi(\boldsymbol{K}_1\boldsymbol{A}\boldsymbol{K}_2) = \Phi(\boldsymbol{K}_1)\boldsymbol{A}\Phi(\boldsymbol{K}_2),$$

for all $\boldsymbol{A} \in \mathcal{A}_0$. Therefore, all the $j$-chain properties are obviously satisfied. Thus, all the links that we find are stable under homogenization.

*Proof.* Let $\Phi$ be the Jordan isomorphism of $\mathrm{End}(\mathbb{R}^2)$. If we restrict $\Phi$ to a 3D subspace $\{x\phi(e^{i\gamma}) + \psi(u) \mid x \in \mathbb{R},\ u \in \mathbb{C}\}$ then we have, according to Proposition 4,

$$\Phi(x\phi(e^{i\gamma}) + \psi(u)) = x\phi(e^{i\alpha}e^{i\gamma}) + \psi(u).$$

Therefore, for every $\gamma \in [0, \pi)$, every $x \in \mathbb{R}$ and $u \in \mathbb{C}$

$$\Phi(x\phi(e^{i\gamma}) + \psi(u)) = x\phi(e^{i\alpha(\gamma)}e^{i\gamma}) + \psi(u).$$

But $\Phi$ is linear and therefore there are complex numbers $p_1$ and $p_2$ such that $\Phi(\phi(z)) = \phi(p_1 z + p_2\overline{z})$. Consequently, $e^{i\alpha(\gamma)}e^{i\gamma} = p_1 e^{i\gamma} + p_2 e^{-i\gamma}$. Thus, for every $\gamma \in [0, \pi)$ we have $|p_1 + p_2 e^{-2i\gamma}| = 1$. This equation says geometrically that the circle centered at $p_1$ with radius $|p_2|$ is a subset of the unit circle. There are only two possibilities: $p_1 = 0$, $|p_2| = 1$ or $p_2 = 0$, $|p_1| = 1$. Thus, either $\Phi(\phi(v) + \psi(u)) = \phi(e^{i\alpha}v) + \psi(u)$ or $\Phi(\phi(v) + \psi(u)) = \phi(e^{i\alpha}\overline{v}) + \psi(u)$. In the matrix notation, either $\Phi(\boldsymbol{K}) = \phi(e^{i\alpha/2})\boldsymbol{K}\phi(e^{i\alpha/2})$ or $\Phi(\boldsymbol{K}) = \phi(e^{i\alpha/2})\boldsymbol{K}^T\phi(e^{i\alpha/2})$, and the first part of the theorem is proved.

We have already described all the Jordan isomorphisms between solutions of (27). It is now a simple matter to check that each one of the Jordan isomorphisms described in Propositions 1–4 is indeed a restriction of one of the Jordan isomorphisms of $\mathrm{End}(\mathbb{R}^2)$. □

We remark that the group of all Jordan isomorphisms of $\mathrm{End}(\mathbb{R}^2)$ is isomorphic to $O(2)$. Another remark is that Theorem 16 saves us a lot of work. All we have to do is to compute the links between uncoupled problems for $\mathrm{End}(\mathbb{R}^2)$. Any other link is just a restriction of the global link to one of the exact relations computed before. We also observe that the Jordan isomorphism $\tau(\boldsymbol{K}) = \boldsymbol{K}^T$ corresponds to the link $(\boldsymbol{L}(\boldsymbol{x})^T)^* = (\boldsymbol{L}^*)^T$, which is a theorem of Murat and Tartar [40] that says that if $\boldsymbol{L}^\epsilon(\boldsymbol{x})$ H-converges to $\boldsymbol{L}^*(\boldsymbol{x})$ then $(\boldsymbol{L}^\epsilon(\boldsymbol{x}))^T$ H-converges to $(\boldsymbol{L}^*(\boldsymbol{x}))^T$. If we combine this link with the links corresponding to $\Phi(\boldsymbol{K}) = \phi(e^{i\alpha})\boldsymbol{K}\phi(e^{i\alpha})$ we will obtain the links corresponding to $\Phi(\boldsymbol{K}) = \phi(e^{i\alpha})\boldsymbol{K}^T\phi(e^{i\alpha})$. All we have to do now is to apply the method of Section 8.7.3 to the subspace

$$\Pi = \{[\boldsymbol{K}, \phi(e^{i\alpha})\boldsymbol{K}\phi(e^{i\alpha})] :\ \boldsymbol{K} \in \mathrm{End}(\mathbb{R}^2)\}.$$

According to the formula (46) we define $L_1' = \sigma_1^{-1/2} L_1 \sigma_1^{-1/2}$, $P_1 = I + \alpha_1 S$ and $L_2' = (1/s)\sigma_1^{-1/2} L_2 \sigma_1^{-1/2}$, $P_2 = I + \alpha_2 S$. Then the link between $L_1$ and $L_2$ has the form

$$\left[(P_2 - L_2')^{-1} - u \otimes u\right]^{-1} = \phi(e^{i\alpha})\left[(P_1 - L_1')^{-1} - u \otimes u\right]^{-1} \phi(e^{i\alpha}). \quad (60)$$

The relation between $L_1$ and $L_2$ does not depend on the choice of the unit vector $u$. Indeed, we can rewrite (60) as

$$(P_2 - L_2')^{-1} = \phi(e^{-i\alpha})(P_1 - L_1')^{-1}\phi(e^{-i\alpha}) + u \otimes u - \phi(e^{-i\alpha})u \otimes \phi(e^{i\alpha})u.$$

Now we use our complex calculus to show that $u \otimes u - \phi(e^{-i\alpha})u \otimes \phi(e^{i\alpha})u$ does not depend on the choice of the unit vector $u$. Indeed, a rank-1 matrix $a \otimes b$ can be written as $(\phi(a\overline{b}) + \psi(ab))/2$. Thus, we have

$$\begin{aligned} u \otimes u - \phi(e^{-i\alpha})u \otimes \phi(e^{i\alpha})u &= (\phi(1) + \psi(u^2))/2 \\ &\quad - (\phi(e^{-i\alpha}u\overline{e^{i\alpha}u}) + \psi(e^{-i\alpha}ue^{i\alpha}u))/2, \end{aligned}$$

and therefore

$$L_2' = P_2 - \left[\phi(e^{-i\alpha})(P_1 - L_1')^{-1}\phi(e^{-i\alpha}) + \sin(\alpha)\phi(ie^{-i\alpha})\right]^{-1}. \quad (61)$$

We examine the right-hand side of (61) with Maple and after a lengthy investigation obtain the link in terms of the conductivity tensors $\sigma$ and the Hall coefficient $r$:

$$\sigma_2 = c_0 \frac{\sigma_1}{(r_0 - r_1)^2 + \det \sigma_1}, \qquad r_2 = c_0 \frac{r_0 - r_1}{(r_0 - r_1)^2 + \det \sigma_1} + q_0, \quad (62)$$

where $c_0 > 0$, $r_0$ and $q_0$ are constants. This link was first obtained by Dykhne [12] in a particular case of an isotropic composite made with two isotropic phases. In its present form the link was first derived by Milton [37]. The links corresponding to the the remaining Jordan isomorphisms can be obtained by combining (62) with the link $L_2 = L_1^T$. In other words, the other set of links has the form (62) but with $r_1$ replaced by $-r_1$:

$$\sigma_2 = c_0 \frac{\sigma_1}{(r_0 + r_1)^2 + \det \sigma_1}, \qquad r_2 = c_0 \frac{r_0 + r_1}{(r_0 + r_1)^2 + \det \sigma_1} + q_0. \quad (63)$$

Following Milton [37], we observe that the trivial exact relation "$r_1(x) = 0$ implies $r_1^* = 0$" transforms into the relation (48) according to the link (62) (or (63)).

Restricting the links to the case of conductivity we obtain the well-known result [36, 39]

$$(\sigma(x)/\det \sigma(x))^* = \sigma^*/\det \sigma^*.$$

It is possible to play further with the links and exact relations, applying them to the case of two-phase composites, polycrystals, and such. We will not pursue this and stop here.

## 8.8. Conclusions

Our theory of exact relations permitted us to derive the following *microstructure independent* results for effective properties of composites in the context of the 2D Hall effect. All of them were previously known (see [12, 14, 37, 47]). In a weak magnetic field the local conducting properties of a 2D composite are described by the nonsymmetric positive definite $2 \times 2$ matrix field $\boldsymbol{L}(\boldsymbol{x}) = \boldsymbol{\sigma}(\boldsymbol{x}) + r(\boldsymbol{x})\boldsymbol{S}$. We have obtained the following *complete* list of results.

1. If $\boldsymbol{L}(\boldsymbol{x})\boldsymbol{a} = \boldsymbol{b}$ for all $\boldsymbol{x} \in Q$ then $\boldsymbol{L}^*\boldsymbol{a} = \boldsymbol{b}$.
2. If $r(\boldsymbol{x}) = r_0$ then $r^* = r_0$.
3. If $\boldsymbol{L}^*$ is the effective tensor for $\boldsymbol{L}(\boldsymbol{x})$ then $(\boldsymbol{L}^*)^T$ is the effective tensor for $\boldsymbol{L}(\boldsymbol{x})^T$.
4. If $(r(\boldsymbol{x}) - r_0)^2 + \det \boldsymbol{\sigma}(\boldsymbol{x}) = c_0$ for all $\boldsymbol{x} \in Q$ then $(r^* - r_0)^2 + \det \boldsymbol{\sigma}^* = c_0$.
5. If $\boldsymbol{L}^* = \boldsymbol{\sigma}^* + r^*\boldsymbol{S}$ is the effective tensor for $\boldsymbol{L}(\boldsymbol{x}) = \boldsymbol{\sigma}(\boldsymbol{x}) + r(\boldsymbol{x})\boldsymbol{S}$ then $\widehat{\boldsymbol{L}}^* = \widehat{\boldsymbol{\sigma}}^* + \widehat{r}^*\boldsymbol{S}$ is the effective tensor for $\widehat{\boldsymbol{L}}(\boldsymbol{x}) = \widehat{\boldsymbol{\sigma}}(\boldsymbol{x}) + \widehat{r}(\boldsymbol{x})\boldsymbol{S}$, where

$$\widehat{\boldsymbol{\sigma}}(\boldsymbol{x}) = c_0 \frac{\boldsymbol{\sigma}(\boldsymbol{x})}{(r_0 - r(\boldsymbol{x}))^2 + \det \boldsymbol{\sigma}(\boldsymbol{x})}, \quad \widehat{r}(\boldsymbol{x}) = c_0 \frac{r_0 - r(\boldsymbol{x})}{(r_0 - r(\boldsymbol{x}))^2 + \det \boldsymbol{\sigma}(\boldsymbol{x})} + q_0$$

and

$$\widehat{\boldsymbol{\sigma}}^* = c_0 \frac{\boldsymbol{\sigma}^*}{(r_0 - r^*)^2 + \det \boldsymbol{\sigma}^*}, \quad \widehat{r}^* = c_0 \frac{r_0 - r^*}{(r_0 - r^*)^2 + \det \boldsymbol{\sigma}^*} + q_0.$$

Technically speaking, there are other results, but all of them are consequences of the ones listed above. Our theory also guarantees that there are no other microstructure-independent equalities.

Throughout this paper we have described several important open questions in the theory. We summarize them here.

- The polycrystalline G-closures possess an important convexity property that was the cornerstone of the present paper. Can it be used to obtain a finite geometric algorithm for constructing G-closures? Such a construction was found by Francfort and Milton [15] for 2D conducting polycrystals, suggesting a positive answer to our question. For more detailed discussion see Section 8.5.
- Equations (27) have to be solved in each physical setting to determine all the exact relations there. So far we have been able to solve these equations for problems of modest size, where the brute-force approach works. Yet, equations (27) say that the subspace $\Pi$ has special structure of the Jordan algebra. Can this structure be used to help solve the equation? In fact, our result [21, Theorem 5.2] suggests that the question is meaningful.
- Finally, there is the question of whether the sufficient conditions for a manifold to be an exact relation are also necessary. See the more detailed formulation of this question at the end of Section 8.6.

**Acknowledgments**

The author wishes to thank Robert Kohn and Graeme Milton for their valuable comments. The author gratefully acknowledges the support of the National Science Foundation through grants DMS-9704813, NSF-0096133, and NSF-0094089.

## References

[1] G. E. Backus, Long-wave elastic anisotropy produced by horizontal layering, *J. Geophys. Res.*, **67** (1962), 4427–4440.

[2] P. Bak, *How Nature Works: The Science of Self-Organized Criticality*, Springer-Verlag, New York, 1996.

[3] P. Bak, C. Tang, and K. Wiesenfeld, Self-organized criticality: An explanation of $1/f$ noise. *Phys. Rev. Lett.*, **59** (1987), 381.

[4] P. Bak, C. Tang, and K. Wiesenfeld, Self-organized criticality, *Phys. Rev.* A, **38** (1988), 364.

[5] A. Bensoussan, J. L. Lions, and G. Papanicolaou, *Asymptotic Analysis of Periodic Structures*, North-Holland, Amsterdam, 1978.

[6] C. Davis, All convex invariant functions of hermitian matrices, *Arch. Math.*, **8** (1957), 276–278.

[7] E. De Giorgi and S. Spagnolo, Sulla convergenza degli integrali dell'energia per operatori ellittici del secondo ordine, *Boll. Un. Mat. Ital.* (4), **8** (1973), 391–411.

[8] G. de Marsily, *Quantitative Hydrogeology*, Academic Press, New York, 1986.

[9] G. F. Dell'Antonio, R. Figari, and E. Orlandi, An approach through orthogonal projections to the study of inhomogeneous random media with linear response, *Ann. Inst. H. Poincaré*, **44** (1986), 1–28.

[10] W. J. Drugan, Micromechanics-based variational estimates for a higher-order nonlocal constitutive equation and optimal choice of effective moduli for elastic composites, *J. Mech. Phys. Solids*, **48**-6–7 (2000) (the J. R. Willis 60th anniversary volume), 1359–1387.

[11] W. J. Drugan and J. R. Willis, A micromechanics-based nonlocal constitutive equation and estimates of representative volume element size for elastic composites, *J. Mech. Phys. Solids*, **44**-4 (1996), 497–524.

[12] A. M. Dykhne, Anomalous plasma resistance in a strong magnetic field, *Zh. Eksp. Teor. Fiz.*, **59** (1970), 641–647 (in Russian); *Sov. Phys. JETP*, **32** (1971), 348–351 (in English).

[13] A. M. Dykhne, Conductivity of a two-dimensional two-phase system, *Sov. Phys. JETP*, **32** (1971), 63–65.

[14] A. M. Dykhne and I. M. Ruzin, On the theory of the fractional quantum Hall effect: The two-phase model, *J. Phys. Rev.* B, **50** (1994), 2369–2379.

[15] G. A. Francfort and G. W. Milton, Optimal bounds for conduction in two-dimensional, multiphase, polycrystalline media, *J. Statist. Phys.*, **46**-1–2 (1987), 161–177.

[16] G. A. Francfort and G. W. Milton, Sets of conductivity and elasticity tensors stable under lamination, *Comm. Pure Appl. Math.*, **47** (1994), 257–279.

[17] G. A. Francfort and F. Murat, Homogenization and optimal bounds in linear elasticity, *Arch. Rational Mech. Anal.*, **94** (1986), 307–334.

[18] K. M. Golden and G. Papanicolaou, Bounds for effective parameters of heterogeneous media by analytic continuation, *Comm. Math. Phys.*, **90** (1983), 473–491.

[19] Y. Grabovsky, The G-closure of two well-ordered anisotropic conductors, *Proc. Roy. Soc. Edinburgh Ser.* A, **123** (1993), 423–432.

[20] Y. Grabovsky and G. W. Milton, Rank one plus a null-lagrangian is an inherited property of two-dimensional compliance tensors under homogenization, *Proc. Roy. Soc. Edinburgh Ser.* A, **128** (1998), 283–299.

[21] Y. Grabovsky, G. W. Milton, and D. S. Sage, Exact relations for effective tensors of polycrystals: Necessary conditions and sufficient conditions, *Comm. Pure. Appl. Math.*, **53**-3 (2000), 300–353.

[22] R. Hill, Elastic properties of reinforced solids: Some theoretical principles, *J. Mech. Phys. Solids*, **11** (1963), 357–372.

[23] R. Hill, Theory of mechanical properties of fibre-strengthened materials I: Elastic behaviour, *J. Mech. Phys. Solids*, **12** (1964), 199–212.

[24] V. V. Jikov, S. M. Kozlov, and O. A. Oleĭnik, *Homogenization of Differential Operators and Integral Functionals*, Springer-Verlag, Berlin, 1994 (translated from the Russian by G. A. Yosifian).

[25] Y. Kantor and D. J. Bergman, Improved rigorous bounds on the effective elastic moduli of a composite material, *J. Mech. Phys. Solids*, **32** (1984), 41–62.

[26] J. B. Keller, A theorem on the conductivity of a composite medium, *J. Math. Phys.*, **5** (1964), 548–549.

[27] P. R. King, The use of renormalization for calculating effective permeability, *Transp. Porous Media*, **4** (1989), 37–58.

[28] W. Kohler and G. C. Papanicolaou, Bounds for effective conductivity of random media, in R. Burridge, S. Childress, and G. Papanicolaou, eds., *Macroscopic Properties of Disordered Media*, Springer-Verlag, Berlin, 1982, 111–130.

[29] R. V. Kohn and R. Lipton, Optimal bounds for the effective energy of a mixture of isotropic, incompressible, elastic materials, *Arch. Rational Mech. Anal.*, **102** (1988), 331–350.

[30] S. M. Kozlov, Averaging of random structures, *Dokl. Akad. Nauk SSSR*, **241**-5 (1978), 1016–1019.

[31] K. A. Lurie and A. V. Cherkaev, $G$-closure of a set of anisotropic conducting media in the case of two dimensions, *Dokl. Akad. Nauk SSSR*, **259**-2 (1981), 328–331 (in Russian).

[32] K. A. Lurie and A. V. Cherkaev, Exact estimates of conductivity of composites formed by two isotropically conducting media taken in prescribed proportion, *Proc. Roy. Soc. Edinburgh Ser.* A, **99** (1984), 71–87.

[33] K. A. Lurie and A. V. Cherkaev, G-closure of some particular sets of admissible material characteristics for the problem of bending of thin plates, *J. Optim. Theory Appl.*, **42** (1984), 305–316.

[34] K. A. Lurie and A. V. Cherkaev, Exact estimates of a binary mixture of isotropic components, *Proc. Roy. Soc. Edinburgh Ser.* A, **104** (1986), 21–38.

[35] K. A. Lurie, A. V. Cherkaev, and A. V. Fedorov, On the existence of solutions to some problems of optimal design for bars and plates, *J. Optim. Theory Appl.*, **42**-2 (1984), 247–281.

[36] K. S. Mendelson, A theorem on the conductivity of two-dimensional heterogeneous medium, *J. Appl. Phys.*, **46** (1975), 4740–4741.

[37] G. W. Milton, Classical hall effect in two-dimensional composites: A characterization of the set of realizable effective conductivity tensors, *Phys. Rev.* B, **38**-16 (1988), 11296–11303.

[38] G. W. Milton, On characterizing the set of possible effective tensors of composites: the variational method and the translation method, *Comm. Pure Appl. Math.*, **43** (1990), 63–125.

[39] G. W. Milton, *The Theory of Composites*, Cambridge University Press, Cambridge, UK, 2001.

[40] F. Murat and L> Tartar, $H$-convergence, in *Topics in the Mathematical Modelling of Composite Materials*, Birkhäuser Boston, Boston, 1997, 21–43.

[41] S. P. Neuman and S. Orr, Prediction of steady state flow in nonuniform geologic media by conditional moments: Exact non-local formalism, effective conductivities and weak approximation, *Water Resources Res.*, **29**-2 (1993), 341–364.

[42] G. Papanicolaou and S. Varadhan, Boundary value problems with rapidly oscillating coefficients, in *Colloquia Mathematica Societatis János Bolyai* 27: *Random Fields* (*Esztergom, Hungary* 1979), North-Holland, Amsterdam, 1982, 835–873.

[43] U. Raitums, On the local representation of G-closure, *Arch. Rational Mech. Anal.*, **158**-3 (2001), 213–234.

[44] A. Saucier, Effective permeability of multifractal porous media, *Phys.* A, **183** (1992), 381–397.

[45] S. Spagnolo, Sulla convergenza di soluzioni di equazioni paraboliche ed ellittiche, *Ann. Scuola Norm. Sup. Pisa* (3), **22** (1968), 571–597; errata, *Ann. Scuola Norm. Sup. Pisa* (3), **22** (1968), 673.

[46] S. Spagnolo, Sul limite delle soluzioni di problemi di Cauchy relativi all'equazione del calore, *Ann. Scuola Norm. Sup. Pisa* (3), **21** (1967), 657–699.

[47] D. Stroud and D. J. Bergman, New exact results for the Hall-coefficient and magnetoresistance of inhomogeneous two-dimensional metals, *Phys. Rev.* B, **30** (1984), 447–449.

[48] L. Tartar, Estimation de coefficients homogénéisés, in *Computing Methods in Applied Sciences and Engineering: Proceedings of the Third International Symposium, Versailles*, 1977, vol. I, Lecture Notes in Mathematics 704, Springer-Verlag, Berlin, 1979, 364–373.

[49] L. Tartar, Estimation fines des coefficients homogénéisés, in P. Kree, ed., *Ennio de Giorgi's Colloquium*, Pitman, London, 1985, 168–187.

[50] J. R. Willis, Elasticity theory of composites, in H. G. Hopkins and M. J. Sewell, eds., *Mechanics of solids*, Pergamon Press, Oxford, New York, 1982, 653–686.

[51] V. V. Zhikov, Estimates for the homogenized matrix and the homogenized tensor, *Russian Math. Surveys*, **46**-3 (1991), 65–136.

# 9. A Multifield Theory for the Modeling of the Macroscopic Behavior of Shape Memory Materials

Davide Bernardini* and Thomas J. Pence†

**Abstract.** The macroscopic behavior of shape memory materials is modeled within the framework of multifield theories. Two scalar fields and a second-order tensor field are used as descriptors of the relevant microstructural phenomena. In this way it is possible to allow for pseudoelasticity and shape memory effect, as well as low and high temperature reorientation of Martensitic variants. The general aspects of the theory are discussed paying special attention to the treatment of balance equations and to the exploitation of the constitutive structure, which is characterized by the prescription of a response function for the entropy production. An example of an explicit model is also given.

## 9.1. Introduction and literature survey

The term shape memory material (SMM) describes a class of materials that exhibit, at the macroscopic scale, the properties of pseudoelasticity and shape memory (see, e.g., Otsuka and Shimizu, 1986, Otsuka and Wayman, 1998). These behaviors arise in materials that transform between different solid phases in a manner that does not require atomic diffusion. These transformations can be induced by mechanical, thermal, or even magnetic, energy supply (Fischer et al., 1994). Among the several examples of materials belonging to this class, binary and ternary metallic alloys like NiTi and CuZnAl are widely used in several fields of engineering, industry and medicine (Melton, 1999). Pseudoelasticity refers to the ability of small changes in stress to generate large, but bounded, changes in strain. Unloading may or may not reverse the deformation and recover the original shape, the determination being dependent on the material temperature. The original shape is recovered on unloading at high temperature, with the reverse deformation taking place at lower stress levels than those associated with the original deformation. The original shape is not recovered at low temperature meaning that a residual deformation remains after unloading. This

*Dipartimento di Ingegneria Strutturale e Geotecnica, Università di Roma "La Sapienza," Via Eudossiana 18, 00184, Rome, Italy, davide.bernardini@uniroma1.it.

†Department of Mechanical Engineering, Michigan State University, East Lansing, MI 48824-1226, pence@egr.msu.edu.

residual deformation can be annihilated by heating, thus returning the material to its original shape and yielding the so-called shape memory effect.

These effects arise because the high temperature phase, called Austenite and abbreviated in this section by A, has a higher crystallographic symmetry than the low temperature phase, called Martensite and abbreviated in this section by M. The Austenite can therefore transform into one or more variants of Martensite that differ mainly by their orientation relation to the Austenite parent. By contrast, all Martensite variants tend to transform into a single common Austenite crystal structure. The different crystallographic structures of Austenite and the Martensite variants account for the large deformations at the macroscopic scale. Transformation from A to M gives rise to variant combinations that are energetically favored in the prevailing stress field, thus providing the high temperature pseudoelasticity. Transformations between different M variant structures are also driven by changes in the stress field, thus providing a low temperature pseudoelastic effect that is called reorientation. There is hysteresis in transformation between A and M and in M variant reorientation. This hysteresis accounts for the absence of shape recovery upon unloading at low temperature and also accounts for the lower stress values associated with shape recovery at high temperature.

In this article we present a framework for describing these materials in the context of multifield theory. In addition to the usual field variables of position $\mathbf{x}$ and temperature $\vartheta$, the existence of both A and M phases motivates additional scalar field variables of Austenite phase fraction $\xi_A$ and Martensite phase fraction $\xi_M$. In addition, different M microstructures must be acknowledged due to the variety of variant combinations within individual crystallite grains. In polycrystals, the grain texture then gives rise at the macroscopic scale to what may be regarded as a continuous distribution of different varieties of Martensite. These are distinguished here by an additional second-order tensor field $\mathbf{M}$ that is characteristic of the effective macroscopic transformation strain of the Martensite. The additional fields $\xi_A$, $\xi_M$, $\mathbf{M}$ are subject to balance principles with the same status as the standard balances of momentum and entropy and participate in the energy balance with the same status as the usual fields. This framework permits a unified description of shape memory, pseudoelasticity, reorientation, and hysteresis as appropriate for a macroscopic model. The framework developed here is limited to quasistatic processes and so is not appropriate for issues of wave propagation and inertial dynamics. Fully dynamic processes could be described by extensions to the present framework. Moreover, the present framework is also limited to smooth processes, although again the associated extensions can be contemplated so as to treat processes with various discontinuities in field variables.

Continuum descriptions for SMM materials are now abundant, although the particular multifield framework presented here is apparently new. The variety of continuum models for SMM are reviewed in various survey papers (e.g., Huo and Müller, 1993; Fischer et al., 1996; Bernardini and Pence, 2002a). For the present discussion we restrict attention to macroscopic constitutive theories for SMM that either employ additional field variables similar to $\xi_A$ and $\xi_M$ or else employ some notion of macroscopic transformation strain similar to $\mathbf{M}$. In the latter case this is typically on the basis of mechanical analysis of variant combinations at finer length scales.

Several models have been posed in the framework of continuum thermodynamics with internal variables (Coleman and Gurtin, 1967; Rice, 1971) where the internal variables are quantities introduced in order to take into account some of the microscopic scale phenomena and which are considered as arguments of the standard response functions. They are constrained by ordinary differential constitutive equations which are, usually, explicit in their rates. In the context of SMM, internal variables typically include one or more phase fractions or macroscopic transformation strains. The evolution of the phase transformations may then be governed by ordinary differential equations in the internal variables rates, the phase transformation *kinetics*, which are prescribed constitutively at the outset.

Early application of the internal variable formalism to SMM focused on accounting for uniaxial isothermal pseudoelasticity through the use a single scalar internal variable equivalent to $\xi_M$ (Tanaka et al., 1986; Liang and Rogers, 1990). This internal variable then evolves in response to changes in temperature and to changes in a single component of stress. This evolution in these early treatments is governed by a separate phase transformation kinetic. The kinetic is designed so as to initiate and conclude the A to M and M to A transformations in appropriate regions of a temperature-stress phase diagram. A limitation of this approach is that it is only able to model a Martensite with fixed transformation strain, and thus unable to account for both positive and negative stress and also unable to model the complete shape memory effect. Even so, the inherent simplicity of this type of model has led to its wide use for several applications exploiting the pseudoelastic effect under positive uniaxial isothermal loading. By incorporating a second scalar internal variable to allow for the distinction between stress-induced (or oriented) Martensite and thermally induced (or unbiased) Martensite, these models can be extended so as to allow for the shape memory effect in the context of uniaxial stress states and isothermal conditions for mechanical loading (Brinson, 1993).

The models mentioned above directly pose stress-strain-temperature relations with little attention to an overarching energetic structure. Such structure can be provided in the context of uniaxial pseudoelasticity via minimization of a macroscopic Helmholtz free energy function with dependence on the Martensite fraction $\xi_M$ as in Müller (1989). The resulting model was among the first to provide explanation for the internal subloops observed under incomplete phase transformations. The approach can in turn be extended to three-dimensional stress states by providing richer expressions for the free energy and formally positing a kinetic for the Martensite fraction $\xi_M$, with further refinement provided by decomposition of $\xi_M$ into stress-induced and thermally induced portions (Raniecki et al., 1992; Raniecki and Lexcellent, 1994). Other developments involve the use of Martensite fractions subject to constitutive assumptions that allow also for reorientation of the Martensite variants (Boyd and Lagoudas, 1996).

Many of the above models focus on isothermal behavior and also on loading paths that involve complete phase transitions. A description of pseudoelastic behavior under arbitrary uniaxial thermomechanical loads can be given through the use of a transformation kinetic for the phase fraction that is based on Duhem-type differential equations (Ivshin and Pence, 1994a, 1994b). This framework allows for a systematic modeling of internal subloops and their shakedown behavior, as well as the temperature change induced during mechanical loading in adiabatic conditions and in conditions involving convective heat transfer. Extensions that account for true shape memory, low temperature reorientation, and tension/compression asymmetry can be accomodated by extending the framework so as to decompose the Martensite phase fraction into two separate fractions representative of different Martensite variants of opposite orientation (Wu and Pence, 1998).

A standard feature of the above mentioned internal variable models is a transformation kinetic for the phase fraction (either $\xi_A$ or $\xi_M$) that is specified at the outset as a direct constitutive assignment. A different approach involves a transformation kinetic that is instead derived from a dissipation response function for the rate of entropy production, after enforcing restrictions induced by the balance equations (Rajagopal and Srinivasa, 1999).[1] Consequently, the constitutive prescription reduces to the specification of a free energy function and a dissipation function. While there are standard forms for free energy in the context of uniaxial stress, standard expressions for the dissipation function are lacking. Alternative forms for the dissipation function can capture different hysteretic effects and provide correlation to various thermodynamic process quantities (Bernardini and Pence, 2002b).

Less developed than the notion of a transformation kinetic for a phase fraction scalar or a collection of phase fraction scalars is the notion of a transformation kinetic for a tensor transformation strain internal variable such as **M**. A notable exception in the context of infinitesimal strains is the model of Bondaryev and Wayman (1988), which is also remarkable for its relatively early appearance. Their model provides an account, in a three-dimensional setting, of both pseudoelasticity and reorientation. It is based on the consideration of phase transformation strain as an internal variable so that phase fractions then arise as multipliers of the corresponding flow rule. Further developments in this direction have been given, by a suitably generalized plasticity theory, in Lubliner and Auricchio (1996) and their subsequent works where also finite element implementation is given.

With regard to the notion of describing a macroscopic tensor transformation strain, much effort to date has focused on its determination on the basis of the current local stress state using an effective continuum approach. Within this framework, each point of the macroscopic continuum is put in correspondence with a representative volume element (RVE) of a multiphase elastic material subjected to local transformation strains that represent the effect of the difference in the crystallographic structure of the phases. At the single crystal level, an Austenite matrix can support a specific finite number of Martensite variants, the specific number based upon the reduction of symmetry in the A to M transformation. Each M variant then has a characteristic

[1] A further feature of the model is a more systematic conception of transformation strain that is acheived by allowing the material to have multiple natural configurations.

transformation strain with respect to a common A parent phase. This by itself does not allow a direct description of macroscopic transformation strain in a polycrystal, nor does it address certain issues related to crystallographic phase interfaces.[2]

Macroscopic descriptions of shape memory material behavior can be acheived by considering averaging procedures upon assemblies of grains in an RVE where each grain is endowed with an appropriate microstructure (Patoor et al., 1988; Siredey et al., 1999). Numerous types of microstructural arrangements and averaging procedures can be contemplated in this framework. These include RVEs consisting of nontextured assemblies of spherical grains with averaging based upon self-consistency. The effect of specific texture can then be simulated using experimentally determined grain orientation distribution functions (Gall and Sehitoglu, 1999). At the level of a single crystal, an A matrix with M inclusions made of groups of self-accomodating variants can be arranged so as to resemble experimentally observed wedge-like microstructures prior to averaging (Gao and Brinson, 2000; Gao et al., 2000). Alternatively, averaging can be performed on polycrystals with an RVE consisting of a collection of grains, each wholly in the A phase or wholly consisting of a single M variant. Although the local transformation strain of each grain is then not strictly crystallographic, the resulting description is suggestive of SMM behavior both in the high and in the low temperaturature ranges (Sun and Hwang, 1994).

Dissipation can be included in such modeling by additional averaging over an internal time scale that is representative of the transformation duration in order to obtain an average dissipation rate and driving force. The phase transformations may then be governed, macroscopically, by overall nucleation and propagation energetic criteria that are obtained after integration of the corresponding local criteria over the RVE and over the propagating interfaces. The final evolution equations then follow by invoking an extremum principle, closely related with a maximum dissipation requirement (Levitas, 1998).

Different elastic properties of the A and M phases can be accommodated in these various averaging techniques, which in turn generates differences in the structure of the macroscopic free energy. In particular, free energies for the above referenced multidimensional treatments can follow from volume averaging over boundary value problems of heterogeneous elastic materials with prescribed eigenstrains (Bernardini, 2001). Macroscopic models then follow when such treatments are augmented with a suitable phase transformation kinetic.

In this work the macroscopic thermomechanical behavior of SMM is modeled within the framework of *multifield theories* (Capriz, 1989; Capriz, 2000; Mariano, 2001 and references therein) using microstructural descriptors consisting of both scalar and second-order tensor fields. This approach does not seem to have been previously pursued with reference to SMM. A specific feature of this approach is that

[2]Although not germane for the purpose of this article, it is worth noting that individual phase interfaces are often modeled with the aid of nonconvex or multiwell stored energy functions. Analysis of individual phase domains with distinguishable phase interfaces can then occur in both static situations (e.g., Ericksen, 1975; Falk, 1980; Ball and James, 1987; Bhattacharya and Kohn, 1997; Truskinovsky and Zanzotto, 2002) and dynamic situtations (e.g., Abeyaratne and Knowles, 1991; Pence, 1992; Rosakis and Tsai, 1995; Truskinovsky, 1997).

the descriptors are regarded as truly observable quantities which therefore directly contribute to the balance equations. Each microstructural descriptor is governed by its own balance which is required to be consistent with the observer invariance principle. The framework also naturally allows for spatial weak nonlocality by virtue of a possible dependence on spatial gradients of the microstructural descriptors. This allows also for refined descriptions of microstructural localization, which is an important issue in modeling SMM. Some connection with certain aspects of the present work can be found in Fried and Gurtin (1994) and in Frémond (1996) where balance equations are also proposed for microstructural descriptors that, in their cases, are restricted to be scalar in nature. Here, in addition to the consideration of a second-order tensor order parameter and certain other basic differences with respect to the general development, we also explore additional issues with respect to the constitutive theory.

Specifically, as remarked above and developed in Section 9.3, the present model takes as descriptors of the underlying microstructure two scalar fields $\xi_A$, $\xi_M$ representing the volume fractions of the Austenite and Martensite phases and a second-order tensor field $\mathbf{M}$ that defines the features of the particular arrangement of Martensite variants. The macroscopic displacement is described by the placement field $\mathbf{x}$ and the overall framework is required to satisfy an observer invariance principle as discussed in Section 9.4. In Section 9.5, scalar balance equations are presented for the fields $\xi_A$, $\xi_M$, and a second-order tensor balance equation is presented for the field $\mathbf{M}$. These accompany the usual balance equations for linear momentum, entropy, and energy. Linear momentum balance is a vector equation that is regarded as the balance associated with placement $\mathbf{x}$, while the entropy balance introduces the temperature field $\vartheta$ into the formulation. The general constitutive structure as elaborated in Section 9.6 involves, as customary in multifield theories, the specification of response functions for the volume and surface densities of the interactions and, in addition, also includes a constitutive function for the entropy production. The fields $\xi_A$, $\xi_M$, $\mathbf{M}$ are typically subject to various constraints that in turn give rise to reactive terms in the response functions. The effect of these constraints, as well as possible constraints on $\mathbf{x}$, are considered in Section 9.7. Then in Section 9.8, restrictions on the constitutive functions are derived from the balance equations. It is shown that the set of response functions is completely determined in terms of two scalar functions: the free energy function and the entropy production function.

In contrast to various other treatments for SMM, phase transformations are not governed by relatively ad hoc criteria but rather, as discussed in Section 9.9, are governed by the response functions for the free energy and entropy production via the microstructural balance equations. The hypothesis of rate independence for SMM transformation introduces natural memory parameters into the constitutive description for the rate of entropy production. This accounts for hysteresis with respect to the fields $\xi_A$, $\xi_M$, $\mathbf{M}$ and so in particular describes incomplete transformation effects and the generation of internal subloops in stress-strain response. The requirement of observer invariance puts additional restrictions on the constitutive theory that identify natural combinations of field variables for the constitutive dependence. An explicit example involving linearized kinematics is presented in Section 9.10, which under appropriate specializations reproduces previous descriptions for pseudoelasticity and the shape memory effect, while at the same time, providing a treatment for reorientation of Martensitic variants at both low and high temperature.

## 9.2. Notation

The following conventions for tensor algebra will be used henceforth (Murdoch, 1978). Let E represent the three-dimensional Euclidean space with translation space $\mathcal{V}_{\mathsf{E}}$ and scalar product $\cdot$ and let $\mathbf{a}, \mathbf{b}, \mathbf{c}, \mathbf{d}, \mathbf{e}, \mathbf{f} \in \mathcal{V}_{\mathsf{E}}$ be six arbitrary vectors. Simple *second-*, *third-*, and *fourth-order tensors* are thus defined via the dyadics in the standard way:

$$\mathbf{a} \otimes \mathbf{b}, \qquad \mathbf{a} \otimes \mathbf{b} \otimes \mathbf{c}, \qquad \mathbf{a} \otimes \mathbf{b} \otimes \mathbf{c} \otimes \mathbf{d}.$$

The corresponding general tensors and all the corresponding operations are defined by linearity on a basis for $\mathcal{V}_{\mathsf{E}}$. The scalar products between tensors of second- and third-order are denoted by the same symbol $\cdot$ used for $\mathcal{V}_{\mathsf{E}}$ and, for simple tensors, reads

$$(\mathbf{a} \otimes \mathbf{b}) \cdot (\mathbf{d} \otimes \mathbf{e}) = (\mathbf{a} \cdot \mathbf{d})(\mathbf{b} \cdot \mathbf{e}),$$
$$(\mathbf{a} \otimes \mathbf{b} \otimes \mathbf{c}) \cdot (\mathbf{d} \otimes \mathbf{e} \otimes \mathbf{f}) = (\mathbf{a} \cdot \mathbf{d})(\mathbf{b} \cdot \mathbf{e})(\mathbf{c} \cdot \mathbf{f}).$$

Various types of contractions may apply; those used in this work are denoted by $\overset{1}{\circ}$, $\overset{2}{\circ}$, $\overset{3}{\circ}$ and are defined, on simple tensors, as follows:

$$(\mathbf{a} \otimes \mathbf{b} \otimes \mathbf{c}) \overset{1}{\circ} (\mathbf{d} \otimes \mathbf{e}) = (\mathbf{c} \cdot \mathbf{d})(\mathbf{a} \otimes \mathbf{b} \otimes \mathbf{e}),$$
$$(\mathbf{a} \otimes \mathbf{b}) \overset{1}{\circ} (\mathbf{c} \otimes \mathbf{d} \otimes \mathbf{e}) = (\mathbf{b} \cdot \mathbf{c})(\mathbf{a} \otimes \mathbf{d} \otimes \mathbf{e}),$$
$$(\mathbf{a} \otimes \mathbf{b} \otimes \mathbf{c}) \overset{2}{\circ} (\mathbf{d} \otimes \mathbf{e}) = (\mathbf{c} \cdot \mathbf{d})(\mathbf{b} \cdot \mathbf{e})\mathbf{a},$$
$$(\mathbf{a} \otimes \mathbf{b}) \overset{2}{\circ} (\mathbf{c} \otimes \mathbf{d} \otimes \mathbf{e}) = (\mathbf{b} \cdot \mathbf{c})(\mathbf{a} \cdot \mathbf{d})\mathbf{e},$$
$$(\mathbf{a} \otimes \mathbf{b} \otimes \mathbf{c} \otimes \mathbf{d}) \overset{2}{\circ} (\mathbf{e} \otimes \mathbf{f}) = (\mathbf{d} \cdot \mathbf{e})(\mathbf{c} \cdot \mathbf{f})(\mathbf{a} \otimes \mathbf{b}),$$
$$(\mathbf{a} \otimes \mathbf{b} \otimes \mathbf{c}) \overset{3}{\circ} (\mathbf{d} \otimes \mathbf{e} \otimes \mathbf{f}) = (\mathbf{c} \cdot \mathbf{d})(\mathbf{b} \cdot \mathbf{e})(\mathbf{a} \cdot \mathbf{f}).$$

Moreover, two types of transposition on third-order tensors are considered

$$(\mathbf{a} \otimes \mathbf{b} \otimes \mathbf{c})^T = (\mathbf{c} \otimes \mathbf{b} \otimes \mathbf{a}), \qquad (\mathbf{a} \otimes \mathbf{b} \otimes \mathbf{c})^{\wedge} = (\mathbf{b} \otimes \mathbf{a} \otimes \mathbf{c}),$$

the other possible transpositions being obtainable by their composition. Transposition operations $^{\sim}$, $^{\wedge}$, and $T$ are defined on fourth-order tensors via

$$(\mathbf{a} \otimes \mathbf{b} \otimes \mathbf{c} \otimes \mathbf{d})^{\sim} = (\mathbf{d} \otimes \mathbf{b} \otimes \mathbf{c} \otimes \mathbf{a}),$$
$$(\mathbf{a} \otimes \mathbf{b} \otimes \mathbf{c} \otimes \mathbf{d})^{\wedge} = (\mathbf{b} \otimes \mathbf{a} \otimes \mathbf{c} \otimes \mathbf{d}),$$
$$(\mathbf{a} \otimes \mathbf{b} \otimes \mathbf{c} \otimes \mathbf{d})^T = (\mathbf{c} \otimes \mathbf{d} \otimes \mathbf{a} \otimes \mathbf{b}).$$

All these tensorial notions can be extended also to the case in which different copies of E are involved by considering combinations of vectors belonging to various translation spaces. For ease of notation, the distinction between the various types of tensors will be explicitly stressed only when strictly necessary. Specifically, this will be done in the case of *two-point second-order tensors* which are defined relative to $\mathsf{E}$ and to its copy $\mathsf{A}$, the latter with translation spaces $\mathcal{V}_{\mathsf{A}}$. If $g \in \mathcal{V}_{\mathsf{A}}$, then both

$$\mathbf{a} \otimes \mathbf{g} \quad \text{and} \quad \mathbf{g} \otimes \mathbf{a}$$

furnish such a two-point tensor. Then one may interpret $\mathbf{a} \otimes \mathbf{g} \in \mathrm{Lin}(\mathcal{V}_{\mathsf{A}}, \mathcal{V}_{\mathsf{E}})$ and $\mathbf{g} \otimes \mathbf{a} \in \mathrm{Lin}(\mathcal{V}_{\mathsf{E}}, \mathcal{V}_{\mathsf{A}})$.

## 9.3. Kinematics

The position in space of a body $\mathfrak{B}$ is idealized by a bijection

$$\widehat{\chi} : \mathfrak{B} \to \mathsf{E}, \tag{1}$$

which associates with each element $\mathfrak{p}$ endowed with a distinct identity a point of $\mathsf{E}$ indicated by $\mathbf{x} = \widehat{\chi}(\mathfrak{p})$. The mapping $\widehat{\chi}$ is said to define an *apparent placement* for $\mathfrak{B}$ if the set $\mathcal{B} = \widehat{\chi}(\mathfrak{B}) \subset \mathsf{E}$, obtained when $\mathfrak{p}$ spans $\mathfrak{B}$, is a Noll–Virga fit region (Noll and Virga, 1988). Material elements $\mathfrak{p}$ are acknowledged to be endowed with an underlying crystallographic structure whose features can potentially affect the macroscopic response of the body. In order to provide an idealization of this circumstance, the logical path of multifield theories[3] is followed and the configuration of each $\mathfrak{p}$ is characterized, besides its position in space, by further attributes, selected via physical considerations, as descriptors of the relevant features of the microstructure.

There exists extensive experimental evidence of the multiphase nature of SMM (Otsuka and Shimizu, 1986; Otsuka and Wayman, 1998) that exhibits arrangements of Austenite and variants of Martensite in microstructures of variable complexity depending on several factors such as, for example, alloy composition, cold work, annealing, and loading history. In the simplest case of single crystals subjected to uniaxial stress states, a single correspondence variant pair of Martensitic variants is usually observed (Gall et al., 1999), whereas for polycrystals under three-dimensional stress states, the microstructure generally appears as an arrangement, within a grain structure, of Austenite regions and many groups of variously twinned Martensite variants (Fischer et al., 1994; Shield, 1995).

For the purpose of the present modeling, material elements are understood as composed of an arrangement of an Austenite phase with a Martensite phase where, in turn, the Martensite phase is endowed by further structure reflecting the possibility of being composed by an assembly of variants. This kind of microstructure is introduced in the theory by means of two kinds of descriptors: the volume fraction of the Austenite and Martensite phases and a tensorial quantity that identifies the particular kind of Martensite under consideration.

[3]The idea to endow material elements with further degrees of freedom is a long-standing one and, historically, emerged in a variety of contexts. In this work we refer to the approach pursued in Capriz (1989), Capriz (2000), Mariano (2001); however, various lines of development of the idea can be found in the literature as, for example, in Eringen (1999) and the references therein.

The volume fractions of Austenite and Martensite are specified by a continuous mapping

$$\widehat{\mathbf{z}} : \mathfrak{B} \to \mathsf{Z} \subseteq \mathbb{R}^2$$

which associates with each $\mathfrak{p}$ an ordered pair of real numbers $(\xi_A, \xi_M) = \widehat{\mathbf{z}}(\mathfrak{p})$. The set $\mathsf{Z}$ collects the physically admissible pairs $(\xi_A, \xi_M)$, also indicated by the bold letter $\boldsymbol{\xi}$. The structure of $\mathsf{Z}$ is specified by equality and inequality constraints, i.e., $\mathsf{Z} = \mathsf{Z}^{\text{eq}} \cap \mathsf{Z}^{\text{in}}$ with

$$\mathsf{Z}^{\text{eq}} := \{\boldsymbol{\xi} \in \mathbb{R}^2 | w_i^{\text{eq}}(\boldsymbol{\xi}) = 0 \ (i = 1, \ldots, M_{\text{eq}})\},$$
$$\mathsf{Z}^{\text{in}} := \{\boldsymbol{\xi} \in \mathbb{R}^2 | w_i^{\text{in}}(\boldsymbol{\xi}) \leq 0 \ (i = 1, \ldots, M_{\text{in}})\},$$

where $w_i^{\text{eq}}$ and $w_i^{\text{in}}$ are suitably smooth, scalar-valued, constraint functions so as to render $\mathsf{Z}^{\text{in}}$ convex. Typical constraint functions that arise naturally from the definition of the phase fractions involve $M_{\text{eq}} = 1$, $M_{\text{in}} = 4$ with

$$w_1^{\text{eq}} = \xi_A + \xi_M - 1, \tag{2}$$

$$\begin{aligned} w_1^{\text{in}} &= \xi_A - 1, & w_2^{\text{in}} &= -\xi_A, \\ w_3^{\text{in}} &= \xi_M - 1, & w_4^{\text{in}} &= -\xi_M. \end{aligned} \tag{3}$$

Each set $\mathcal{Z} = \widehat{\mathbf{z}}(\mathfrak{B})$ of elements belonging to $\mathsf{Z}$ is said to be a *phase distribution* for $\mathfrak{B}$, while the mapping $\widehat{\mathbf{z}}$ is a *phase assignment* for $\mathfrak{B}$.

The variant assembly within the Martensite phase is described by a continuous mapping

$$\widehat{\mathbf{m}} : \mathfrak{B} \to \mathsf{M} \subseteq \text{Lin}(\mathcal{V}_\mathsf{A}, \mathcal{V}_\mathsf{E}) \tag{4}$$

which associates with each $\mathfrak{p}$ a two-point second-order tensor $\mathbf{M} = \widehat{\mathbf{m}}(\mathfrak{p})$. The set $\mathsf{M}$ of the physically admissible values of $\mathbf{M}$ is given by $\mathsf{M} = \mathsf{M}^{\text{eq}} \cap \mathsf{M}^{\text{in}}$ with

$$\begin{aligned} \mathsf{M}^{\text{eq}} &:= \{\mathbf{M} \in \text{Lin}(\mathcal{V}_\mathsf{A}, \mathcal{V}_\mathsf{E}) \mid g_j^{\text{eq}}(\mathbf{M}) = 0 \ (j = 1, \ldots, N_{\text{eq}})\}, \\ \mathsf{M}^{\text{in}} &:= \{\mathbf{M} \in \text{Lin}(\mathcal{V}_\mathsf{A}, \mathcal{V}_\mathsf{E}) \mid g_j^{\text{in}}(\mathbf{M}) \leq 0 \ (j = 1, \ldots, N_{\text{in}})\}, \end{aligned} \tag{5}$$

for suitably smooth, scalar-valued, constraint functions $g_j^{\text{eq}}$ and $g_j^{\text{in}}$ so as to render $\mathsf{M}^{\text{in}}$ convex. In (5) $\mathcal{V}_\mathsf{A}$ and $\mathcal{V}_\mathsf{E}$ are the translation spaces of two copies, $\mathsf{A}$ and $\mathsf{E}$, of the three-dimensional Euclidean space such that $\mathsf{E}$ is the space in which the actual position of $\mathfrak{B}$ is represented (see (1)) whereas $\mathsf{A}$ is the representation space for the position of $\mathfrak{B}$ when it is in a fully Austenitic state. The tensor $\mathbf{M}$ may then be interpreted, locally, as arising from a homogeneous microscale transplacement that would convert an entirely Austenitic patch into another one composed completely of the Martensite under consideration.[4] In particular, $\mathbf{M} = \mathbf{I}$ is associated with a Martensite that yields no macroscopic deformation (i.e., self-accommodated), whereas $\mathbf{M} \neq \mathbf{I}$ implies that the Martensite is associated with a collection of variants giving rise to

[4]This interpretation is an attempt to introduce a purely kinematic description of the effect of the microstructure on the macroscopic behavior. This task can be formalized rigorously by following, in a more explicit way, the approach developed in Del Piero and Owen (1993).

a net macroscopic deformation with respect to the Austenite. As regards constraints upon $\mathsf{M}$, an example motivated by the case in which microscale transplacement is volume preserving and limited by crystallographic lattice shifts, is given by $N_{\text{eq}} = 1$, $N_{\text{in}} = 1$ with

$$g_1^{\text{eq}}(\mathbf{M}) = \det \mathbf{M} - 1, \tag{6}$$

$$g_1^{\text{in}}(\mathbf{M}) = (\mathbf{M}^T\mathbf{M} - \mathbf{I}) \cdot (\mathbf{M}^T\mathbf{M} - \mathbf{I}) - (g^*)^2, \tag{7}$$

where $g^*$ is characteristic of the maximum "effective" lattice shift between the particular Martensite under consideration and the Austenite. Each set $\mathcal{M} = \widehat{\mathfrak{m}}(\mathfrak{B})$ of elements belonging to $\mathsf{M}$ is a *Martensite distribution* for $\mathfrak{B}$, while the mapping $\widehat{\mathfrak{m}}$ is a *Martensite assignment* for $\mathfrak{B}$.

The quantities $\boldsymbol{\xi}$, $\mathbf{M}$ are therefore regarded as a complete representation of the microstructure underlying each material element. They are independent since, at each point, independent variation of $\boldsymbol{\xi}$ and $\mathbf{M}$ may occur reflecting physically meaningful different phase transformations. Specifically, a variation of $\boldsymbol{\xi}$ corresponds to a *proper Martensitic transformation* (i.e., conversion of Austenite into a fixed kind of Martensite or vice versa), while a variation of $\mathbf{M}$ indicates that the kind of Martensite is changed and so provides the other typical transformation observed in SMM, which is usually called *Martensite reorientation* (Bernardini and Pence, 2002a). Martensite reorientation provides the only transformational change when the material is in its low temperature range.

Summarizing, the assignment of a *configuration* of the body is defined by the action of the mappings $\widehat{\chi}, \widehat{\mathfrak{z}}, \widehat{\mathfrak{m}}$ which assign to every $\mathfrak{p} \in \mathfrak{B}$ a triple

$$\mathcal{C}(\mathfrak{p}) := \{\widehat{\chi}(\mathfrak{p}), \widehat{\mathfrak{z}}(\mathfrak{p}), \widehat{\mathfrak{m}}(\mathfrak{p})\} \in \mathsf{E} \times \mathsf{Z} \times \mathsf{M}.$$

Clearly, the configuration of the whole $\mathfrak{B}$ is the triple of subsets $\{\mathcal{B}, \mathcal{Z}, \mathcal{M}\}$, respectively, of $\mathsf{E}$, $\mathsf{Z}$, and $\mathsf{M}$. Such a configuration for $\mathfrak{B}$ is said to be *admissible* if $\mathcal{B}$ is a fit region.

Given any two apparent placements $\mathcal{B}_1$ and $\mathcal{B}_2$, the apparent *transplacement* from $\mathcal{B}_1$ to $\mathcal{B}_2$ is the bijective mapping $\chi : \mathcal{B}_1 \to \mathcal{B}_2$ such hat

$$\chi(\mathcal{B}_1) = \mathcal{B}_2. \tag{8}$$

Therefore, taking any apparent placement $\mathcal{R}$ as reference, a material point labeling can be established so as to yield mappings $\chi, \mathbf{z}, \mathbf{m}$ from $\mathcal{R}$ to any admissible configuration $\mathcal{B}$, $\mathcal{Z}$, $\mathcal{M}$ in such a way that

$$\mathcal{B} = \chi(\mathcal{R}), \quad \mathcal{Z} = \mathbf{z}(\mathcal{R}), \quad \mathcal{M} = \mathbf{m}(\mathcal{R}).$$

In what follows, the apparent placement $\mathcal{R}$ associated with undeformed Austenite will be considered as reference. The elements of $\mathcal{R}$ will be denoted by $\mathbf{X}$.

A *motion* is a suitably regular time-parameterized family of configuration mappings

$$\chi(\mathbf{X}, t), \qquad \mathbf{z}(\mathbf{X}, t), \qquad \mathbf{m}(\mathbf{X}, t)$$

such that $\mathbf{x} = \chi(\mathbf{X}, t)$, $\boldsymbol{\xi} = \mathbf{z}(\mathbf{X}, t)$, and $\mathbf{M} = \mathbf{m}(\mathbf{X}, t)$ denote the place and the microstructure at time $t$ of the material point that occupies the place $\mathbf{X}$ in $\mathcal{R}$. Recall also that, more explicitly, $\boldsymbol{\xi}$ stands for

$$\xi_A = z_A(\mathbf{X}, t), \qquad \xi_M = z_M(\mathbf{X}, t).$$

Provided suitable regularity is available, space and time derivatives of motions can be evaluated (at least almost everywhere) giving rise to velocities

$$\mathbf{v} := \frac{\partial \chi(\mathbf{X}, t)}{\partial t}, \quad v^{\xi_A} := \frac{\partial z_A(\mathbf{X}, t)}{\partial t}, \quad v^{\xi_M} := \frac{\partial z_M(\mathbf{X}, t)}{\partial t}, \quad \mathbf{v}^{\mathbf{M}} := \frac{\partial \mathbf{m}(\mathbf{X}, t)}{\partial t}$$

and gradients

$$\mathbf{F} := \frac{\partial \chi(\mathbf{X}, t)}{\partial \mathbf{X}}, \quad \mathbf{F}^{\xi_A} := \frac{\partial z_A(\mathbf{X}, t)}{\partial \mathbf{X}}, \quad \mathbf{F}^{\xi_M} := \frac{\partial z_M(\mathbf{X}, t)}{\partial \mathbf{X}}, \quad \mathbf{F}^{\mathbf{M}} := \frac{\partial \mathbf{m}(\mathbf{X}, t)}{\partial \mathbf{X}}, \tag{9}$$

where $v^{\xi_A}$, $v^{\xi_M}$ are scalars; $\mathbf{v}$, $\mathbf{F}^{\xi_A}$, $\mathbf{F}^{\xi_M}$ are vectors; $\mathbf{v}^{\mathbf{M}}$, $\mathbf{F}$ are two-point second-order tensors and $\mathbf{F}^{\mathbf{M}}$ is a third-order tensor. The standard condition

$$\det \mathbf{F} > 0$$

follows from the invertibility of $\chi$ upon excluding material inversion. Additional equality constraints upon the placement may also apply in the form

$$\begin{aligned} C_k^{\mathrm{eq}}(\mathbf{F}) &= 0 \quad (k = 1, \ldots, P_{\mathrm{eq}}), \\ C_k^{\mathrm{in}}(\mathbf{F}) &\leq 0 \quad (k = 1, \ldots, P_{\mathrm{in}}) \end{aligned} \tag{10}$$

for suitably smooth constraint functions $C_k^{\mathrm{eq}}$ and $C_k^{\mathrm{in}}$. For example, a requirement of isochoric deformation involves $P_{\mathrm{eq}} = 1$, $P_{\mathrm{in}} = 0$ with

$$C_1^{\mathrm{eq}}(\mathbf{F}) = \det \mathbf{F} - 1. \tag{11}$$

## 9.4. Observer dependence and invariance requirement

The above introduced kinematic fields, as well as the fields to be introduced in what follows, are all susceptible to different representations by different observers. The term *observer* here is understood as synonymous with the term *frame* used in Capriz and Virga (1990). Henceforth, attention will be restricted to observers that agree upon the representation of spatial distance between points belonging to the same instantaneous apparent placement. This is consistent not only with the Capriz–Virga family of rigidly related frames but also with the interpretation in Murdoch (1982). For the sake of simplicity, it will be assumed that all observers agree also on the representation of time.

Specifically, let the position of a generic material point $\mathfrak{p}$ at time $t$ be described by the places $\mathbf{x}(t)$ and $\mathbf{x}^+(t)$ for two separate observers. Since, by the above definition, observers agree upon the simultaneous distance between points, there exists a vector $\mathbf{c}(t)$ and a proper orthogonal tensor $\mathbf{Q}(t)$ such that

$$\mathbf{x}^+(t) = \mathbf{c}(t) + \mathbf{Q}(t)\mathbf{x}(t). \tag{12}$$

Similarly, the placement of $\mathfrak{p}$ in the reference configuration is represented as $\mathbf{X}^+$ and $\mathbf{X}$ by the two observers, respectively, with

$$\mathbf{X}^+ = \mathbf{c}_0 + \mathbf{Q}_0\mathbf{X} \tag{13}$$

for vector $\mathbf{c}_0$ and proper orthogonal tensor $\mathbf{Q}_0$ (Murdoch, 2000). The different representations of the apparent transplacement gradient are then related by

$$\mathbf{F}^+ = \mathbf{Q}\mathbf{F}\mathbf{Q}_0^T.$$

According to Capriz and Virga (1990), it is assumed that the possible observer's representations of the microstructural descriptors are related by a set of transformations that form a continuous group that is the homomorphic image of the set of proper orthogonal second-order tensors under a suitable mapping that is to be specified on the basis of the physical significance of the descriptors. This assumption embodies the idea that the microstructural phenomena under consideration are unaltered by observer's translations.

Taking into account the nature of the phase fractions and of the tensor $\mathbf{M}$, it is stipulated that the microstructural descriptors are observer-independent (objective) in the sense that the relations among the different representations given by the observers obey the standard tensorial transformations (Ogden, 1997)

$$(\xi_A)^+ = \xi_A, \qquad (\xi_M)^+ = \xi_M, \qquad \mathbf{M}^+ = \mathbf{Q}\mathbf{M}\mathbf{Q}_0^T. \tag{14}$$

A basic postulate underlying the whole theory is that every observer agrees on the physical behavior of the body, in the sense defined by the following:

> *Observer invariance principle*: All observers must agree on the judgment as to the validity of any statement involving quantities susceptible to different representations by different observers.

Extensive usage of this principle will be made in the following with respect to balance equations, constitutive equations and internal constraints.

## 9.5. Balance equations and entropy inequality

Balance equations for mass, linear momentum, energy, and entropy are assumed to hold as for standard continua (Green and Naghdi, 1977; Müller, 1985; Šilhavý, 1997). In addition, according to Capriz (1989), the variations of configuration associated with the change in the microstructural descriptors are assumed to give rise to balanced microscopic linear momenta. The specific form of the balances is summarized in the following:

> *Balance principle*: Let $\mathcal{P}$ be an open subset, with suitably regular boundary $\partial\mathcal{P}$, of the reference apparent placement $\mathcal{R}$ of $\mathfrak{B}$ and let $\mathcal{T}$ be some open time interval of interest. Then there exists an observer relative to which the equations

$$\beta_{(\circ)}(\mathcal{P}, \partial\mathcal{P}, t) = 0$$

hold for every $\mathcal{P} \subseteq \mathcal{R}$ and almost every time $t \in \mathcal{T}$, where

$$\beta_{(\circ)}(\mathcal{P}, \partial\mathcal{P}, t) := \dot{C}_{(\circ)}(\mathcal{P}, t) - \Phi_{(\circ)}(\partial\mathcal{P}, t) - S_{(\circ)}(\mathcal{P}, t) - P_{(\circ)}(\mathcal{P}, t) \qquad (15)$$

and $(\circ)$ spans $\{\rho, \mathbf{p}, p^{\xi_A}, p^{\xi_M}, \mathbf{p}^{\mathbf{M}}, e, \eta\}$, respectively, denoting mass, macroscopic linear momentum, microscopic linear momenta associated with $\xi_A$, with $\xi_M$, with $\mathbf{M}$, energy, and entropy.[5]

The various symbols introduced in (15) have the following physical meaning relative to the field $(\circ)$: $C_{(\circ)}$ denotes the *content* in $\mathcal{P}$ at time $t$; $\Phi_{(\circ)}$ denotes the rate of *exchange through the surface* $\partial\mathcal{P}$ with the external environment at time $t$; $S_{(\circ)}(\mathcal{P}, t)$ denotes the rate of *supply* to $\mathcal{P}$ by the external environment at time $t$; and $P_{(\circ)}(\mathcal{P}, t)$ denotes the rate of *production* to $\mathcal{P}$ at time $t$. Moreover, $\Phi_{(\circ)}$, $S_{(\circ)}$, and $P_{(\circ)}$ are assumed positive if the corresponding exchanges tend to increase the content of $(\circ)$. Each of the four quantities involved in the balance equations are assumed to be absolutely continuous with respect to the volume and surface measures in the reference placement so that

$$C_{(\circ)}(\mathcal{P}, t) := \int_{\mathcal{P}} \overline{C}_{(\circ)}(\mathbf{X}, t) dV, \qquad \Phi_{(\circ)}(\partial\mathcal{P}, t) := \int_{\partial\mathcal{P}} \overline{\Phi}_{(\circ)}(\mathbf{X}, \mathbf{n}(\mathbf{X}), t) dA,$$

$$S_{(\circ)}(\mathcal{P}, t) := \int_{\mathcal{P}} \overline{S}_{(\circ)}(\mathbf{X}, t) dV, \qquad P_{(\circ)}(\partial\mathcal{P}, t) := \int_{\mathcal{P}} \overline{P}_{(\circ)}(\mathbf{X}, t) dV,$$

where $\mathbf{n}(\mathbf{X})$ is the outward unit normal to the reference surface $\partial\mathcal{P}$ at $\mathbf{X}$, $\overline{C}_{(\circ)}$ are the *field densities*, $\overline{\Phi}_{(\circ)}$ are the *surface exchange rates*, $\overline{S}_{(\circ)}$ are the *bulk supply rates*, and $\overline{P}_{(\circ)}$ are the *bulk production rates*. For the various balances, these densities are assumed to be represented as follows.

Mass is a field that involves neither a surface exchange nor a bulk supply nor a bulk production; i.e.,

$$\overline{\Phi}_{\rho}(\mathbf{X}, \mathbf{n}, t) := 0, \qquad \overline{S}_{\rho}(\mathbf{X}, t) := 0, \qquad \overline{P}_{\rho}(\mathbf{X}, t) := 0.$$

Setting $\overline{C}_{\rho} := \rho$, the arbitrariness of $\mathcal{P}$ gives a local form of the mass balance as

$$\dot{\rho}(\mathbf{X}, t) = 0. \qquad (16)$$

This property permits the remaining fields of linear momenta, energy, and entropy to be specified in terms of mass density rather than volume density.

Concerning macroscopic and microscopic linear momenta the assumptions are summarized in the table

| $(\circ)$ | $\overline{\Phi}_{(\circ)}$ | $\overline{S}_{(\circ)}$ | $\overline{P}_{(\circ)}$ |
|---|---|---|---|
| $\mathbf{p}$ | $\mathbf{t}$ | $\rho\mathbf{b}$ | $\rho\mathbf{f}$ |
| $p^{\xi_A}$ | $t^{\xi_A}$ | $\rho b^{\xi_A}$ | $\rho\Pi^{\xi_A}$ |
| $p^{\xi_M}$ | $t^{\xi_M}$ | $\rho b^{\xi_M}$ | $\rho\Pi^{\xi_M}$ |
| $\mathbf{p}^{\mathbf{M}}$ | $\mathbf{t}^{\mathbf{M}}$ | $\rho\mathbf{b}^{\mathbf{M}}$ | $\rho\mathbf{\Pi}^{\mathbf{M}}$ |

,

[5]Clearly, the symbol $\beta_{(\circ)}$ is assumed to accommodate different tensorial orders as appropriate to the nature of the fields $(\circ)$.

where $\mathbf{t}$ is the nominal traction and $\mathbf{b}$ is the body force per unit mass in the reference placement as is standard in describing balance of macroscopic linear momentum with respect to a reference placement. Also $t^{\xi_A}$, $t^{\xi_M}$, $\mathbf{t}^{\mathbf{M}}$ are interpreted as generalized tractions; $b^{\xi_A}$, $b^{\xi_M}$, $\mathbf{b}^{\mathbf{M}}$ as generalized body forces per unit mass; and $\Pi^{\xi_A}$, $\Pi^{\xi_M}$, $\mathbf{\Pi}^{\mathbf{M}}$ as generalized self-forces per unit mass. In addition, suitable expressions must be given for the content quantities $\overline{C}_{(\circ)}$. Mass density gives the macroinertia associated with positional change and the standard expression $\overline{\mathbf{C}}_{\mathbf{p}} := \rho\mathbf{v}$ applies. Corresponding expressions are needed, in general, to describe inertia associated with the change of microstructure (Capriz, 1989). As our purpose here is limited to a treatment of quasistatic material behavior, all of these fields will be regarded as vanishingly small, i.e., $\overline{\mathbf{C}}_{\mathbf{p}} = \mathbf{0}$, $\overline{C}_{p^{\xi_A}} = \overline{C}_{p^{\xi_M}} = 0$, $\overline{\mathbf{C}}_{\mathbf{p}^{\mathbf{M}}} = \mathbf{0}$. Summarizing, the above assumptions yield

$$
\begin{aligned}
\boldsymbol{\beta}_{\mathbf{p}} &= -\int_{\partial\mathcal{P}} \mathbf{t}\,dA - \int_{\mathcal{P}} \rho(\mathbf{b}+\mathbf{f})\,dV,\\
\beta_{p^{\xi_A}} &= -\int_{\partial\mathcal{P}} t^{\xi_A}\,dA - \int_{\mathcal{P}} \rho(b^{\xi_A} + \Pi^{\xi_A})\,dV,\\
\beta_{p^{\xi_M}} &= -\int_{\partial\mathcal{P}} t^{\xi_M}\,dA - \int_{\mathcal{P}} \rho(b^{\xi_M} + \Pi^{\xi_M})\,dV,\\
\boldsymbol{\beta}_{\mathbf{p}^{\mathbf{M}}} &= -\int_{\partial\mathcal{P}} \mathbf{t}^{\mathbf{M}}\,dA - \int_{\mathcal{P}} \rho(\mathbf{b}^{\mathbf{M}} + \mathbf{\Pi}^{BM})\,dV.
\end{aligned}
$$

By assuming the continuity of the dependence of the surface exchange rates upon the normal, the balanced nature of the interactions ensures that a generalized Cauchy theorem holds also for the microstructural fields (Capriz and Virga, 1990). This, together with suitable transformations to the reference placement, provides the existence of referential macro- and microstresses such that

$$
\mathbf{t} = \mathbf{S}\mathbf{n}, \qquad t^{\xi_A} = \mathbf{s}^{\xi_A}\cdot\mathbf{n}, \qquad t^{\xi_M} = \mathbf{s}^{\xi_M}\cdot\mathbf{n}, \quad \mathbf{t}^{\mathbf{M}} = \mathbf{S}^{\mathbf{M}}\mathbf{n},
$$

where $\mathbf{S}$ is the first Piola–Kirchhoff stress tensor, while the vectors $\mathbf{s}^{\xi_A}$ and $\mathbf{s}^{\xi_M}$ and the third-order tensor $\mathbf{S}^{\mathbf{M}}$ are the generalized microstresses with respect to the reference placement.

As regards the energy balance, energy is taken to be a field that does not involve bulk production; i.e.,

$$
\overline{P}_e = 0.
$$

In the present context, this is regarded as the appropriate statement of the first law of thermodynamics. The remaining energy fields $\overline{C}_e$, $\overline{S}_e$, $\overline{\Phi}_e$ are all decomposed into a part that involves the macrorate $\mathbf{v}$, the microrates $v^{\xi_A}$, $v^{\xi_M}$, $\mathbf{v}_{\mathbf{M}}$, and a part that is independent of these resolved rates. The latter are referred to as the thermal part. Energy exchange through a surface and energy supply through the bulk are associated with configurational changes involving not only change of place but also change of microstructure. Hence, both the surface exchange density rate and the bulk supply density rate are given by the sum of the contributions associated with each possible type of configurational change. Accordingly, the energy content density is

$$\overline{C}_e := \rho e + K + K^{\xi_A} + K^{\xi_M} + K^{\mathbf{M}}$$

where the field $e$ is the *internal energy* per unit mass and $K$, $K^{\xi_A}$, $K^{\xi_M}$, $K^{\mathbf{M}}$ are the *kinetic energy* fields associated with rates of position change, phase change, and Martensite orientation change, respectively. These kinetic energy fields involve the notions of macro- and microinertia alluded to above, which play no part in a quasistatic description and so are also neglected in what follows; i.e., $K = K^{\xi_A} = K^{\xi_M} = K^{\mathbf{M}} = 0$. Energy exchange through a surface and energy supply through the bulk are again assumed to be given by the sum of contributions associated with each possible type of configurational change as well as a thermal part, i.e.,

$$\overline{S}_e := \rho(\mathbf{b}\cdot\mathbf{v} + b^{\xi_A}v^{\xi_A} + b^{\xi_M}v^{\xi_M} + \mathbf{b}^{\mathbf{M}}\cdot\mathbf{v}^{\mathbf{M}} + r),$$
$$\overline{\Phi}_e := \mathbf{t}\cdot\mathbf{v} + t^{\xi_A}v^{\xi_A} + t^{\xi_M}v^{\xi_M} + \mathbf{t}^{\mathbf{M}}\cdot\mathbf{v}^{\mathbf{M}} + h,$$

where $r$ and $h$ are the thermal energy bulk supply and surface exchange, respectively. In combination this gives

$$\beta_e = \frac{d}{dt}\int_{\mathcal{P}} \rho e dV - \int_{\partial\mathcal{P}} \mathbf{t}\cdot\mathbf{v} + t^{\xi_A}v^{\xi_A} + t^{\xi_M}v^{\xi_M} + \mathbf{t}^{\mathbf{M}}\cdot\mathbf{v}^{\mathbf{M}} + h dA$$
$$- \int_{\mathcal{P}} \rho(\mathbf{b}\cdot\mathbf{v} + b^{\xi_A}v^{\xi_A} + b^{\xi_M}v^{\xi_M} + \mathbf{b}^{\mathbf{M}}\cdot\mathbf{v}^{\mathbf{M}} + r)dV$$

The dependence of $h$ upon normal $\mathbf{n}$ is again determined by a Cauchy theorem as

$$h = -\mathbf{q}\cdot\mathbf{n},$$

where $\mathbf{q}$ is called the heat flux vector.

As regards the entropy balance, the surface exchange $\overline{\Phi}_\eta$, and the bulk supply $\overline{S}_\eta$ are assumed to be proportional, respectively, to $h$ and $r$ which are the analogous thermal quantities in the energy balance.[6] The inverse of the common positive proportionality factor is called the *absolute temperature* $\vartheta$, and thus

$$\overline{\Phi}_\eta := \frac{h}{\vartheta}, \qquad \overline{S}_\eta := \rho\frac{r}{\vartheta}.$$

Further, defining $\eta$ as the specific entropy content and $\Gamma$ as the specific entropy production, the content and the production terms are

$$\overline{C}_\eta := \rho\eta, \qquad \overline{P}_\eta := \rho\dot{\Gamma},$$

so that finally

$$\beta_\eta = \frac{d}{dt}\int_{\mathcal{P}} \rho\eta dV - \int_{\partial\mathcal{P}} \frac{h}{\vartheta} dA - \int_{\mathcal{P}} \rho\frac{r}{\vartheta} + \rho\dot{\Gamma} dV.$$

In addition to the above balance equations, the second law of thermodynamics gives an entropy inequality requiring, as customary, a nonnegative rate of entropy

[6] According to Müller (1967), different assumptions can be made in this respect.

production; i.e.,

$$P_\eta = \int_{\mathcal{P}} \rho\dot{\Gamma} dV \geq 0. \tag{17}$$

Here and in what follows it is assumed that all fields are sufficiently regular to support the analytical tools, explicitly or implictly, invoked. Thus, the arbitrariness of $\mathcal{P}$ yields the following local statements

$$\begin{aligned}
&\mathrm{Div}(\mathbf{S}^T\mathbf{v} + \mathbf{s}^{\xi_A}\upsilon^{\xi_A} + \mathbf{s}^{\xi_M}\upsilon^{\xi_M} + (\mathbf{S}^{\mathbf{M}})^T \overset{2}{\circ} \mathbf{v}^{\mathbf{M}} - \mathbf{q}) \\
&\qquad + \rho(\mathbf{b}\cdot\mathbf{v} + b^{\xi_A}\upsilon^{\xi_A} + b^{\xi_M}\upsilon^{\xi_M} + \mathbf{b}^{\mathbf{M}}\cdot\mathbf{v}^{\mathbf{M}} + r) = \rho\dot{e}, \\
&\mathrm{Div}\,\mathbf{S} + \rho(\mathbf{b}+\mathbf{f}) = \mathbf{0}, \qquad \mathrm{Div}\,\mathbf{S}^{\mathbf{M}} + \rho(\mathbf{b}^{\mathbf{M}} + \boldsymbol{\Pi}^{\mathbf{M}}) = \mathbf{0}, \\
&\mathrm{Div}\,\mathbf{s}^{\xi_A} + \rho(b^{\xi_A} + \Pi^{\xi_A}) = 0, \qquad \mathrm{Div}\,\mathbf{s}^{\xi_M} + \rho(b^{\xi_M} + \Pi^{\xi_M}) = 0, \\
&-\mathrm{Div}\left(\frac{\mathbf{q}}{\vartheta}\right) + \rho\left(\frac{r}{\vartheta} + \dot{\Gamma}\right) = \rho\dot{\eta}, \qquad \rho\dot{\Gamma} \geq 0.
\end{aligned} \tag{18}$$

After inspection of the previous relations, it is recognized that 23 fields have been introduced. In order to facilitate the next developments, any ordered 23-ple of fields defined on $\mathcal{R}\times\mathcal{T}$ and belonging to a suitably smooth function space $\mathsf{Y}$ is said to be a *possible process* for $\mathfrak{B}$ and it is denoted as $\mathsf{y} := \{\mathsf{x}, \mathsf{c}, \mathsf{f}, \mathsf{p}, \mathsf{s}\}$ where the notation emphasizes the different role played by the fields within the theory. In particular, omitting the dependence upon $(\mathbf{X}, t)$,

| | |
|---|---|
| configuration fields: | $\mathsf{x} := \{\mathbf{x}, \vartheta, \xi_A, \xi_M, \mathbf{M}\}$, |
| contents: | $\mathsf{c} := \{\rho, e, \eta\}$, |
| fluxes: | $\mathsf{f} := \{\mathbf{S}, \mathbf{q}, \mathbf{s}^{\xi_A}, \mathbf{s}^{\xi_M}, \mathbf{S}^{\mathbf{M}}\}$, |
| productions: | $\mathsf{p} := \{\mathbf{f}, \Gamma, \Pi^{\xi_A}, \Pi^{\xi_M}, \boldsymbol{\Pi}^{\mathbf{M}}\}$, |
| sources: | $\mathsf{s} := \{\mathbf{b}, r, b^{\xi_A}, b^{\xi_M}, \mathbf{b}^{\mathbf{M}}\}$. |

The set of possible processes $\mathsf{Y}$ collects all the 23-ples of fields whose elements, individually, can be realized in some circumstance. However, in general, while the realization of each single field is possible, not all combinations $\mathsf{y}$ describe real situations. The processes which describe physically meaningful circumstances are called *admissible thermomechanical processes* and belong to a set $\mathsf{Y}_{\mathbb{A}} \subset \mathsf{Y}$ which will be singled out from $\mathsf{Y}$ after enforcing various requirements: balance equations, constitutive equations, internal constraints, and observer invariance. To this end, the effect of balance equations on processes can be emphasized by writing compactly $\mathbb{B}^{(\circ)}(\mathsf{y}) = \mathbf{0}$, with appropriate $(\circ)$, for the balance equations in (18), and $\mathbb{B}^{\Gamma}(\mathsf{y}) \geq 0$ for the entropy inequality again in (18).

The principle of observer invariance is now applied to the balance equations. This will yield restrictions on the possible processes so as to ensure that all observers agree on the validity of the balances.

Specifically, let the balance equations associated with the observers measuring placement $\mathbf{x}^+$ and $\bar{\mathbf{x}}$ be written, respectively, as $\beta^+_{(\circ)} = 0$ and $\bar{\beta}_{(\circ)} = 0$, where symbols are defined in the obvious way in terms of the individual fields as measured by the observers, for example,

$$\bar{\beta}_e = \int_{\mathcal{P}} \bar{\rho}(\dot{\bar{e}} - \bar{\mathbf{b}} \cdot \bar{\mathbf{v}} - \bar{b}^{\xi_A}\bar{v}^{\xi_A} - \bar{b}^{\xi_M}\bar{v}^{\xi_M} - \bar{\mathbf{b}}^{\mathbf{M}} \cdot \bar{\mathbf{v}}^{\mathbf{M}} - \bar{r})dV$$
$$- \int_{\partial\mathcal{P}} \bar{\mathbf{t}} \cdot \bar{\mathbf{v}} + \bar{t}^{\xi_A}\bar{v}^{\xi_A} + \bar{t}^{\xi_M}\bar{v}^{\xi_M} + \bar{\mathbf{t}}^{\mathbf{M}} \cdot \bar{\mathbf{v}}^{\mathbf{M}} + \bar{h}dA.$$

As regards the balance equations for the quasistatic theory, the observer invariance principle thus requires that for every pair of observers,

$$\beta^+_{(\circ)} = 0 \Leftrightarrow \bar{\beta}_{(\circ)} = 0. \tag{19}$$

Transformation rules that augment (12)–(14) for the additional fields appearing in (19) are derived from the assumption of their objectivity. Material time differentiation of the transformation rules for the configuration fields yield the corresponding rules for the various velocity terms

$$\mathbf{v}^+ = \mathbf{Q}(\bar{\mathbf{v}} + \mathbf{u} + \boldsymbol{\Omega}\bar{\mathbf{x}}), \qquad (v^{\xi_A})^+ = \bar{v}^{\xi_A},$$
$$(v^{\xi_M})^+ = \bar{v}^{\xi_M}, \qquad (\mathbf{v}^{\mathbf{M}})^+ = \mathbf{Q}(\boldsymbol{\Omega}\bar{\mathbf{M}} + \bar{\mathbf{v}}^{\mathbf{M}})\mathbf{Q}_0^T,$$

where $\mathbf{u} := \mathbf{Q}^T\dot{\mathbf{c}}$ and $\boldsymbol{\Omega} := \mathbf{Q}^T\dot{\mathbf{Q}}$, a skew second-order tensor. The scalar field quantities $\vartheta$, $\rho$, $h$, $r$, $e$, $\eta$, $\Gamma$ share the same representation

$$\vartheta^+ = \bar{\vartheta}, \quad \rho^+ = \bar{\rho}, \quad h^+ = \bar{h}, \quad r^+ = \bar{r}, \quad e^+ = \bar{e}, \quad \eta^+ = \bar{\eta}, \quad \Gamma^+ = \bar{\Gamma}, \tag{20}$$

while recalling that the theory is quasistatic, the standard and generalized tractions, body forces, and self forces obey

$$\mathbf{t}^+ = \mathbf{Q}\bar{\mathbf{t}}, \quad (t^{\xi_A})^+ = \bar{t}^{\xi_A}, \quad (t^{\xi_M})^+ = \bar{t}^{\xi_M}, \quad (\mathbf{t}^{\mathbf{M}})^+ = \mathbf{Q}\bar{\mathbf{t}}^{\mathbf{M}}\mathbf{Q}_0^T,$$
$$\mathbf{b}^+ = \mathbf{Q}\bar{\mathbf{b}}, \quad (b^{\xi_A})^+ = \bar{b}^{\xi_A}, \quad (b^{\xi_M})^+ = \bar{b}^{\xi_M}, \quad (\mathbf{b}^{\mathbf{M}})^+ = \mathbf{Q}\bar{\mathbf{b}}^{\mathbf{M}}\mathbf{Q}_0^T,$$
$$\mathbf{f}^+ = \mathbf{Q}\bar{\mathbf{f}}, \quad (\Pi^{\xi_A})^+ = \bar{\Pi}^{\xi_A}, \quad (\Pi^{\xi_M})^+ = \bar{\Pi}^{\xi_M}, \quad (\boldsymbol{\Pi}^{\mathbf{M}})^+ = \mathbf{Q}\bar{\boldsymbol{\Pi}}^{\mathbf{M}}\mathbf{Q}_0^T.$$

It is immediate from the above expressions that

$$\beta^+_\rho = \bar{\beta}_\rho, \qquad \beta^+_{p^{\xi_A}} = \bar{\beta}_{p^{\xi_A}}, \qquad \beta^+_{p^{\xi_M}} = \bar{\beta}_{p^{\xi_M}},$$
$$\boldsymbol{\beta}^+_{\mathbf{p}} = \mathbf{Q}\bar{\boldsymbol{\beta}}_{\mathbf{p}}, \quad \boldsymbol{\beta}^+_{\mathbf{p}^{\mathbf{M}}} = \mathbf{Q}\bar{\boldsymbol{\beta}}_{\mathbf{p}^{\mathbf{M}}}\mathbf{Q}_0^T, \quad \beta^+_\eta = \bar{\beta}_\eta,$$

whereupon the balances associated with $\rho$, $\mathbf{p}$, $p^A$, $p^M$, $\mathbf{p}^{\mathbf{M}}$, $\eta$ are automatically observer invariant. As regards observer invariance of the energy balance, it is noted if $\mathbf{Q}_0 = \mathbf{I}$ that

$$\mathbf{t}^+ \cdot \mathbf{v}^+ = \bar{\mathbf{t}} \cdot \bar{\mathbf{v}} + \bar{\mathbf{t}} \cdot \mathbf{u} + \bar{\mathbf{t}} \cdot \boldsymbol{\Omega}\bar{\mathbf{x}},$$
$$\mathbf{b}^+ \cdot \mathbf{v}^+ = \bar{\mathbf{b}} \cdot \bar{\mathbf{v}} + \bar{\mathbf{b}} \cdot \mathbf{u} + \bar{\mathbf{b}} \cdot \boldsymbol{\Omega}\bar{\mathbf{x}},$$
$$(\mathbf{t}^{\mathbf{M}})^+ \cdot (\mathbf{v}^{\mathbf{M}})^+ = \bar{\mathbf{t}}^{\mathbf{M}} \cdot \bar{\mathbf{v}}^{\mathbf{M}} + \bar{\mathbf{t}}^{\mathbf{M}} \cdot \boldsymbol{\Omega}\bar{\mathbf{M}},$$
$$(\mathbf{b}^{\mathbf{M}})^+ \cdot (\mathbf{v}^{\mathbf{M}})^+ = \bar{\mathbf{b}}^{\mathbf{M}} \cdot \bar{\mathbf{v}}^{\mathbf{M}} + \bar{\mathbf{b}}^{\mathbf{M}} \cdot \boldsymbol{\Omega}\bar{\mathbf{M}},$$

whereupon elementary manipulations give

$$\beta_e^+ = \bar{\beta}_e - \int_{\partial\mathcal{P}} \bar{\mathbf{t}} \cdot \mathbf{u} + \bar{\mathbf{t}} \cdot \boldsymbol{\Omega}\bar{\mathbf{x}} + \bar{\mathbf{t}}^{\mathbf{M}} \cdot \boldsymbol{\Omega}\bar{\mathbf{M}} dA$$
$$- \int_{\mathcal{P}} \bar{\rho}(\bar{\mathbf{b}} \cdot \mathbf{u} + \bar{\mathbf{b}} \cdot \boldsymbol{\Omega}\bar{\mathbf{x}} + \bar{\mathbf{b}}^{\mathbf{M}} \cdot \boldsymbol{\Omega}\bar{\mathbf{M}}) dV.$$

Terms involving $\Omega$ can be rewritten as

$$\bar{\mathbf{t}} \cdot \boldsymbol{\Omega}\bar{\mathbf{x}} = \boldsymbol{\Omega} \cdot (\bar{\mathbf{t}} \otimes \bar{\mathbf{x}}), \qquad \bar{\mathbf{b}} \cdot \boldsymbol{\Omega}\bar{\mathbf{x}} = \boldsymbol{\Omega} \cdot (\bar{\mathbf{b}} \otimes \bar{\mathbf{x}}),$$
$$\bar{\mathbf{t}}^{\mathbf{M}} \cdot \boldsymbol{\Omega}\bar{\mathbf{M}} = \boldsymbol{\Omega} \cdot (\bar{\mathbf{t}}^{\mathbf{M}}\bar{\mathbf{M}}^T), \qquad \bar{\mathbf{b}}^{\mathbf{M}} \cdot \boldsymbol{\Omega}\bar{\mathbf{M}} = \boldsymbol{\Omega} \cdot (\bar{\mathbf{b}}^{\mathbf{M}}\bar{\mathbf{M}}^T),$$

which in turn gives

$$\beta_e^+ - \overline{\beta}_e = -\mathbf{u} \cdot \left[ \int_{\partial\mathcal{P}} \bar{\mathbf{t}} dA + \int_{\mathcal{P}} \bar{\rho} dV \right]$$
$$- \boldsymbol{\Omega} \cdot \left[ \int_{\partial\mathcal{P}} \bar{\mathbf{t}} \otimes \bar{\mathbf{x}} + \bar{\mathbf{t}}^{\mathbf{M}}\bar{\mathbf{M}}^T dA + \int_{\mathcal{P}} \bar{\rho}(\bar{\mathbf{b}} \otimes \bar{\mathbf{x}} + \bar{\mathbf{b}}^{\mathbf{M}}\bar{\mathbf{M}}^T) dV \right]. \tag{21}$$

The arbitrariness in the choice of the observer measuring placement $\mathbf{x}^+$ implies that the expression on the right of the equality in (21) must vanish for arbitrary vectors $\mathbf{u}$ and arbitrary skew tensors $\boldsymbol{\Omega}$. This gives that the first bracketed quantity (a vector) must vanish identically and that the second bracketed quantity (a second-order tensor) must be symmetric. Further, the arbitrariness in the choice of the observer measuring placement $\bar{\mathbf{x}}$ implies that all conclusions with respect to these bracketed quantities must hold with the overbar removed. This gives

$$\int_{\partial\mathcal{P}} \mathbf{t} dA + \int_{\mathcal{P}} \rho \mathbf{b} dV = \mathbf{0},$$

$$\underbrace{\operatorname{skw}\left[ \int_{\partial\mathcal{P}} \mathbf{t} \otimes \mathbf{x} dA + \int_{\mathcal{P}} \rho \mathbf{b} \otimes \mathbf{x} dV \right]}_{\mathcal{W}}$$
$$+ \underbrace{\operatorname{skw}\left[ \int_{\partial\mathcal{P}} \mathbf{t}^{\mathbf{M}}\mathbf{M}^T dA + \int_{\mathcal{P}} \rho \mathbf{b}^{\mathbf{M}}\mathbf{M}^T dV \right]}_{\mathcal{Z}} = \mathbf{0}.$$

Now, by virtue of $\mathbf{t} = \mathbf{S}\mathbf{n}$, the divergence theorem provides

$$\int_{\mathcal{P}} \operatorname{Div} \mathbf{S} + \rho \mathbf{b} dV = \mathbf{0} \tag{22}$$

and

$$\mathcal{W} = \operatorname{skw} \int_{\mathcal{P}} \mathbf{S}\mathbf{F}^T + (\operatorname{Div} \mathbf{S} + \rho \mathbf{b}) \otimes \mathbf{x} dV.$$

Similarly, by virtue of $\mathbf{t}^{\mathbf{M}} = \mathbf{S}^{\mathbf{M}}\mathbf{n}$, the divergence theorem gives

$$\mathcal{Z} = \operatorname{skw} \int_{\mathcal{P}} \mathbf{S}^{\mathbf{M}} \overset{2}{\circ} (\mathbf{F}^{\mathbf{M}})^T + (\operatorname{Div} \mathbf{S}^{\mathbf{M}} + \rho \mathbf{b}^{\mathbf{M}})\mathbf{M}^T dV,$$

hence

$$\operatorname{skw}\int_{\mathcal{P}} \mathbf{S}\mathbf{F}^T + (\operatorname{Div}\mathbf{S} + \rho\mathbf{b}) \otimes \mathbf{x} + \mathbf{S}^{\mathbf{M}} \overset{2}{\circ} (\mathbf{F}^{\mathbf{M}})^T + (\operatorname{Div}\mathbf{S}^{\mathbf{M}} + \rho\mathbf{b}^{\mathbf{M}})\mathbf{M}^T dV = \mathbf{0}. \tag{23}$$

Note that the entropy inequality (17) is automatically observer invariant by virtue of (20). Under suitable regularity of the various fields involved in (22) and (23), the arbitrariness of $\mathcal{P}$ leads to the local statements

$$\begin{aligned} \operatorname{Div}\mathbf{S} + \rho\mathbf{b} = \mathbf{0}, \\ \operatorname{skw}[\mathbf{S}\mathbf{F}^T + (\operatorname{Div}\mathbf{S} + \rho\mathbf{b}) \otimes \mathbf{x} + \mathbf{S}^{\mathbf{M}} \overset{2}{\circ} (\mathbf{F}^{\mathbf{M}})^T + (\operatorname{Div}\mathbf{S}^{\mathbf{M}} + \rho\mathbf{b}^{\mathbf{M}})\mathbf{M}^T] = \mathbf{0}. \end{aligned} \tag{24}$$

Thus the relations (24) express the restrictions induced on the processes $\mathsf{y}$ by the request of observer invariance of the balance equations. As a companion to the compact notation $\mathbb{B}^{(\circ)}(\mathsf{y}) = \mathbf{0}$ for the balance equations in (18), the observer invariance requirements (24) will be referred to as $\mathbb{B}^{\mathbf{u}}(\mathsf{y}) = \mathbf{0}$ and $\mathbb{B}^{\mathbf{\Omega}}(\mathsf{y}) = \mathbf{0}$. It is now possible to summarize the effect of balance equations on the possible processes in the following.

**Definition 1.** A process $\mathsf{y} \in \mathsf{Y}$ is said to be *consistent with balance equations* $\mathbb{B}$ or, equivalently, $\mathsf{y} \in \mathsf{Y}_{\mathbb{B}} \subset \mathsf{Y}$, if, after substitution of $\mathsf{y}$ into (18), the resulting relations

$$\mathbb{B}^{(\circ)}(\mathsf{y}) = 0, \qquad \mathbb{B}^{\Gamma}(\mathsf{y}) \geq 0$$

hold true for every $(\circ)$ spanning the fields $\{\rho, \mathbf{p}, p^{\xi_A}, p^{\xi_M}, \mathbf{p}^{\mathbf{M}}, e, \eta\}$ and almost every $(\mathbf{X}, t) \in \mathcal{B} \times \mathcal{T}$. If, in addition,

$$\mathbb{B}^{\mathbf{u}}(\mathsf{y}) = \mathbf{0}, \qquad \mathbb{B}^{\mathbf{\Omega}}(\mathsf{y}) = \mathbf{0},$$

then $\mathsf{y}$ is said to be *objectively consistent with balance equations* $\mathbb{B}$, or $\mathsf{y} \in \mathsf{Y}_{\overline{\mathbb{B}}} \subset \mathsf{Y}_{\mathbb{B}}$.

## 9.6. Constitutive assumptions

Not all processes $\mathsf{y} \in \mathsf{Y}_{\mathbb{B}}$ are relevant for modeling the behavior of shape memory materials, since the above definition applies equally to all materials whose microstructure can be described by two scalars and an objective two-point second-order tensor. In order to characterize specific materials and environments, constitutive equations are added via relations between the configuration fields $\mathsf{x}$ and

$$\begin{aligned} \mathsf{r} &:= \{\underbrace{\psi, \eta,}_{\mathsf{c}} \ \underbrace{\mathbf{S}, \mathbf{q}, \mathbf{s}^{\xi_A}, \mathbf{s}^{\xi_M}, \mathbf{S}^{\mathbf{M}},}_{\mathsf{f}} \ \underbrace{\mathbf{f}, \Gamma, \Pi^{\xi_A}, \Pi^{\xi_M}, \mathbf{\Pi}^{\mathbf{M}}}_{\mathsf{p}}\}, \\ \mathsf{s} &= \{\mathbf{b}, r, b^{\xi_A}, b^{\xi_M}, \mathbf{b}^{\mathbf{M}}\}. \end{aligned} \tag{25}$$

Here and in what follows, the definition of $\mathsf{c}$ is updated so as to eliminate mass density $\rho$, which is completely determined by (16), and to replace internal energy in favor of the Helmholtz free energy

$$\psi := e - \vartheta\eta. \tag{26}$$

According to Noll (1973), constitutive equations for $\mathsf{r}$ and $\mathsf{s}$ are called, respectively, *internal* if they characterize the constitution of the material and *external* if they characterize the interactions of the body with the environment. Both of them are specified according to the following:

> *Constitutive principle*: Let the symbol $\alpha$ span the fields in $\{\mathsf{r}, \mathsf{s}\}$. Then, relative to a given observer, there exist suitably regular constitutive functions[7] $\mathbb{C}_\alpha$ such that
>
> $$\alpha(\mathbf{X}, t) = \mathbb{C}_\alpha[\mathcal{X}_\alpha(\mathsf{x})]; \tag{27}$$
>
> i.e., the value of the field $\alpha$ at $(\mathbf{X}, t)$ equals the value of the constitutive function $\mathbb{C}_\alpha$ at $\mathcal{X}_\alpha(\mathsf{x})$, where $\mathcal{X}_\alpha(\mathsf{x})$ denotes the result of some operation that extracts the arguments of the $\mathbb{C}_\alpha$s from the fields $\mathsf{x}$.

In the most general case, the $\mathcal{X}_\alpha$s might be identified with functions that extract from $\mathsf{x}$ the past history up to the time $t$ so that the constitutive equations would be affected by the past history of all the fields $\mathsf{x}$ all over the whole body. However, usually, this level of generality is not manageable nor actually needed and therefore the $\mathcal{X}_\alpha$s are restricted by further assumptions that reflect the specific degree of space and time nonlocality of the theory. When specialized to $\mathsf{r}$, the above principle requires the assignment of 5 constitutive functions

$$\mathbb{S} := \{\widehat{\mathbf{b}}, \widehat{r}, \widehat{b}^{\xi_A}, \widehat{b}^{\xi_M}, \widehat{\mathbf{b}}^{\mathbf{M}}\}$$

that characterize an *environment*, while relative to $\mathsf{s}$, it requires the assignment of 12 constitutive functions, also called *response functions*,

$$\mathbb{R} := \{\widehat{\psi}, \widehat{\eta}, \widehat{\mathbf{S}}, \widehat{\mathbf{q}}, \widehat{s}^{\xi_A}, \widehat{s}^{\xi_M}, \widehat{\mathbf{S}}^{\mathbf{M}}, \widehat{\mathbf{f}}, \widehat{\Gamma}, \widehat{\Pi}^{\xi_A}, \widehat{\Pi}^{\xi_M}, \widehat{\mathbf{\Pi}}^{\mathbf{M}}\} \tag{28}$$

that characterize a *material*. Two different observers will, in general, give different representations of the constitutive functions, say $\mathbb{C}^+_\alpha$ and $\overline{\mathbb{C}}_\alpha$ (Murdoch, 1982). However, such representations must obey the observer invariance principle in the sense that

$$\alpha^+ = \mathbb{C}^+_\alpha[\mathcal{X}_\alpha(\mathsf{x}^+)] \Leftrightarrow \overline{\alpha} = \overline{\mathbb{C}}_\alpha[\mathcal{X}_\alpha(\overline{\mathsf{x}})], \tag{29}$$

where $\mathsf{x}^+$, $\alpha^+$ and $\overline{\mathsf{x}}$, $\overline{\alpha}$ are the different representations of the relevant fields as given by the two observers ($\alpha \in \{\mathsf{r}, \mathsf{s}\}$).

**Definition 2.** A process $\mathsf{y} \in \mathsf{Y}$ is said to be *consistent with the constitutive functions* $\mathbb{C}$ or, equivalently, $\mathsf{y} \in \mathsf{Y}_\mathbb{C} \subset \mathsf{Y}$ if after substitution of $\mathsf{y}$ in (27) the resulting relations

$$\alpha(\mathbf{X}, t) = \mathbb{C}_\alpha[\mathcal{X}_\alpha(\mathsf{x})]$$

hold true for every $\alpha \in \{\mathsf{r}, \mathsf{s}\}$ and almost every $(\mathbf{X}, t) \in \mathcal{B} \times \mathcal{T}$. If, in addition, $\mathsf{y}$ satisfies (29) for every pair of observers, then $\mathsf{y}$ is said to be *objectively consistent with the constitutive functions* $\mathbb{C}$, or $\mathsf{y} \in \mathsf{Y}_{\overline{\mathbb{C}}} \subset \mathsf{Y}_\mathbb{C}$.

[7]The term *function* is used throughout without attempting to pursue the distinctions implied by other terms like *function, functional, operator*. The meaning can be inferred from the nature of the domain and codomain.

If $\mathsf{X}_{\mathbb{C}}$ is the set, assumed to be nonempty, of fields $\mathsf{x}$ such that all constitutive functions $\mathbb{C}$ are simultaneously defined, then any process in $\mathsf{Y}_{\mathbb{C}}$ has the form

$$\mathsf{y} = \{\mathsf{x}, \mathbb{R}[\mathcal{X}(\mathsf{x})], \mathbb{S}[\mathcal{X}(\mathsf{x})]\} \quad \text{for some } \mathsf{x} \in \mathsf{X}_{\mathbb{C}}$$

and it is said that $\mathsf{y}$ is *induced* by the configuration fields $\mathsf{x}$ through $\mathbb{C}$. If, furthermore, the external constitutive functions $\mathbb{S}$ are such that the linear momenta and entropy balances are fulfilled for every $\mathsf{x}$ and $\mathbb{R}[\mathcal{X}(\mathsf{x})]$, then $\mathbb{S}$ are said to *sustain* the process $\mathsf{y}$.

The particular type of constitutive assignment to be considered here motivates the notation

$$\mathsf{e} := \{\mathbf{F}, \vartheta\}, \qquad \mathsf{m} := \{\xi_A, \xi_M, \mathbf{M}, \mathbf{F}^{\xi_A}, \mathbf{F}^{\xi_M}, \mathbf{F}^{\mathbf{M}}\},$$

which distinguishes between the relevant macroscopic fields, collected in $\mathsf{e}$, from the microstructural fields and their gradients, collected in $\mathsf{m}$. Moreover, the *referential temperature gradient* $g(\mathbf{X}, t)$ shall denote the derivative of the temperature field with respect to $\mathbf{X}$; i.e.,

$$g = \frac{\partial \vartheta(\mathbf{X}, t)}{\partial \mathbf{X}}.$$

A further notational tool is required to distinguish the values of the fields at $(\mathbf{X}, t)$,

$$\underline{\mathsf{e}} := \{\mathsf{e}|_{(\mathbf{X},t)}\}, \qquad \underline{\mathsf{m}} := \{\mathsf{m}|_{(\mathbf{X},t)}\}, \qquad \underline{g} := g|_{(\mathbf{X},t)},$$

from the *difference history* up to the time $t$ at $\mathbf{X}$ of the fields $\mathsf{m}$ (Coleman, 1964), which is indicated as

$$\mathsf{m}_d(s) := \{(\xi_A, \xi_M, \mathbf{M}, \mathbf{F}^{\xi_A}, \mathbf{F}^{\xi_M}, \mathbf{F}^{\mathbf{M}})|_d(\mathbf{X}, s)\},$$

where, for example,

$$\mathbf{M}_d(\mathbf{X}, s) := \mathbf{M}(\mathbf{X}, t - s) - \mathbf{M}(\mathbf{X}, t) \quad \text{for } s \geq 0.$$

Summarizing, henceforth the underscore will mean field-evaluation at $(\mathbf{X}, t)$ while juxtaposition of $_d(s)$ will mean difference history up to the time $t$ at $\mathbf{X}$.

Taking advantage of this notation, it is now easy to indicate which particular $\mathcal{X}_\alpha$s are adopted here. Sources are specified by a direct field assignment, equivalent to 15 scalar fields, and $\mathcal{X}_\alpha$s degenerate into

$$\mathcal{X}_\alpha := \{\mathbf{X}, t\} \quad \text{for } \alpha \in \mathsf{s} = \{\mathbf{b}, r, b^{\xi_A}, b^{\xi_M}, \mathbf{b}^{\mathbf{M}}\}.$$

The response function for the entropy production $\Gamma$ is assumed to depend on the actual value of $\mathsf{e}$, $\mathbf{g}$, $\mathsf{m}$, as well as on the past history of $\mathsf{m}$; i.e.,

$$\mathcal{X}_\Gamma := \{\underline{\mathsf{e}}, \underline{\mathbf{g}}, \underline{\mathsf{m}}, \mathsf{m}(s); \mathbf{X}\}.$$

All other response functions, equivalent to 61 scalar functions, are assumed to depend only on the present values of $\mathsf{e}, g, \mathsf{m}$; i.e.,

$$\mathcal{X}_\alpha := \{\underline{\mathsf{e}}, \underline{\mathbf{g}}, \underline{\mathsf{m}}; X\} \quad \text{for } \alpha = \{\psi, \eta, \mathbf{S}, \mathbf{q}, \mathbf{s}^{\xi_A}, \mathbf{s}^{\xi_M}, \mathbf{S}^{\mathbf{M}}, \mathbf{f}, \Pi^{\xi_A}, \Pi^{\xi_M}, \mathbf{\Pi}^{\mathbf{M}}\}. \quad (30)$$

The dependence on the past history of the microstructural descriptors is introduced in order to model the hysteresis typically observed in SMM, while the fact that this dependence here applies only to $\Gamma$ has been introduced for the sake of simplicity.

In view of the central role that the entropy production function will play in the theory, additional structure is now attributed to $\Gamma$ to reflect some basic physical assumptions. It is assumed that it is possible to resolve additively two uncoupled contributions to the entropy production. One contribution $\Gamma_{HC}$ is due to the heat conduction. The other contribution $\Gamma_{PT}$ is due to the dissipative mechanisms associated with the phase transformations and Martensite reorientation; i.e.,

$$\Gamma(\mathbf{X}, t) = \Gamma_{HC}(\mathbf{X}, t) + \Gamma_{PT}(\mathbf{X}, t).$$

Moreover, the dependence on the past history of the microstructural fields applies only to the latter contribution so that

$$\begin{cases} \Gamma_{HC}(\mathbf{X}, t) = \widehat{\Gamma}_{HC}(\underline{\mathsf{e}}, \underline{\mathbf{g}}, \underline{\mathsf{m}}; \mathbf{X}), \\ \Gamma_{PT}(\mathbf{X}, t) = \widehat{\Gamma}_{PT}(\underline{\mathsf{e}}, \underline{\mathsf{m}}, \mathsf{m}(s); \mathbf{X}). \end{cases}$$

Since $\Gamma$ enters the theory, not directly, but only through its time derivative at time $t$, the constitutive functions $\widehat{\Gamma}_{HC}$ and $\widehat{\Gamma}_{PT}$ are now further restricted by two hypotheses that characterize the present values of their time derivative. First, it is assumed that the function $\widehat{\Gamma}_{HC}$ is such that

$$\dot{\Gamma}_{HC}(\mathbf{X}, t) = \frac{1}{\underline{\vartheta}^2}\underline{\mathbf{g}} \cdot \widehat{\mathbf{K}}(\underline{\mathsf{e}}, \underline{\mathbf{g}}, \underline{\mathsf{m}}; \mathbf{X})\underline{\mathbf{g}}, \tag{31}$$

where $\widehat{\mathbf{K}}$ is a positive-definite, symmetric, second-order, tensor-valued constitutive function (the heat conduction tensor).

Second, it is assumed that the function $\widehat{\Gamma}_{PT}$ is *rate independent* in the sense that it is unaltered by time rescalings. More precisely (Owen and Williams, 1968), for any increasing continuous function $\varphi$ of the elapsed time $s$ that maps $[0, \infty]$ into itself, the function $\widehat{\Gamma}_{PT}$ satisfies

$$\widehat{\Gamma}_{PT}[\ldots, \mathsf{m}_d(\varphi(s))] = \widehat{\Gamma}_{PT}[\ldots, \mathsf{m}_d(s)]. \tag{32}$$

The hypothesis (32) stems from experimental observations that, to within effects induced by the thermomechanical coupling, the quasistatic progress of SMM phase transformation is practically insensitive to rate of mechanical load application and rate of induced temperature change (Otsuka and Shimizu, 1986).[8] This observation is often regarded as due to the fact that atomic diffusion is not required for Austenite–Martensite phase transformation or for Martensite reorientation (Fischer et al., 1994). Hypothesis (32) imparts special significance to instants of time at which $\mathsf{m}(s)$ undergoes reversal. For scalar-valued fields this notion is straightforward and can be described with the aid of a monotonicity partition of the past into a class of intervals during which the field either increases, remains constant, or decreases. When a suitable order relation and a notion of piecewise-monotonicity are available for vector- and tensor-valued fields, the set of *switching instants* $\{t_i^\alpha\}_{i \in J^\alpha}$ ($J^\alpha$ is an index set and $\alpha$ spans the fields in $\mathsf{m}$) can be defined as those instants that partition the past

[8] Time scales introduced by thermomechanical coupling (e.g., mechanical loading in a convective environment) will however confer rate dependence in this setting and so are consistent with this notion of rate independence (Ivshin and Pence, 1994b).

into subintervals where the fields preserve their monotonicity. The set of the values attained by the microstructural fields at the switching instants is denoted

$$\underline{\mathsf{m}}_i := \{\alpha|_{t_i^\alpha} \quad \text{for every } \alpha \in \mathsf{m} \quad \text{and} \quad i \in J^\alpha\}.$$

Under suitable regularity assumptions, a time derivative at $t$ of the function $\widehat{\Gamma}_{PT}$ can be defined according to Owen and Williams (1969) and it is denoted as $\dot{\Gamma}_{PT}$, after omission of the caret for the sake of simplicity. The following representation for $\dot{\Gamma}_{PT}$ can then be obtained.

**Theorem 1.** *Let* $\dot{\underline{\mathsf{m}}} := \{\underline{\upsilon}^{\xi_A}, \underline{\upsilon}^{\xi_M}, \underline{\mathbf{v}}^{\mathbf{M}}, \dot{\underline{\mathbf{F}}}^{\xi_A}, \dot{\underline{\mathbf{F}}}^{\xi_M}, \dot{\underline{\mathbf{F}}}^{\mathbf{M}}\}$ *and* $\dot{\underline{\mathsf{e}}} := \{\dot{\underline{\mathbf{F}}}, \dot{\underline{\vartheta}}\}$. *Suppose that* $\widehat{\Gamma}_{PT}$ *is rate independent in the above sense and* $\mathsf{m}(s)$ *is piecewise monotone. Then there exist two continuous functions,* $\mathbf{\Lambda}^{\mathsf{m}}$ *and* $\mathbf{\Lambda}^{\mathsf{e}}$, *of* $(\underline{\mathsf{e}}, \underline{\mathsf{m}}, \underline{\mathsf{m}}_i)$ *such that*

$$\dot{\Gamma}_{PT}[\underline{\mathsf{e}}, \underline{\mathsf{m}}, \mathsf{m}(s)] = \mathbf{\Lambda}^{\mathsf{m}}(\underline{\mathsf{e}}, \underline{\mathsf{m}}, \underline{\mathsf{m}}_i) \cdot \dot{\underline{\mathsf{m}}} + \mathbf{\Lambda}^{\mathsf{e}}(\underline{\mathsf{e}}, \underline{\mathsf{m}}, \underline{\mathsf{m}}_i) \cdot \dot{\underline{\mathsf{e}}}, \tag{33}$$

*where the notation* $\cdot$ *denotes the appropriate scalar products.*

The theorem can be proved by a combination of arguments from Owen and Williams (1969) and Brokate and Sprekels (1995) and so provides theoretical justification for the introduction of the *discrete memory parameters*, here denoted by $\underline{\mathsf{m}}_i$, which in the SMM literature have often been introduced in order to describe internal subloops (see, e.g., Ivshin and Pence, 1994b; Rajagopal and Srinivasa, 1999). More explicitly and for later use, equation (33) can also be written as

$$\begin{aligned}\dot{\Gamma}_{PT} = {} & \Lambda^{\xi_A}\dot{\xi}_A + \Lambda^{\xi_M}\dot{\xi}_M + \mathbf{\Lambda}^{\mathbf{M}} \cdot \mathbf{v}^{\mathbf{M}} + \mathbf{\Lambda}^{\mathbf{F}^{\xi_A}} \cdot \dot{\mathbf{F}}^{\xi_A} \\ & + \mathbf{\Lambda}^{F^{\xi_M}} \cdot \dot{\mathbf{F}}^{\xi_M} + \mathbf{\Lambda}^{\mathbf{F}^{\mathbf{M}}} \cdot \dot{\mathbf{F}}^{\mathbf{M}} + \mathbf{\Lambda}^{\mathbf{F}} \cdot \dot{\mathbf{F}} + \Lambda^{\vartheta}\dot{\vartheta},\end{aligned} \tag{34}$$

where it is remarked that all the rates are evaluated at $(\mathbf{X}, t)$.

## 9.7. Internal constraints

Various internal constraints have already been introduced in order to delimit the physically admissible values of the microstructural fields and of the apparent transplacement gradient. For convenience, they are recalled below

$$\begin{aligned} & w_i^{\mathrm{eq}}(\boldsymbol{\xi}) = 0, \quad g_j^{\mathrm{eq}}(\mathbf{M}) = 0, \quad C_k^{\mathrm{eq}}(\mathbf{F}) = 0, \quad (i, j, k) = 1, \ldots, (M_{\mathrm{eq}}, N_{\mathrm{eq}}, P_{\mathrm{eq}}), \\ & w_i^{\mathrm{in}}(\boldsymbol{\xi}) \le 0, \quad g_j^{\mathrm{in}}(\mathbf{M}) \le 0, \quad C_k^{\mathrm{in}}(\mathbf{F}) \le 0, \quad (i, j, k) = 1, \ldots, (M_{\mathrm{in}}, N_{\mathrm{in}}, P_{\mathrm{in}}), \end{aligned} \tag{35}$$

which, compactly, can also be written as

$$\mathbb{D}^{\mathrm{eq}}(\mathsf{x}) = 0, \qquad \mathbb{D}^{\mathrm{in}}(\mathsf{x}) \le 0. \tag{36}$$

The observer invariance principle applied to (35) requires that all the constraint functions satisfy, for all proper orthogonal tensors $\mathbf{Q}(t)$ and $\mathbf{Q}_0$,

$$
\begin{aligned}
&g_j^{\text{eq}}(\mathbf{M}) = g_j^{\text{eq}}[\mathbf{Q}(t)\mathbf{M}\mathbf{Q}_0^T], \qquad C_k^{\text{eq}}(\mathbf{F}) = C_k^{\text{eq}}[\mathbf{Q}(t)\mathbf{F}\mathbf{Q}_0^T],\\
&\quad (j,k) = 1,\ldots,(N_{\text{eq}}, P_{\text{eq}}),\\
&g_j^{\text{in}}(\mathbf{M}) = g_j^{\text{in}}[\mathbf{Q}(t)\mathbf{M}\mathbf{Q}_0^T], \qquad C_k^{\text{in}}(\mathbf{F}) = C_k^{\text{in}}[\mathbf{Q}(t)\mathbf{F}\mathbf{Q}_0^T],\\
&\quad (j,k) = 1,\ldots,(N_{\text{in}}, P_{\text{in}}).
\end{aligned}
\tag{37}
$$

It is noted that examples (6), (11) satisfy (37).

The effect of the constraints on the possible processes is summarized in the following.

**Definition 3.** A process $\mathsf{y} \in \mathsf{Y}$ is said to be *consistent with the constraints* $\mathbb{D}$ or, equivalently, $\mathsf{y} \in \mathsf{Y}_{\mathbb{D}} \subset \mathsf{Y}$ if after substitution of $\mathsf{y}$ into (36) the resulting relations hold true for almost every $(\mathbf{X}, t) \in \mathcal{B} \times \mathcal{T}$. If, in addition, the constraint functions satisfy (37), $\mathsf{y}$ is said to be *objectively consistent with the constraints*, or $\mathsf{y} \in \mathsf{Y}_{\overline{\mathbb{D}}} \subset \mathsf{Y}_{\mathbb{D}}$.

For the next developments it is necessary to obtain explicit expressions of the effect of the constraints on the present values of the rates of the constrained quantities. This is provided by the following.

**Lemma 1.** *Let* $\mathsf{y}$ *be a process consistent with the constraints* (35) *for suitably regular constraint functions. Then for almost every* $(\mathbf{X}, t) \in \mathcal{B} \times \mathcal{T}$,

$$
\begin{aligned}
\mathbf{S}_0 \cdot \dot{\mathbf{F}} &= 0,\\
\mathbf{\Pi}_0^{\mathbf{M}} \cdot \mathbf{v}^{\mathbf{M}} + \mathbf{S}_0^{\mathbf{M}} \cdot \dot{\mathbf{F}}^{M} &= 0,\\
\Pi_0^{\xi_A} v^{\xi_A} + \Pi_0^{\xi_M} v^{\xi_M} + \mathbf{s}_0^{\xi_A} \cdot \dot{\mathbf{F}}^{\xi_A} + \mathbf{s}_0^{\xi_M} \cdot \dot{\mathbf{F}}^{\xi_M} &= 0,
\end{aligned}
\tag{38}
$$

*for any*

$$
\begin{aligned}
\Pi_0^{\xi_A} &:= \Pi_{\text{eq}}^{\xi_A} + \Pi_{\text{in}}^{\xi_A}, & \mathbf{s}_0^{\xi_A} &:= \mathbf{s}_{\text{eq}}^{\xi_A} + \mathbf{s}_{\text{in}}^{\xi_A},\\
\Pi_0^{\xi_M} &:= \Pi_{\text{eq}}^{\xi_M} + \Pi_{\text{in}}^{\xi_M}, & \mathbf{s}_0^{\xi_M} &:= \mathbf{s}_{\text{eq}}^{\xi_M} + \mathbf{s}_{\text{in}}^{\xi_M},\\
\mathbf{\Pi}_0^{\mathbf{M}} &:= \mathbf{\Pi}_{\text{eq}}^{\mathbf{M}} + \mathbf{\Pi}_{\text{in}}^{\mathbf{M}}, & \mathbf{S}_0^{\mathbf{M}} &:= \mathbf{S}_{\text{eq}}^{\mathbf{M}} + \mathbf{S}_{\text{in}}^{\mathbf{M}},\\
& & \mathbf{S}_0 &:= \mathbf{S}_{\text{eq}} + \mathbf{S}_{\text{in}},
\end{aligned}
\tag{39}
$$

*such that*

$$
\begin{aligned}
\Pi_{\text{eq}}^{\xi_A} &:= \sum_{i=1}^{M_{\text{eq}}} \left[ \bar{a}_i^{\text{eq}} \frac{\partial w_i^{\text{eq}}}{\partial \xi_A} + \frac{\partial^2 w_i^{\text{eq}}}{\partial \xi_A^2} \mathbf{F}^{\xi_A} \cdot \bar{\mathbf{c}}_i^{\text{eq}} \right],\\
\Pi_{\text{eq}}^{\xi_M} &:= \sum_{i=1}^{M_{\text{eq}}} \left[ \bar{a}_i^{\text{eq}} \frac{\partial w_i^{\text{eq}}}{\partial \xi_M} + \frac{\partial^2 w_i^{\text{eq}}}{\partial \xi_M^2} \mathbf{F}^{\xi_M} \cdot \bar{\mathbf{c}}_i^{\text{eq}} \right],\\
\mathbf{\Pi}_{\text{eq}}^{\mathbf{M}} &:= \sum_{j=1}^{N_{\text{eq}}} \left[ a_j^{\text{eq}} \frac{\partial g_j^{\text{eq}}}{\partial \mathbf{M}} + \frac{\partial^2 g_j^{\text{eq}}}{\partial \mathbf{M}^2} \overset{2}{\circ} (\mathbf{F}^{\mathbf{M}} \overset{1}{\circ} \mathbf{c}_j^{\text{eq}})^T \right],
\end{aligned}
\tag{40}
$$

*and*

$$\mathbf{s}^{\xi_A}_{\text{eq}} := \sum_{i=1}^{M_{\text{eq}}} \left[ \frac{\partial w_i^{\text{eq}}}{\partial \xi_A} \bar{\mathbf{c}}_i^{\text{eq}} \right], \qquad \mathbf{s}^{\xi_M}_{\text{eq}} := \sum_{i=1}^{M_{\text{eq}}} \left[ \frac{\partial w_i^{\text{eq}}}{\partial \xi_M} \bar{\mathbf{c}}_i^{\text{eq}} \right],$$
$$\mathbf{S}^{\mathbf{M}}_{\text{eq}} := \sum_{j=1}^{N_{\text{eq}}} \left[ \frac{\partial g_j^{\text{eq}}}{\partial \mathbf{M}} \otimes \mathbf{c}_j^{\text{eq}} \right], \qquad \mathbf{S}_{\text{eq}} := \sum_{k=1}^{P_{\text{eq}}} \left[ p_k^{\text{eq}} \frac{\partial C_k^{\text{eq}}}{\partial \mathbf{F}} \right], \tag{41}$$

*with arbitrary scalar fields* $a_j^{\text{eq}}$, $\bar{a}_i^{\text{eq}}$, $p_k^{\text{eq}}$, *arbitrary vector fields* $\mathbf{c}_j^{\text{eq}}$, $\bar{\mathbf{c}}_i^{\text{eq}}$. *Moreover, the quantities* $\Pi^{\xi_A}_{\text{in}}$, $\Pi^{\xi_M}_{\text{in}}$, $\mathbf{\Pi}^{\mathbf{M}}_{\text{in}}$, $\mathbf{s}^{\xi_A}_{\text{in}}$, $\mathbf{s}^{\xi_M}_{\text{in}}$, $\mathbf{S}^{\mathbf{M}}_{\text{in}}$, $\mathbf{S}_{\text{in}}$ *have the same structure as* (40) *and* (41) *but involve arbitrary nonpositive scalar fields* $a_j^{\text{in}} \le 0$, $\bar{a}_i^{\text{in}} \le 0$, $p_k^{\text{in}} \le 0$ *and arbitrary componentwise nonpositive vectors fields* $\mathbf{c}_j^{\text{in}} \le 0$, $\bar{\mathbf{c}}_i^{\text{in}} \le 0$ *whenever the associated inequality constraint in* (37) *is satisfied with strict equality.*

*Proof.* The contribution of the equality constraints to (38)–(41) follow after differentiation of the constraints, combination of the derivatives by arbitrary fields $a_j^{\text{eq}}$, $\bar{a}_i^{\text{eq}}$, $p_k^{\text{eq}}$, $\mathbf{c}_j^{\text{eq}}$, $\bar{\mathbf{c}}_i^{\text{eq}}$, and subsequent application of the chain rule. To obtain the contribution of the inequality constraints, consider that whenever $w_i^{\text{in}}[\boldsymbol{\xi}(\mathbf{X}, t)] < 0$, $g_j^{\text{in}}[\mathbf{M}(\mathbf{X}, t)] < 0$, $C_k^{\text{in}}[\mathbf{F}(\mathbf{X}, t)] < 0$, then each constraint can either decrease, increase, or remain constant. However, whenever one or more constraints become active, i.e., satisfied as equalities, then the right time derivatives of the constraints are restricted to be nonpositive. The same applies to each single component of the spatial gradient of the active constraints. Accordingly, standard complementarity relations follow,

$$\sum_{k=1}^{P_{\text{in}}} p_k^{\text{in}} \frac{\partial C_k^{\text{in}}}{\partial t} = 0,$$
$$\sum_{j=1}^{N_{\text{in}}} a_j^{\text{in}} \frac{\partial g_j^{\text{in}}}{\partial t} + \mathbf{c}_j^{\text{in}} \cdot \frac{\partial g_j^{\text{in}}}{\partial \mathbf{X}} = 0,$$
$$\sum_{i=1}^{M_{\text{in}}} \bar{a}_i \frac{\partial w_i^{\text{in}}}{\partial t} + \bar{\mathbf{c}}_i^{\text{in}} \cdot \frac{\partial w_i^{\text{in}}}{\partial \mathbf{X}} = 0,$$
$$a_j^{\text{in}} \le 0, \qquad \bar{a}_i^{\text{in}} \le 0, \qquad p_k^{\text{in}} \le 0, \qquad \mathbf{c}_j^{\text{in}} \le \mathbf{0}, \qquad \bar{\mathbf{c}}_i^{\text{in}} \le \mathbf{0},$$

where the time derivatives are understood from the right and the nonpositiveness of the vectors $\mathbf{c}_j^{\text{in}}$ and $\bar{\mathbf{c}}_i^{\text{in}}$ is componentwise. The relevant part of (38)–(41) then follows by application of the chain rule. □

This lemma provides a characterization of the effect of the constraints on the present values of the rates of the constrained quantities, i.e., $v^{\xi_A}$, $v^{\xi_M}$, $\mathbf{v}^{\mathbf{M}}$, $\dot{\mathbf{F}}^{\xi_A}$, $\dot{\mathbf{F}}^{\xi_M}$, $\dot{\mathbf{F}}^{\mathbf{M}}$, $\dot{\mathbf{F}}$. The arbitrary fields $a_j^{\text{eq}}$, $\bar{a}_i^{\text{eq}}$, $p_k^{\text{eq}}$, $\mathbf{c}_j^{\text{eq}}$, $\bar{\mathbf{c}}_i^{\text{eq}}$, $a_j^{\text{in}}$, $\bar{a}_i^{\text{in}}$, $p_k^{\text{in}}$, $\mathbf{c}_j^{\text{in}}$, $\bar{\mathbf{c}}_i^{\text{in}}$ are referred to as the *reactions* of the constraints and are assumed to exist for every choice of the configurational fields $\mathsf{x}$. A standard physical interpretation for these reactions is in terms of productions and stresses that enforce the constraints. The reaction associated with an inequality constraint vanishes when the inequality is strict. Under these circumstances the inequality constraint is said to be inactive. In contrast, equality constraints are always active.

For the case (11), the definitions (41) give

$$\mathbf{S}_{\text{eq}} = p_1^{\text{eq}} \mathbf{F}^{-T}$$

so that $p_1^{\text{eq}}$ is the hydrostatic pressure. For the case (2), the definitions (40) and (41) give reactive productions and reactive microstresses for phase transformation in the form

$$\Pi_{\text{eq}}^{\xi_A} = \Pi_{\text{eq}}^{\xi_M} = \bar{a}_1^{\text{eq}}, \qquad \mathbf{s}_{\text{eq}}^{\xi_A} = \mathbf{s}_{\text{eq}}^{\xi_M} = \bar{\mathbf{c}}_1^{\text{eq}}.$$

For the case (6), the definitions (40) and (41) give reactive productions and reactive microstresses for Martensite reorientation in the form

$$\begin{aligned} \mathbf{\Pi}_{\text{eq}}^{\mathbf{M}} &= a_1^{\text{eq}} \mathbf{M}^{-T} + [(\mathbf{M}^{-T} \otimes \mathbf{M}^{-T}) - (\mathbf{M}^{-T} \otimes \mathbf{M}^{-T})^{\hat{}\tilde{}\hat{}}] \overset{2}{\circ} (\mathbf{F}^{\mathbf{M}} \overset{1}{\circ} \mathbf{c}_1^{\text{eq}})^T, \\ \mathbf{S}_{\text{eq}}^{M} &= \mathbf{M}^{-T} \otimes \mathbf{c}_1^{\text{eq}}. \end{aligned}$$

## 9.8. Characterization of admissible processes

Combining Definitions 1, 2, and 3, it is finally possible to identify the processes actually relevant for the theory.

**Definition 4.** A process $\mathsf{y} \in \mathsf{Y}$ is said to be an *admissible thermomechanical process* with respect to balance equations $\mathbb{B}$, constitutive functions $\mathbb{C}$, and internal constraints $\mathbb{D}$ if

$$\mathsf{y} \in \mathsf{Y}_{\overline{\mathbb{B}}} \cap \mathsf{Y}_{\overline{\mathbb{C}}} \cap \mathsf{Y}_{\overline{\mathbb{D}}} := \mathsf{Y}_{\mathbb{A}} \subset \mathsf{Y}. \tag{42}$$

The aim of the next developments is to find a procedure that can ultimately lead to boundary value problems whose solutions describe admissible processes. To this end, it is assumed that, for every choice of the configuration fields $\mathsf{x}$, there exist constitutive functions $\mathbb{C}_{\mathsf{x}}$ such that the process induced by $\mathsf{x}$ through $\mathbb{C}_{\mathsf{x}}$ is consistent with balance equations (Dunn and Serrin, 1985). It follows therefore that the characterization of $\mathsf{Y}_{\mathbb{A}}$ can be reduced to the search for suitable constitutive functions.

The next lemma provides conditions on the constitutive functions that ensure the simultaneous consistency of a process with both balance and constitutive equations.

**Lemma 2.** *Let* $\mathsf{y} \in \mathsf{Y}_{\mathbb{C}} \cap \mathsf{Y}_{\mathbb{D}}$ *be a process consistent with the constitutive functions* $\mathbb{C} = \{\mathbb{R}, \mathbb{S}\}$ *and with the constraints* $\mathbb{D}$. *Then* $\mathsf{y} \in \mathsf{Y}_{\mathbb{B}}$ *if and only if the external constitutive functions* $\mathbb{S}$ *sustain* $\mathsf{y}$ *and the response functions* $\mathbb{R}$ *satisfy*

$$
\begin{aligned}
&\widehat{\eta} = -\frac{\partial \widehat{\psi}}{\partial \vartheta} - \vartheta \Lambda^{\vartheta}, && \widehat{\mathbf{S}} = \rho \frac{\partial \widehat{\psi}}{\partial \mathbf{F}} + \rho \vartheta \boldsymbol{\Lambda}^{\mathbf{F}} + \mathbf{S}_0, \\
&\widehat{\mathbf{f}} = \mathbf{0}, \qquad \frac{\partial \widehat{\psi}}{\partial \mathbf{g}} = \mathbf{0}, && \widehat{\mathbf{q}} = -\rho \widehat{\mathbf{K}} \mathbf{g}, \\
&\widehat{\Pi}^{\xi_A} = -\frac{\partial \widehat{\psi}}{\partial \xi_A} - \vartheta \Lambda^{\xi_A} + \Pi_0^{\xi_A}, && \widehat{\mathbf{s}}^{\xi_A} = \rho \frac{\partial \widehat{\psi}}{\partial \mathbf{F}^{\xi_A}} + \rho \vartheta \boldsymbol{\Lambda}^{\mathbf{F}^{\xi_A}} - \rho \mathbf{s}_0^{\xi_A}, \\
&\widehat{\Pi}^{\xi_M} = -\frac{\partial \widehat{\psi}}{\partial \xi_M} - \vartheta \Lambda^{\xi_M} + \Pi_0^{\xi_M}, && \widehat{\mathbf{s}}^{\xi_M} = \rho \frac{\partial \widehat{\psi}}{\partial \mathbf{F}^{\xi_M}} + \rho \vartheta \boldsymbol{\Lambda}^{\mathbf{F}^{\xi_M}} - \rho \mathbf{s}_0^{\xi_M}, \\
&\widehat{\boldsymbol{\Pi}}^{\mathbf{M}} = -\frac{\partial \widehat{\psi}}{\partial \mathbf{M}} - \vartheta \boldsymbol{\Lambda}^{\mathbf{M}} + \boldsymbol{\Pi}_0^{\mathbf{M}}, && \widehat{\mathbf{S}}^{\mathbf{M}} = \rho \frac{\partial \widehat{\psi}}{\partial \mathbf{F}^{\mathbf{M}}} + \rho \vartheta \Lambda^{\mathbf{F}^{\mathbf{M}}} - \rho \mathbf{S}_0^{\mathbf{M}},
\end{aligned}
\tag{43}
$$

*for some choice of the arbitrary fields in* (40), (41).

*Proof.* To prove necessity, assume that $\mathsf{y} = \{\mathsf{x}, \mathsf{r}, \mathsf{s}\}$ is consistent with the balance equations. Then the balance equations for the linear momenta and entropy are satisfied; i.e.,

$$
\mathbb{B}^{\mathbf{p}}(\mathsf{y}) = \mathbf{0}, \quad \mathbb{B}^{p^{\xi_A}}(\mathsf{y}) = \mathbb{B}^{p^{\xi_M}}(\mathsf{y}) = \mathbb{B}^{\eta}(\mathsf{y}) = 0, \quad \mathbb{B}^{\mathbf{p}^{\mathbf{M}}}(\mathsf{y}) = \mathbf{0}. \tag{44}
$$

Now since (44) are affine and decoupled in the sources, one can solve these for $\mathsf{s}$ and substitute the result into $\mathbb{B}^e(\mathsf{y}) = 0$ which, after using (26) and the assignments (28), gives

$$
\begin{aligned}
&-\rho \widehat{\mathbf{f}} \cdot \mathbf{v} - \rho \widehat{\Pi}^{\xi_A} v^{\xi_A} - \rho \widehat{\Pi}^{\xi_M} v^{\xi_M} - \rho \widehat{\boldsymbol{\Pi}}^{\mathbf{M}} \cdot \mathbf{v}^{\mathbf{M}} \\
&+ \widehat{\mathbf{S}} \cdot \dot{\mathbf{F}} + \widehat{\mathbf{s}}^{\xi_A} \cdot \dot{\mathbf{F}}^{\xi_A} + \widehat{\mathbf{s}}^{\xi_M} \cdot \dot{\mathbf{F}}^{\xi_M} + \widehat{\mathbf{S}}^{\mathbf{M}} \cdot \dot{\mathbf{F}}^{\mathbf{M}} \\
&- \frac{\widehat{\mathbf{q}}}{\vartheta} \cdot \mathbf{g} - \rho \widehat{\eta} \dot{\vartheta} - \rho \dot{\psi} - \rho \vartheta \dot{\Gamma} = 0.
\end{aligned}
\tag{45}
$$

with $\dot{\psi} = d\widehat{\psi}/dt$ and $\dot{\Gamma} = d\widehat{\Gamma}/dt$. Since $\mathsf{y} \in \mathsf{Y}_{\mathbb{D}}$, the previous lemma applies and the three equations in (38) multiplied by the respective factors $-1$, $\rho$, $\rho$ can be added to the left-hand side of (45) without altering its value. By using the chain rule for $\dot{\psi}$ and the representations of $\dot{\Gamma}$ in (31) and (34) it follows that (45) becomes

$$
\begin{aligned}
&\left(\widehat{\mathbf{S}} - \rho\frac{\partial\widehat{\psi}}{\partial\mathbf{F}} - \rho\vartheta\mathbf{\Lambda}^{\mathbf{F}} - \sum_{k=1}^{P_1} p_k\frac{\partial C_k}{\partial\mathbf{F}}\right)\cdot\dot{\mathbf{F}} \\
&\quad -\rho\left(\widehat{\eta} + \frac{\partial\widehat{\psi}}{\partial\vartheta} + \vartheta\Lambda^{\vartheta}\right)\dot{\vartheta} - \rho\widehat{\mathbf{f}}\cdot\mathbf{v} - \rho\frac{\partial\widehat{\psi}}{\partial\mathbf{g}}\cdot\dot{\mathbf{g}} - \frac{1}{\vartheta}\left(\widehat{\mathbf{q}} + \rho\widehat{\mathbf{K}}\mathbf{g}\right)\cdot\mathbf{g} \\
&\quad -\rho\left(\widehat{\Pi}^{\xi_A} + \frac{\partial\widehat{\psi}}{\partial\xi_A} + \vartheta\Lambda^{\xi_A} - \sum_{i=1}^{M_1}\frac{\partial w_i}{\partial\xi_A}\bar{a}_i + \frac{\partial^2 w_i}{\partial\xi_A^2}\mathbf{F}^{\xi_A}\cdot\bar{\mathbf{c}}_i\right)\upsilon^{\xi_A} \\
&\quad -\rho\left(\widehat{\Pi}^{\xi_M} + \frac{\partial\widehat{\psi}}{\partial\xi_M} + \vartheta\Lambda^{\xi_M} - \sum_{i=1}^{M_1}\frac{\partial w_i}{\partial\xi_M}\bar{a}_i + \frac{\partial^2 w_i}{\partial\xi_M^2}\mathbf{F}^{\xi_M}\cdot\bar{\mathbf{c}}_i\right)\upsilon^{\xi_M} \\
&\quad -\rho\left(\widehat{\mathbf{\Pi}}^{\mathbf{M}} + \frac{\partial\widehat{\psi}}{\partial\mathbf{M}} + \vartheta\mathbf{\Lambda}^{\mathbf{M}} - \sum_{j=1}^{N_1} a_j\frac{\partial g_j}{\partial\mathbf{M}} + \frac{\partial^2 g_j}{\partial\mathbf{M}^2}\overset{2}{\circ}\left(\mathbf{F}^{\mathbf{M}}\overset{1}{\circ}\mathbf{c}_j\right)^T\right)\cdot\mathbf{v}^{\mathbf{M}} \\
&\quad +\left(\widehat{\mathbf{S}}^{\xi_A} - \rho\frac{\partial\widehat{\psi}}{\partial\mathbf{F}^{\xi_A}} - \rho\vartheta\mathbf{\Lambda}^{\mathbf{F}^{\xi_A}} + \rho\sum_{i=1}^{M_1}\frac{\partial w_i}{\partial\xi_A}\bar{\mathbf{c}}_i\right)\cdot\dot{\mathbf{F}}^{\xi_A} \\
&\quad +\left(\widehat{\mathbf{S}}^{\xi_M} - \rho\frac{\partial\widehat{\psi}}{\partial\mathbf{F}^{\xi_M}} - \rho\vartheta\mathbf{\Lambda}^{\mathbf{F}^{\xi_M}} + \rho\sum_{i=1}^{M_1}\frac{\partial w_i}{\partial\xi_M}\bar{\mathbf{c}}_i\right)\cdot\dot{\mathbf{F}}^{\xi_M} \\
&\quad +\left(\widehat{\mathbf{S}}^{\mathbf{M}} - \rho\frac{\partial\widehat{\psi}}{\partial\mathbf{F}^{\mathbf{M}}} - \rho\vartheta\mathbf{\Lambda}^{\mathbf{F}^{\mathbf{M}}} + \rho\sum_{j=1}^{N_1}\frac{\partial g_j}{\partial\mathbf{M}}\otimes\mathbf{c}_j\right)\cdot\dot{\mathbf{F}}^{\mathbf{M}} = 0.
\end{aligned}
\tag{46}
$$

For the sake of readability, the effects of the equality and inequality constraints in (39) have been condensed into the same symbol. The actual meaning should be clear from (39)–(41). The left-hand side of (46) depends on $(\mathbf{X}, t)$ via substitution of the configuration fields $\mathsf{x}$ into the response functions. Since $\mathsf{y} \in \mathsf{Y}_{\mathbb{C}}$, it is induced by some configuration field $\mathsf{x}$. Since $\mathsf{y}$ is arbitrary in $\mathsf{Y}_{\mathbb{C}}$, due to the assumption following Definition 3, (46) holds for every choice of $\mathsf{x}$. Then $\mathbb{S}$ sustains $\mathsf{y}$. Taking into account the arbitrariness of the fields $\mathbf{c}_h^{\text{eq}}, \bar{\mathbf{c}}_k^{\text{eq}}, p_i^{\text{eq}}, a_h^{\text{eq}}, \bar{a}_k^{\text{eq}}$ and $\mathbf{c}_h^{\text{in}}, \bar{\mathbf{c}}_k^{\text{in}}, p_i^{\text{in}}, a_h^{\text{in}}, \bar{a}_k^{\text{in}}$, the rates $\upsilon^{\xi_A}, \upsilon^{\xi_M}, \mathbf{v}^{\mathbf{M}}, \dot{\mathbf{F}}^{\xi_A}, \dot{\mathbf{F}}^{\xi_M}, \dot{\mathbf{F}}^{\mathbf{M}}, \dot{\mathbf{F}}$ can take arbitrary values at $(\mathbf{X}, t)$ independently from $\underline{\mathsf{x}}$. Hence the parenthesized terms, which in general can depend on $\underline{\mathsf{x}}$ and on the past history of the microstructural fields but not on the present values of the above rates, vanish, hence establishing (43). Sufficiency follows by direct substitution. □

The principle benefit of this lemma is to reduce the constitutive assignment for the response functions $\mathbb{R}$ to the specification of two functions, namely $\widehat{\psi}$ and $\widehat{\Gamma}$.

Following Murdoch (1982), objectivity conditions for $\widehat{\psi}$ and $\widehat{\Gamma}$ can be enforced by the requirement that

$$
\begin{aligned}
\widehat{\psi}(\ldots, \mathbf{F}, \mathbf{M}, \mathbf{F}^{\mathbf{M}}) &= \widehat{\psi}_{\mathbf{Q}} \\
\widehat{\Gamma}[\ldots, \mathbf{F}, \mathbf{M}, \mathbf{F}^{\mathbf{M}}, \mathbf{M}(s), \mathbf{F}^{\mathbf{M}}(s)] &= \widehat{\Gamma}_{\mathbf{Q}}
\end{aligned}
\tag{47}
$$

with

$$
\begin{aligned}
&\widehat{\psi}_{\mathbf{Q}} := \widehat{\psi}(\ldots, \mathbf{QF}, \mathbf{QM}, \mathbf{Q} \overset{1}{\circ} \mathbf{F^M}) \\
&\widehat{\Gamma}_{\mathbf{Q}} := \widehat{\Gamma}[\ldots, \mathbf{QF}, \mathbf{QM}, \mathbf{Q} \overset{1}{\circ} \mathbf{F^M}, \mathbf{Q}(s)\mathbf{M}(s), \mathbf{Q}(s) \overset{1}{\circ} \mathbf{F^M}(s)]
\end{aligned}
\tag{48}
$$

for every proper orthogonal tensor $\mathbf{Q}$ and every proper orthogonal tensor history $\mathbf{Q}(s)$ such that $\mathbf{Q}(0) = \mathbf{Q}$. It follows by direct calculation on (43) using (47), (48), (37) that the remaining response functions in $\mathbb{R}$ are then also objective.

**Definition 5.** A set of constitutive functions $\mathbb{C} = \{\mathbb{R}, \mathbb{S}\}$ is said to be an *admissible constitutive assignment* if, for every $\mathsf{y} \in \mathsf{Y}_{\mathbb{C}}$: (i) $\mathbb{R}$ obey (43) and (47), (ii) $\mathbb{B}^{\Gamma}(\mathsf{y}) \geq 0$, (iii) $\mathbb{S}$ sustains $\mathsf{y}$.

We are now in position to characterize admissible processes.

**Theorem 2.** *Let $\mathbb{C}$ be an admissible constitutive assignment, then a process $\mathsf{y} \in \mathsf{Y}$ is an* admissible thermomechanical process *for $\mathcal{B}$ over $\mathcal{T}$ if and only if*

$$
\mathbb{B}^{\mathbf{p}}(\mathsf{y}) = \mathbf{0}, \quad \mathbb{B}^{p^{\xi_A}}(\mathsf{y}) = \mathbf{0}, \quad \mathbb{B}^{p^{\xi_M}}(\mathsf{y}) = \mathbf{0}, \quad \mathbb{B}^{\mathbf{p^M}}(\mathsf{y}) = \mathbf{0}, \quad \mathbb{B}^{\eta}(\mathsf{y}) = 0 \tag{49}
$$

*for almost every $(\mathbf{X}, t) \in \mathcal{B} \times \mathcal{T}$.*

*Proof.* Necessity is immediate. As regards sufficiency, there is only the need to prove consistency with the energy balance and observer invariance. The observer invariance of constitutive equations is ensured by hypothesis via Definition 5. The condition $\mathbb{B}^{e}(\mathsf{y}) = 0$ follows from (43) in conjunction with the subsequent discussion. Due to $\widehat{\mathbf{f}} = \mathbf{0}$ and $\mathbf{f} = \widehat{\mathbf{f}}$ the condition $\mathbb{B}^{\mathbf{p}}(\mathsf{y}) = \mathbf{0}$ gives $\mathbb{B}^{\mathbf{u}}(\mathsf{y}) = \mathbf{0}$. Therefore, it remains only to show that $\mathbb{B}^{\mathbf{\Omega}}(\mathsf{y}) = \mathbf{0}$. To this end observe from $\mathbb{B}^{\mathbf{u}}(\mathsf{y}) = \mathbf{0}$, $\mathbb{B}^{\mathbf{p^M}}(\mathsf{y}) = \mathbf{0}$ that $\mathbb{B}^{\mathbf{\Omega}}(\mathsf{y}) = \mathbf{0}$ is equivalent to

$$
\operatorname{skw}[\widehat{\mathbf{S}}\mathbf{F}^{T} + \widehat{\mathbf{S}}^{\mathbf{M}} \overset{2}{\circ} (\mathbf{F^M})^{T} - \rho \widehat{\mathbf{\Pi}}^{\mathbf{M}} \mathbf{M}^{T}] = \mathbf{0}.
$$

Substituting from the expressions for $\widehat{\mathbf{S}}$, $\widehat{\mathbf{S}}^{\mathbf{M}}$, and $\widehat{\mathbf{\Pi}}^{\mathbf{M}}$ gives, after rearrangement,

$$
\begin{aligned}
&\rho\,\mathrm{skw}\left[\frac{\partial\widehat{\psi}}{\partial\mathbf{F}}\mathbf{F}^T+\frac{\partial\widehat{\psi}}{\partial\mathbf{F}^{\mathbf{M}}}\overset{2}{\circ}\left(\mathbf{F}^{\mathbf{M}}\right)^T+\frac{\partial\widehat{\psi}}{\partial\mathbf{M}}\mathbf{M}^T\right]\\
&\quad+\rho\vartheta\,\mathrm{skw}\left[\boldsymbol{\Lambda}^{\mathbf{F}}\mathbf{F}^T+\boldsymbol{\Lambda}^{\mathbf{F}^{\mathbf{M}}}\overset{2}{\circ}\left(\mathbf{F}^{\mathbf{M}}\right)^T+\boldsymbol{\Lambda}^{\mathbf{M}}\mathbf{M}^T\right]\\
&\quad+\sum_{k=1}^{P_{\mathrm{eq}}}p_k^{\mathrm{eq}}\,\mathrm{skw}\left[\frac{\partial C_k^{\mathrm{eq}}}{\partial\mathbf{F}}\mathbf{F}^T\right]-\rho\sum_{j=1}^{N_{\mathrm{eq}}}a_j^{\mathrm{eq}}\,\mathrm{skw}\left[\frac{\partial g_j^{\mathrm{eq}}}{\partial\mathbf{M}}\mathbf{M}^T\right]\\
&\quad-\rho\sum_{j=1}^{N_{\mathrm{eq}}}\mathrm{skw}\left[\left(\frac{\partial g_j^{\mathrm{eq}}}{\partial\mathbf{M}}\otimes\mathbf{c}_j^{\mathrm{eq}}\right)\overset{2}{\circ}\left(\mathbf{F}^{\mathbf{M}}\right)^T+\left(\frac{\partial^2 g_j^{\mathrm{eq}}}{\partial\mathbf{M}^2}\overset{2}{\circ}\left(\mathbf{F}^{\mathbf{M}}\overset{1}{\circ}\mathbf{c}_j^{\mathrm{eq}}\right)^T\right)\mathbf{M}^T\right]\\
&\quad+\sum_{k=1}^{P_{\mathrm{in}}}p_k^{\mathrm{in}}\,\mathrm{skw}\left[\frac{\partial C_k^{\mathrm{in}}}{\partial\mathbf{F}}\mathbf{F}^{\mathbf{T}}\right]-\rho\sum_{j=1}^{N_{\mathrm{in}}}a_j^{\mathrm{in}}\,\mathrm{skw}\left[\frac{\partial g_j^{\mathrm{in}}}{\partial\mathbf{M}}\mathbf{M}^T\right]\\
&\quad-\rho\sum_{j=1}^{N_{\mathrm{in}}}\mathrm{skw}\left[\left(\frac{\partial g_j^{\mathrm{in}}}{\partial\mathbf{M}}\otimes\mathbf{c}_j^{\mathrm{in}}\right)\overset{2}{\circ}\left(\mathbf{F}^{\mathbf{M}}\right)^T+\left(\frac{\partial^2 g_j^{\mathrm{in}}}{\partial\mathbf{M}^2}\overset{2}{\circ}\left(\mathbf{F}^{\mathbf{M}}\overset{1}{\circ}\mathbf{c}_j^{\mathrm{in}}\right)^T\right)\mathbf{M}^T\right]=\mathbf{0}.
\end{aligned}
\tag{50}
$$

It can now be shown that each of the bracketed second-order tensors in (50) is individually symmetric. To show this for the first one consider the observer invariance on $\widehat{\psi}$ for a parametrized smooth family of proper orthogonal tensors $\mathbf{Q}(\lambda)$ so that

$$\widehat{\psi}[\ldots,\mathbf{Q}(\lambda)\mathbf{F},\mathbf{Q}(\lambda)\mathbf{M},\mathbf{Q}(\lambda)\overset{1}{\circ}\mathbf{F}^{\mathbf{M}}]=\widehat{\psi}(\ldots,\mathbf{F},\mathbf{M},\mathbf{F}^{\mathbf{M}}). \tag{51}$$

For an arbitrary skew tensor $\mathbf{W}$ pick $\mathbf{Q}(\lambda)$ such that

$$\mathbf{Q}(0)=\mathbf{I},\qquad \frac{\partial\mathbf{Q}(\lambda)}{\partial\lambda}|_{\lambda=0}=\mathbf{W}.$$

Differentiating (51) with respect to $\lambda$ and setting $\lambda=0$ yields

$$\left[\frac{\partial\widehat{\psi}}{\partial\mathbf{F}}\mathbf{F}^T+\frac{\partial\widehat{\psi}}{\partial\mathbf{F}^{\mathbf{M}}}\overset{2}{\circ}(\mathbf{F}^{\mathbf{M}})^T+\frac{\partial\widehat{\psi}}{\partial\mathbf{M}}\mathbf{M}^T\right]\cdot\mathbf{W}=0. \tag{52}$$

Since (52) holds for arbitrary skew $\mathbf{W}$ it follows that the bracketed tensor is symmetric. Similar arguments using the objectivity of $\widehat{\Gamma}_{PT}$ from (47), (48), and the observer invariance of the constraints (37) give the symmetry of the other bracketed tensors in (50). □

## 9.9. Statement of specific problems

After the above discussion of the general aspects of the theory, specific problems can now be stated for a given admissible constitutive assignment. Here "*to state a specific problem*" is understood in the sense "*to give a specific choice of the sources and determine the corresponding admissible thermomechanical process.*"

The input of a problem thus involves a set of source fields $\mathbb{S}$ that characterizes the specific environment under consideration. Substitution from the constitutive assignments and gradient definitions (9) into the balance equations (49) then provides the equivalent of 15 scalar PDE's in the unknown fields $\mathsf{x}$ which, given suitable boundary conditions, describe the quasistatic behavior of shape memory materials. If one or more solutions of the system are found,[9] then the previous theorem ensures that they induce an admissible process via back-substitution within the constitutive assignments. Since the energy balance has already been used to find restrictions (43), it does not contribute to the final system. If constraints are present, the system is enlarged to include (35) and the unknown fields are augmented by the associated reactions.

The equations arising from macroscopic linear momentum and entropy balance are standard in structure whereas those arising from the balances of microstructural linear momenta are peculiar for the description of the shape memory materials behavior as they govern the evolution of the microstructure. Specifically, the two scalar equations

$$\begin{aligned}
&\underbrace{\left(\rho\frac{\partial\widehat{\psi}}{\partial\xi_A} - \operatorname{Div}\rho\frac{\partial\widehat{\psi}}{\partial\mathbf{F}^{\xi_A}}\right)} + \underbrace{\left(\rho\vartheta\Lambda^{\xi_A} - \operatorname{Div}\rho\vartheta\boldsymbol{\Lambda}^{\mathbf{F}^{\xi_A}}\right)} \\
&\qquad + \underbrace{(-\rho\Pi_0^{\xi_A} + \operatorname{Div}\rho\mathbf{s}_0^{\xi_A})} = \rho\widehat{b}^{\xi_A}, \\
&\underbrace{\left(\rho\frac{\partial\widehat{\psi}}{\partial\xi_M} - \operatorname{Div}\rho\frac{\partial\widehat{\psi}}{\partial\mathbf{F}^{\xi_M}}\right)} + \underbrace{(\rho\vartheta\Lambda^{\xi_M} - \operatorname{Div}\rho\vartheta\boldsymbol{\Lambda}^{\mathbf{F}^{\xi_M}})} \\
&\qquad + \underbrace{(-\rho\Pi_0^{\xi_M} + \operatorname{Div}\rho\mathbf{s}_0^{\xi_M})} = \rho\widehat{b}^{\xi_M}
\end{aligned} \tag{53}$$

provide *Martensite transformation balances* that govern the conversion between Austenite and Martensite, whereas the tensorial equation

$$\begin{aligned}
&\underbrace{\left(\rho\frac{\partial\widehat{\psi}}{\partial\mathbf{M}} - \operatorname{Div}\rho\frac{\partial\widehat{\psi}}{\partial\mathbf{F}^{\mathbf{M}}}\right)} + \underbrace{\left(\rho\vartheta\boldsymbol{\Lambda}^{\mathbf{M}} - \operatorname{Div}\rho\vartheta\boldsymbol{\Lambda}^{\mathbf{F}^{\mathbf{M}}}\right)} \\
&\qquad + \underbrace{\left(-\rho\boldsymbol{\Pi}_0^{\mathbf{M}} + \operatorname{Div}\rho\mathbf{S}_0^{\mathbf{M}}\right)} = \rho\widehat{\mathbf{b}}^{\mathbf{M}}
\end{aligned}$$

can be regarded as a *Martensite reorientation balance* and governs the transformation among different types of Martensite. In the above balances four groups of terms are identified: one, depending on the free energy $\widehat{\psi}$, gives the *internal driving force* of the phase transformations; a second one, arising from the dissipation function $\Gamma_{PT}$ via the various $\Lambda$s, provides the *internal resistance* to the phase transformations; a third one associated with the *reactions* of the internal constraints; and, finally, a source term that describes the *external actions* directly influencing the phase transformations. Simplified versions of the Martensite transformation balances were derived within an internal variable framework in Bernardini and Pence (2002b).

[9] We do not discuss existence and uniqueness issues.

The case of (2) permits elimination of the phase fraction reactions between the two phase transformation balances (53) when the inequality constraints are inactive. This gives

$$
\begin{aligned}
&\rho\left(\frac{\partial\widehat{\psi}}{\partial\xi_A}-\frac{\partial\widehat{\psi}}{\partial\xi_M}\right)+\rho\vartheta(\Lambda^{\xi_A}-\Lambda^{\xi_M})\\
&=\operatorname{Div}\rho\left(\frac{\partial\widehat{\psi}}{\partial\mathbf{F}^{\xi_A}}-\frac{\partial\widehat{\psi}}{\partial\mathbf{F}^{\xi_M}}\right)+\operatorname{Div}\rho\vartheta(\mathbf{\Lambda}^{\mathbf{F}^{\xi_A}}-\mathbf{\Lambda}^{\mathbf{F}^{\xi_M}})+\rho(\widehat{b}^{\xi_A}-\widehat{b}^{\xi_M}).
\end{aligned}
\tag{54}
$$

As described in the introduction, other models for shape memory materials are based on *phase transformation kinetics* that are either posed directly or obtained by an alternative development. In the context of the present development as embodied in (54), kinetic equations follow as time derivatives of the transformation balances and therefore, besides the fundamentally different derivation, exhibit a richer structure due to the presence of the divergence term and the source term.

In the statement of specific problems the issue arises as to the explicit dependence of the constitutive functions and constraints upon $\mathbf{F}$, $\mathbf{M}$, $\mathbf{F}^{\mathbf{M}}$ so as to provide necessary and sufficient conditions for the observer invariance of the constraints and constitutive equations. With respect to (37) this issue is addressed by the following well-known result (see, e.g., Ogden, 1997) which follows from the polar decomposition theorem of nonsingular linear transformations.

**Lemma 3.** *Let $\mathcal{W}$ be a vector space. If a function $z : \mathrm{Lin}\,\mathcal{V}_{\mathsf{E}} \to \mathcal{W}$ satisfies $z(\mathbf{A}) = z(\mathbf{QA})$ for all proper orthogonal $\mathbf{Q}$ and nonsingular $\mathbf{A}$, then there exists a companion function $z_2$ such that*

$$
z(\mathbf{A}) = z_2(\mathbf{A}^T\mathbf{A}). \tag{55}
$$

It follows from this lemma that the constraint functions present in (37) can thus be expressed in the form

$$
\begin{aligned}
&\widehat{g}_j^{\mathrm{eq}}(\mathbf{C}^{\mathbf{M}}) = 0, \qquad \widehat{C}_k^{\mathrm{eq}}(\mathbf{C}) = 0, \quad (j,k) = 1,\ldots,(N_{\mathrm{eq}}, P_{\mathrm{eq}}),\\
&\widehat{g}_j^{\mathrm{in}}(\mathbf{C}^{\mathbf{M}}) \le 0, \qquad \widehat{C}_k^{\mathrm{in}}(\mathbf{C}) \le 0, \quad (j,k) = 1,\ldots,(N_{\mathrm{in}}, P_{\mathrm{in}}).
\end{aligned}
\tag{56}
$$

with

$$
\mathbf{C} := \mathbf{F}^T\mathbf{F}, \qquad \mathbf{C}^{\mathbf{M}} := \mathbf{M}^T\mathbf{M}. \tag{57}
$$

Under (12), (13) with $\mathbf{Q}_0 = \mathbf{I}$ the appropriate observer representation transformation rules are

$$
\mathbf{C}^+ = \mathbf{C}, \qquad (\mathbf{C}^{\mathbf{M}})^+ = \mathbf{C}^{\mathbf{M}}, \tag{58}
$$

whereupon it follows that no additional conditions are required on the functions $\widehat{g}_j^{\mathrm{eq}}$, $\widehat{C}_k^{\mathrm{eq}}$, $\widehat{g}_j^{\mathrm{in}}$, $\widehat{C}_k^{\mathrm{in}}$ to ensure observer invariance. The reactive stress $\mathbf{S}_0$ given in (39) can thus be expressed in terms of derivatives of the functions $\widehat{C}_k^{\mathrm{eq}}$, $\widehat{C}_k^{\mathrm{in}}$ with respect to $\mathbf{C}$. Similarly, the reactive stress $\mathbf{S}_0^{\mathbf{M}}$ and reactive production $\mathbf{\Pi}_0^{\mathbf{M}}$ given in (39)–(40) can be expressed in terms of derivatives of the functions $\widehat{g}_j^{\mathrm{eq}}$, $\widehat{g}_j^{\mathrm{in}}$ with respect to $\mathbf{C}^{\mathbf{M}}$.

The functions $\widehat{g}_j^{\text{eq}}$, $\widehat{C}_k^{\text{eq}}$, $\widehat{g}_j^{\text{in}}$, $\widehat{C}_k^{\text{in}}$ can always be taken to have symmetric dependence on their symmetric tensor argument of either $\mathbf{C}$ or $\mathbf{C}^{\mathbf{M}}$, which in turn simplifies the resulting derivatives so as to give

$$\mathbf{S}_{\text{eq}} = 2\sum_{i=1}^{P_{\text{eq}}} p_i^{\text{eq}} \mathbf{F} \frac{\partial \widehat{C}_i^{\text{eq}}}{\partial \mathbf{C}},$$

$$\mathbf{S}_{\text{eq}}^{\mathbf{M}} = 2\sum_{j=1}^{N_{\text{eq}}} \mathbf{M} \frac{\partial \widehat{g}_j^{\text{eq}}}{\partial \mathbf{C}^{\mathbf{M}}} \otimes \mathbf{c}_j^{\text{eq}},$$

$$\boldsymbol{\Pi}_{\text{eq}}^{\mathbf{M}} = 2\sum_{j=1}^{N_{\text{eq}}} a_j^{\text{eq}} \mathbf{M} \frac{\partial \widehat{g}_j^{\text{eq}}}{\partial \mathbf{C}^{\mathbf{M}}} + (\mathbf{F}^{\mathbf{M}} \overset{1}{\circ} \mathbf{c}_j^{\text{eq}}) \frac{\partial \widehat{g}_j^{\text{eq}}}{\partial \mathbf{C}^{\mathbf{M}}} + 2\mathbf{M}\left[\frac{\partial^2 \widehat{g}_j^{\text{eq}}}{\partial \mathbf{C}^{\mathbf{M}} \partial \mathbf{C}^{\mathbf{M}}} \overset{2}{\circ} ((\mathbf{F}^{\mathbf{M}} \overset{1}{\circ} \mathbf{c}_j^{\text{eq}})^T \mathbf{M})\right].$$

Analagous expressions hold for $\mathbf{S}_{\text{in}}$, $\mathbf{S}_{\text{in}}^{M}$, $\boldsymbol{\Pi}_{\text{in}}^{\mathbf{M}}$.

Turning to the explicit dependence of the constitutive functions upon $\mathbf{F}, \mathbf{M}, \mathbf{F}^{\mathbf{M}}$ the appropriate extension of (55) is given by the following.

**Lemma 4.** *Let $\mathcal{W}$ be a vector space. If a function $z : \text{Lin}\,\mathcal{V}_{\mathsf{E}} \times \text{Lin}\,\mathcal{V}_{\mathsf{E}} \times \text{Lin}(\mathcal{V}_{\mathsf{E}}, \text{Lin}(\mathcal{V}_{\mathsf{E}})) \to \mathcal{W}$ satisfies $z(\mathbf{A}, \mathbf{B}, \mathbf{G}) = z(\mathbf{QA}, \mathbf{QB}, \mathbf{Q} \overset{1}{\circ} \mathbf{G})$ for all proper orthogonal second-order $\mathbf{Q}$, nonsingular second-order $\mathbf{A}$, second-order $\mathbf{B}$, and third-order $\mathbf{G}$, then there exists a companion function $z_2$ such that*

$$z(\mathbf{A}, \mathbf{B}, \mathbf{G}) = z_2(\mathbf{A}^T\mathbf{A}, \mathbf{A}^{\mathbf{T}}\mathbf{B}, \mathbf{A}^{\mathbf{T}} \overset{1}{\circ} \mathbf{G}). \tag{59}$$

*Proof.* The definition

$$z_1(\mathbf{A}, \mathbf{D}, \mathbf{H}) := z(\mathbf{A}, \mathbf{A}^{-T}\mathbf{D}, \mathbf{A}^{-T} \overset{1}{\circ} \mathbf{H})$$

gives $z(\mathbf{A}, \mathbf{B}, \mathbf{G}) = z_1(\mathbf{A}, \mathbf{A}^T\mathbf{B}, \mathbf{A}^T \overset{1}{\circ} \mathbf{G})$ and

$$z_1(\mathbf{QA}, \mathbf{D}, \mathbf{H}) = z[\mathbf{QA}, \mathbf{Q}(\mathbf{A}^{-T}\mathbf{D}), \mathbf{Q} \overset{1}{\circ} (\mathbf{A}^{-T} \overset{1}{\circ} \mathbf{H})] = z_1(\mathbf{A}, \mathbf{D}, \mathbf{H})$$

for all proper orthogonal second-order $\mathbf{Q}$, whereupon the previous lemma gives the existence of $z_2$ such that $z_1(\mathbf{A}, \mathbf{D}, \mathbf{H}) = z_2(\mathbf{A}^T\mathbf{A}, \mathbf{D}, \mathbf{H})$, whence

$$z(\mathbf{A}, \mathbf{B}, \mathbf{G}) = z_1(\mathbf{A}, \mathbf{A}^T\mathbf{B}, \mathbf{A}^T \overset{1}{\circ} \mathbf{G}) = z_2(\mathbf{A}^T\mathbf{A}, \mathbf{A}^T\mathbf{B}, \mathbf{A}^T \overset{1}{\circ} \mathbf{G}). \qquad \square$$

As a consequence of (59) it follows from (47) and (48) that there exist suitably smooth functions $\overline{\psi}$, $\overline{\mathbf{K}}$, $\overline{\Gamma}_{PT}$ such that

$$\widehat{\psi}(\underline{\mathsf{e}}, \underline{\mathsf{m}}; \mathbf{X}) = \overline{\psi}(\underline{\mathsf{e}}^{\text{ob}}, \underline{\mathsf{m}}^{\text{ob}}; \mathbf{X}),$$
$$\widehat{\mathbf{K}}(\underline{\mathsf{e}}, \underline{\mathsf{g}}, \underline{\mathsf{m}}; \mathbf{X}) = \overline{\mathbf{K}}(\underline{\mathsf{e}}^{\text{ob}}, \underline{\mathsf{g}}, \underline{\mathsf{m}}^{\text{ob}}; \mathbf{X}),$$
$$\widehat{\Gamma}_{PT}(\underline{\mathsf{e}}, \underline{\mathsf{m}}, \mathsf{m}(s); \mathbf{X}) = \overline{\Gamma}_{PT}(\underline{\mathsf{e}}^{\text{ob}}, \underline{\mathsf{m}}^{\text{ob}}, \mathsf{m}^{\text{ob}}(s); \mathbf{X}),$$

where the modified set of arguments read as

$$\underline{\mathsf{e}}^{\text{ob}} := \{\mathbf{C}, \vartheta\}, \qquad \underline{\mathsf{m}}^{\text{ob}} := \left\{\xi_A, \xi_M, \mathbf{H}^{\mathbf{M}}, \mathbf{F}^{\xi_A}, \mathbf{F}^{\xi_M}, \mathbf{H}^{\mathbf{F}^{\mathbf{M}}}\right\}$$

with

$$\mathbf{H}^{\mathbf{M}} := \mathbf{F}^T\mathbf{M}, \qquad \mathbf{H}^{\mathbf{F}^{\mathbf{M}}} := \mathbf{F}^T \overset{1}{\circ} \mathbf{F}^{\mathbf{M}}. \tag{60}$$

Since (43) gives expressions involving derivatives of $\widehat{\psi}$ with respect to $\mathbf{F}$, $\mathbf{M}$, $\mathbf{F}^{\mathbf{M}}$, the following connections to $\overline{\psi}$ are noted:

$$\frac{\partial\widehat{\psi}}{\partial\mathbf{F}} = 2\mathbf{F}\frac{\partial\overline{\psi}}{\partial\mathbf{C}} + \mathbf{M}\left(\frac{\partial\overline{\psi}}{\partial\mathbf{H}^{\mathbf{M}}}\right)^T + \mathbf{F}^{\mathbf{M}} \overset{2}{\circ} \left(\frac{\partial\overline{\psi}}{\partial\mathbf{H}^{\mathbf{F}^{\mathbf{M}}}}\right)^T, \tag{61}$$

$$\frac{\partial\widehat{\psi}}{\partial\mathbf{M}} = \mathbf{F}\frac{\partial\overline{\psi}}{\partial\mathbf{H}^{\mathbf{M}}}, \tag{62}$$

$$\frac{\partial\widehat{\psi}}{\partial\mathbf{F}^{\mathbf{M}}} = \mathbf{F} \overset{1}{\circ} \frac{\partial\overline{\psi}}{\partial\mathbf{H}^{\mathbf{F}^{\mathbf{M}}}}, \tag{63}$$

where, in (61), $\overline{\psi}$ is required, without loss of generality, to have symmetric dependence on $\mathbf{C}$. Similarly, from (34) it follows that

$$\mathbf{\Lambda}^{\mathbf{F}} = 2\mathbf{F}\frac{\partial\overline{\Gamma}_{PT}}{\partial\mathbf{C}} + \mathbf{M}\left(\frac{\partial\overline{\Gamma}_{PT}}{\partial\mathbf{H}^{\mathbf{M}}}\right)^T + \mathbf{F}^{\mathbf{M}} \overset{2}{\circ} \left(\frac{\partial\overline{\Gamma}_{PT}}{\partial\mathbf{H}^{\mathbf{F}^{\mathbf{M}}}}\right)^T, \tag{64}$$

$$\mathbf{\Lambda}^{\mathbf{M}} = \mathbf{F}\frac{\partial\overline{\Gamma}_{PT}}{\partial\mathbf{H}^{\mathbf{M}}}, \tag{65}$$

$$\mathbf{\Lambda}^{\mathbf{F}^{\mathbf{M}}} = \mathbf{F} \overset{1}{\circ} \frac{\partial\overline{\Gamma}_{PT}}{\partial\mathbf{H}^{\mathbf{F}^{\mathbf{M}}}}, \tag{66}$$

where the dependence of $\overline{\Gamma}_{PT}$ on $\mathbf{C}$ is also symmetric and the derivatives are understood, if needed, in a generalized sense. The response function specification in (43) thus holds under the replacement $\{\widehat{\psi}, \widehat{\mathbf{K}}\} \to \{\overline{\psi}, \overline{\mathbf{K}}\}$, where (61)–(66) aid in the expression of $\mathbf{S}$, $\mathbf{\Pi}^{\mathbf{M}}$, and $\mathbf{S}^{\mathbf{M}}$. Under (12), (13) with $\mathbf{Q}_0 = \mathbf{I}$ the appropriate observer invariance transformation rules for (60) are

$$\left(\mathbf{H}^{\mathbf{M}}\right)^+ = \mathbf{H}^{\mathbf{M}}, \qquad \left(\mathbf{H}^{\mathbf{F}^{\mathbf{M}}}\right)^+ = \mathbf{H}^{\mathbf{F}^{\mathbf{M}}},$$

whereupon it follows with the aid of (58) that *no additional conditions are required on the functions* $\overline{\psi}$, $\overline{\Gamma}_{PT}$, $\overline{\mathbf{K}}$ *to ensure observer invariance*. This in turn confers observer invariance on the full set of response functions $\mathbb{R}$ in (28).

## 9.10. An infinitesimal strain theory

### 9.10.1. General development

On the basis of the previous developments, an infinitesimal strain model can be constructed for the case where

$$(\mathbf{F}-\mathbf{I})\cdot(\mathbf{F}-\mathbf{I}) \ll 1, \qquad (\mathbf{M}-\mathbf{I})\cdot(\mathbf{M}-\mathbf{I}) \ll 1$$

such that $(\mathbf{F}-\mathbf{I})\cdot(\mathbf{F}-\mathbf{I})$ and $(\mathbf{M}-\mathbf{I})\cdot(\mathbf{M}-\mathbf{I})$ are allowed be of similar magnitude. Internal constraints are assumed to be given by (2), (3) and (6), (7). To this end, it is useful to introduce the displacement gradient $\boldsymbol{\delta}^{\mathbf{F}} := \mathbf{F} - \mathbf{I}$ and the corresponding quantity $\boldsymbol{\delta}^{\mathbf{M}} := \mathbf{M} - \mathbf{I}$. Then the following linearizations are obtained:

$$\begin{aligned} \mathbf{C} &= \mathbf{I} + \boldsymbol{\delta}^{\mathbf{F}} + (\boldsymbol{\delta}^{\mathbf{F}})^T + o(\delta^2), & \mathbf{C}^{\mathbf{M}} &= \mathbf{I} + \boldsymbol{\delta}^{\mathbf{M}} + (\boldsymbol{\delta}^{\mathbf{M}})^T + o(\delta^2), \\ \mathbf{H}^{\mathbf{M}} &= \mathbf{I} + \boldsymbol{\delta}^{\mathbf{M}} + (\boldsymbol{\delta}^{\mathbf{F}})^T + o(\delta^2), & \mathbf{H}^{\mathbf{F}^{\mathbf{M}}} &= \mathbf{F}^{\mathbf{M}} + o(\delta). \end{aligned} \tag{67}$$

Equations (67) thus motivate the definitions

$$\boldsymbol{\varepsilon} := \frac{1}{2}[\boldsymbol{\delta}^{\mathbf{F}} + (\boldsymbol{\delta}^{\mathbf{F}})^T], \qquad \boldsymbol{\alpha} := \frac{1}{2}[\boldsymbol{\delta}^{\mathbf{M}} + (\boldsymbol{\delta}^{\mathbf{M}})^T], \qquad \boldsymbol{\mu} := \frac{1}{2}[\boldsymbol{\delta}^{\mathbf{M}} + (\boldsymbol{\delta}^{\mathbf{F}})^T],$$

where $\boldsymbol{\varepsilon}$ is the conventional infinitesimal strain tensor and $\boldsymbol{\alpha}$ can be regarded as the corresponding Martensite transformation strain. Accordingly, in this infinitesimal theory, the fields $\{\boldsymbol{\varepsilon}, \boldsymbol{\alpha}, \boldsymbol{\mu}, \mathbf{F}^{\mathbf{M}}\}$ replace $\{\mathbf{C}, \mathbf{C}^{\mathbf{M}}, \mathbf{H}^{\mathbf{M}}, \mathbf{H}^{\mathbf{F}^{\mathbf{M}}}\}$. Direct connections between these fields are given by

$$\boldsymbol{\varepsilon} + \boldsymbol{\alpha} = \boldsymbol{\mu} + \boldsymbol{\mu}^T, \qquad \mathbf{F}^{\mathbf{M}} + (\mathbf{F}^{\mathbf{M}})^{\wedge} = 2\frac{\partial \boldsymbol{\alpha}}{\partial \mathbf{X}}.$$

The constraint functions (6) and (7) expressed in terms of $\boldsymbol{\alpha}$ give, to leading order,

$$g_1^{\mathrm{eq}}(\boldsymbol{\alpha}) = \boldsymbol{\alpha} \cdot \mathbf{I}, \qquad g_1^{\mathrm{in}}(\boldsymbol{\alpha}) = 4\boldsymbol{\alpha} \cdot \boldsymbol{\alpha} - (g^*)^2,$$

where $g^*$ is necessarily small. Thus the reactive stresses and productions associated with the equality constraints are now

$$\begin{aligned} \Pi_{\mathrm{eq}}^{\xi_A} &= \Pi_0^{\xi_M} = \bar{a}_1^{\mathrm{eq}}, & \mathbf{s}_0^{\xi_A} &= \mathbf{s}_0^{\xi_M} = \bar{\mathbf{c}}_1^{\mathrm{eq}}, \\ \boldsymbol{\Pi}_{\mathrm{eq}}^{\mathbf{M}} &= a_1^{\mathrm{eq}}\mathbf{I} + [(\mathbf{I} \otimes \mathbf{I}) - (\mathbf{I} \otimes \mathbf{I})^{\hat{}\tilde{}\hat{}}] \overset{2}{\circ} (\mathbf{F}^{\mathbf{M}} \overset{1}{\circ} \mathbf{c}_1^{\mathrm{eq}})^T, & \mathbf{S}_{\mathrm{eq}}^{\mathbf{M}} &= \mathbf{I} \otimes \mathbf{c}_1^{\mathrm{eq}}. \end{aligned}$$

The constitutive assignments in the infinitesimal treatment are inherited from the finite strain treatment simply by replacing $\{\mathbf{C}, \mathbf{H}^{\mathbf{M}}, \mathbf{H}^{\mathbf{F}^{\mathbf{M}}}\}$ with $\{\mathbf{I}+2\boldsymbol{\varepsilon}, \mathbf{I}+2\boldsymbol{\mu}, \mathbf{F}^{\mathbf{M}}\}$ in the response functions $\overline{\psi}, \overline{\Gamma}_{PT}, \overline{\mathbf{K}}$ after retaining only leading-order terms in $\{\boldsymbol{\varepsilon}, \boldsymbol{\mu}\}$. This entails the existence of suitably smooth functions $\widetilde{\psi}, \widetilde{\mathbf{K}}, \widetilde{\Gamma}_{PT}$ such that

$$\begin{aligned} \psi(\mathbf{X}, t) &= \widetilde{\psi}(\underline{\mathsf{e}}^{\mathrm{inf}}, \underline{\mathsf{m}}^{\mathrm{inf}}; \mathbf{X}), \\ \mathbf{K}(\mathbf{X}, t) &= \widetilde{\mathbf{K}}(\underline{\mathsf{e}}^{\mathrm{inf}}, \underline{\mathbf{g}}, \underline{\mathsf{m}}^{\mathrm{inf}}; \mathbf{X}), \\ \Gamma_{PT}(\mathbf{X}, t) &= \widetilde{\Gamma}_{PT}(\underline{\mathsf{e}}^{\mathrm{inf}}, \underline{\mathsf{m}}^{\mathrm{inf}}, \mathsf{m}^{\mathrm{inf}}(s); \mathbf{X}), \end{aligned}$$

where the modified set of arguments read as

$$\underline{\mathsf{e}}^{\mathrm{inf}} := \{\boldsymbol{\varepsilon}, \vartheta\}, \qquad \underline{\mathsf{m}}^{\mathrm{inf}} := \{\xi_A, \xi_M, \boldsymbol{\mu}, \mathbf{F}^{\xi_A}, \mathbf{F}^{\xi_M}, \mathbf{F}^{\mathbf{M}}\}.$$

The application of (33) to $\widetilde{\Gamma}_{PT}$ then leads to functions $\widetilde{\Lambda}$ analogous with those already introduced in (34). The remaining reduced constitutive equation fields follow from the previous development after substitution of the modified response functions

$$\mathbf{S} = \rho\frac{\partial\widetilde{\psi}}{\partial\boldsymbol{\varepsilon}} + \frac{1}{2}\rho\left(\frac{\partial\widetilde{\psi}}{\partial\boldsymbol{\mu}}\right)^T + \rho\mathbf{F}^{\mathbf{M}} \overset{2}{\circ} \left(\frac{\partial\widetilde{\psi}}{\partial\mathbf{F}^{\mathbf{M}}}\right)^T + \rho\vartheta\widetilde{\boldsymbol{\Lambda}}^{\mathbf{F}},$$
$$\eta = -\frac{\partial\widetilde{\psi}}{\partial\vartheta} - \vartheta\widetilde{\Lambda}^{\vartheta}, \qquad \mathbf{q} = -\rho\widetilde{\mathbf{K}}\mathbf{g}, \tag{68}$$

$$\Pi^{\xi_A} = -\frac{\partial\widetilde{\psi}}{\partial\xi_A} - \vartheta\widetilde{\Lambda}^{\xi_A} + \Pi_0^{\xi_A}, \qquad \mathbf{s}^{\xi_A} = \rho\frac{\partial\widetilde{\psi}}{\partial\mathbf{F}^{\xi_A}} + \rho\vartheta\widetilde{\boldsymbol{\Lambda}}^{\mathbf{F}^{\xi_A}} - \rho\mathbf{s}_0^{\xi_A},$$
$$\Pi^{\xi_M} = -\frac{\partial\widetilde{\psi}}{\partial\xi_M} - \vartheta\widetilde{\Lambda}^{\xi_M} + \Pi_0^{\xi_M}, \qquad \mathbf{s}^{\xi_M} = \rho\frac{\partial\widetilde{\psi}}{\partial\mathbf{F}^{\xi_M}} + \rho\vartheta\widetilde{\boldsymbol{\Lambda}}^{\mathbf{F}^{\xi_M}} - \rho\mathbf{s}_0^{\xi_M}, \tag{69}$$
$$\boldsymbol{\Pi}^{\mathbf{M}} = -\frac{1}{2}\frac{\partial\widetilde{\psi}}{\partial\boldsymbol{\mu}} - \vartheta\widetilde{\boldsymbol{\Lambda}}^{\mathbf{M}} + \boldsymbol{\Pi}_0^{\mathbf{M}}, \qquad \mathbf{S}^{M} = \rho\frac{\partial\widetilde{\psi}}{\partial\mathbf{F}^{\mathbf{M}}} + \rho\vartheta\widetilde{\boldsymbol{\Lambda}}^{\mathbf{F}^{\mathbf{M}}} - \rho\mathbf{S}_0^{\mathbf{M}}$$

with

$$\widetilde{\boldsymbol{\Lambda}}^{\mathbf{F}} = \frac{\partial\widetilde{\Gamma}_{PT}}{\partial\boldsymbol{\varepsilon}} + \frac{1}{2}\frac{\partial\widetilde{\Gamma}_{PT}}{\partial\boldsymbol{\mu}}^T + \mathbf{F}^{\mathbf{M}} \overset{2}{\circ} \left(\frac{\partial\widetilde{\Gamma}_{PT}}{\partial\mathbf{F}^{\mathbf{M}}}\right)^T,$$
$$\widetilde{\Lambda}^{\xi_A} = \frac{\partial\widetilde{\Gamma}_{PT}}{\partial\xi_A}, \qquad \widetilde{\Lambda}^{\xi_M} = \frac{\partial\widetilde{\Gamma}_{PT}}{\partial\xi_M}, \qquad \widetilde{\boldsymbol{\Lambda}}^{\mathbf{M}} = \frac{1}{2}\frac{\partial\widetilde{\Gamma}_{PT}}{\partial\boldsymbol{\mu}},$$
$$\widetilde{\boldsymbol{\Lambda}}^{\mathbf{F}^{\xi_A}} = \frac{\partial\widetilde{\Gamma}_{PT}}{\partial\mathbf{F}^{\xi_A}}, \qquad \widetilde{\boldsymbol{\Lambda}}^{\mathbf{F}^{\xi_M}} = \frac{\partial\widetilde{\Gamma}_{PT}}{\partial\mathbf{F}^{\xi_M}}, \qquad \widetilde{\boldsymbol{\Lambda}}^{\mathbf{F}^{\mathbf{M}}} = \frac{\partial\widetilde{\Gamma}_{PT}}{\partial\mathbf{F}^{\mathbf{M}}}, \tag{70}$$
$$\widetilde{\Lambda}^{\vartheta} = \frac{\partial\widetilde{\Gamma}_{PT}}{\partial\vartheta},$$

where the dependence of both $\widetilde{\psi}$ and $\widetilde{\Gamma}_{PT}$ upon $\boldsymbol{\varepsilon}$ is taken to be given by a symmetric representation and, if needed, the derivatives have to be understood in a generalized sense.

A useful special case occurs if the dependence of $\widetilde{\psi}$ upon $\boldsymbol{\mu}$ is mediated solely through sym$[\boldsymbol{\mu}]$. Then, since sym$[\boldsymbol{\mu}] = \frac{1}{2}(\boldsymbol{\alpha} + \boldsymbol{\varepsilon})$, it follows that

$$\widetilde{\psi}(\underline{\mathsf{e}}^{\text{inf}}, \underline{\mathsf{m}}^{\text{inf}}; \mathbf{X}) = \breve{\psi}(\underline{\mathsf{e}}^{\text{inf}}, \breve{\underline{\mathsf{m}}}^{\text{inf}}; \mathbf{X})$$

for a suitably smooth function $\breve{\psi}$, where

$$\breve{\mathsf{m}}^{\text{inf}} := \{\xi_A, \xi_M, \boldsymbol{\alpha}, \mathbf{F}^{\xi_A}, \mathbf{F}^{\xi_M}, \mathbf{F}^{\mathbf{M}}\}.$$

Taking a symmetric representation for the dependence of $\widetilde{\psi}$ on $\boldsymbol{\varepsilon}$ and a symmetric representation for the dependence of $\breve{\psi}$ on both $\boldsymbol{\varepsilon}$ and $\boldsymbol{\alpha}$ gives the connections

$$\frac{\partial\widetilde{\psi}}{\partial\boldsymbol{\varepsilon}} = \frac{\partial\breve{\psi}}{\partial\boldsymbol{\varepsilon}} - \frac{\partial\breve{\psi}}{\partial\boldsymbol{\alpha}}, \qquad \frac{\partial\widetilde{\psi}}{\partial\boldsymbol{\mu}} = 2\frac{\partial\breve{\psi}}{\partial\boldsymbol{\alpha}},$$

whereupon (68), (69) gives that $\mathbf{S}$ and $\mathbf{\Pi}^{\mathbf{M}}$ take the form

$$\mathbf{S} = \rho \frac{\partial \breve{\psi}}{\partial \boldsymbol{\varepsilon}} + \rho \mathbf{F}^{\mathbf{M}} \overset{2}{\circ} \widetilde{\left( \frac{\partial \breve{\psi}}{\partial \mathbf{F}^{\mathbf{M}}} \right)} + \rho \vartheta \widetilde{\mathbf{\Lambda}}^{\mathbf{F}},$$

$$\mathbf{\Pi}^{\mathbf{M}} = -\frac{\partial \breve{\psi}}{\partial \boldsymbol{\alpha}} - \vartheta \widetilde{\mathbf{\Lambda}}^{\mathbf{M}} + \mathbf{\Pi}^{\mathbf{M}}_0 .$$

Similar conclusions apply to $\widetilde{\mathbf{\Lambda}}^{\mathbf{F}}$, $\widetilde{\mathbf{\Lambda}}^{\mathbf{M}}$ in the event that the dependence of $\widetilde{\Gamma}_{PT}$ upon $\boldsymbol{\mu}$ is solely through sym$[\boldsymbol{\mu}]$.

### 9.10.2. Example

A simple example is described in the following for the purpose of showing certain features regarding the structure of the resulting model. To this end, the following specialization of the basic response functions are considered:

$$\begin{aligned} \psi &= \breve{\psi}(\boldsymbol{\varepsilon}, \boldsymbol{\alpha}, \vartheta, \xi_A, \xi_M), \\ \Gamma_{PT} &= \widetilde{\Gamma}_{PT}[\xi_A, \xi_M, \{\xi_A, \xi_M\}_i], \\ \mathbf{K} &= \kappa \mathbf{I}, \quad \kappa > 0, \end{aligned} \tag{71}$$

where, as prescribed by Theorem 1, $\{\xi_A, \xi_M\}_i$ are the values attained by the phase fractions at the switching instants which allow for the modeling of internal pseudoelastic subloops. The above constitutive dependence for $\Gamma_{PT}$ has been introduced for the sake of simplicity and it provides for hysteresis in the transformations between Austenite and Martensite but not in the Martensite reorientation. The latter hysteresis is associated with Martensite reorientation that lags $\mathbf{S}$ and can be accommodated directly in this framework by considering $\widetilde{\Gamma}_{PT}$ with dependence also on $\boldsymbol{\alpha}$.

Taking $(\boldsymbol{\varepsilon}, \boldsymbol{\alpha}, \vartheta, \xi_A, \xi_M) = (\mathbf{0}, \boldsymbol{\alpha}, \vartheta_0, 1, 0)$ for any $\boldsymbol{\alpha}$ obeying $g_1^{\text{in}}(\boldsymbol{\alpha}) \le 0$ as reference state for Austenite and $(\boldsymbol{\varepsilon}, \boldsymbol{\alpha}, \vartheta, \xi_A, \xi_M) = (\mathbf{0}, \mathbf{0}, \vartheta_0, 0, 1)$ as reference state for Martensite, the following expression is considered for the free energy,

$$\begin{aligned} \breve{\psi}(\boldsymbol{\varepsilon}, \boldsymbol{\alpha}, \vartheta, \xi_A, \xi_M) &= \frac{1}{2}(\boldsymbol{\varepsilon} - \xi_M \boldsymbol{\alpha}) \overset{2}{\circ} \mathsf{C} \overset{2}{\circ} (\boldsymbol{\varepsilon} - \xi_M \boldsymbol{\alpha}) \\ &\quad + c \left[ \vartheta - \vartheta_0 - \vartheta \ln \frac{\vartheta}{\vartheta_0} \right] - \eta_0 (\vartheta - \vartheta_0) + \psi_0 \\ &\quad + \xi_M [-\Delta \eta_0 (\vartheta - \vartheta_0) + \psi^{\text{int}}_{MM}(\boldsymbol{\alpha})] + \psi^{\text{int}}_{AM}(\xi_A, \xi_M), \end{aligned} \tag{72}$$

where

$$\psi^{\text{int}}_{AM}(\xi_A, \xi_M) := \Omega \xi_A \xi_M, \qquad \psi^{\text{int}}_{MM}(\boldsymbol{\alpha}) := \frac{1}{2} \boldsymbol{\alpha} \overset{2}{\circ} \mathsf{B} \overset{2}{\circ} \boldsymbol{\alpha}, \tag{73}$$

respectively, provide expressions for an Austenite–Martensite and a Martensite–Martensite interaction energy. The scalar $\Omega$ and the fourth-order tensor $\mathsf{B}$ characterize the details of the underlying model for the phase arrangement and, for the sake of simplicity, are assumed to be constants. They can be either derived experimentally or computed by micromechanical analyses (Bernardini, 2001). Also, $\mathsf{C}$ is the elastic modulus fourth-order tensor, $c$ the specific heat, $\eta_0$ the entropy at an Austenite reference state, $\eta_0 + \Delta\eta_0$ is the entropy at the Martensite reference state, and $\psi_0$ is the free energy at both an Austenite and the Martensite reference state. Again for the sake of simplicity the difference in the material parameters between Austenite and Martensite have been neglected but it can be taken into account by following, e.g., Bernardini (2001). Also, piezocaloric terms associated with, say, thermal expansion are here omitted. The tensors $\mathsf{C}$ and $\mathsf{B}$ are assumed to obey the symmetries

$$\mathsf{C} = \mathsf{C}^{\wedge} = \mathsf{C}^{T} = \mathsf{C}^{T\wedge T}, \qquad \mathsf{B} = \mathsf{B}^{\wedge} = \mathsf{B}^{T} = \mathsf{B}^{T\wedge T}, \tag{74}$$

and are also taken to be invertible with respective inverse tensors $\mathsf{S}$ and $\mathsf{A}$. By observing that

$$\breve{\psi}(\mathbf{0}, \boldsymbol{\alpha}, \vartheta, 1, 0) - \breve{\psi}(\mathbf{0}, \mathbf{0}, \vartheta, 0, 1) = \Delta\eta_0(\vartheta - \vartheta_0),$$

it follows if $\Delta\eta_0 \neq 0$ that $\vartheta_0$ is a unique reference temperature in the sense that $\vartheta \neq \vartheta_0$ gives

$$\breve{\psi}(\mathbf{0}, \boldsymbol{\alpha}, \vartheta, 1, 0) \neq \breve{\psi}(\mathbf{0}, \mathbf{0}, \vartheta, 0, 1).$$

As a consequence of the above assumptions, it now follows from (68), (69) that

$$\begin{aligned} &\mathbf{S} = \rho \mathsf{C} \overset{2}{\circ} (\boldsymbol{\varepsilon} - \xi_M \boldsymbol{\alpha}), \qquad \eta = c \ln \frac{\vartheta}{\vartheta_0} + \eta_0 + \xi_M \Delta\eta_0, \\ &\mathbf{s}^{\xi_A} = -\rho \bar{\mathbf{c}}_1, \qquad \mathbf{s}^{\xi_M} = -\rho \bar{\mathbf{c}}_1, \qquad \mathbf{S}^{\mathbf{M}} = -\rho \mathbf{I} \otimes \mathbf{c}_1, \end{aligned} \tag{75}$$

and

$$\begin{aligned} \Pi^{\xi_A} &= -\Omega \xi_M + \bar{a}_1 - \vartheta \widetilde{\Lambda}^{\xi_A}, \\ \Pi^{\xi_M} &= -\Omega \xi_A + \Delta\eta_0(\vartheta - \vartheta_0) + \boldsymbol{\alpha} \overset{2}{\circ} \mathsf{C} \overset{2}{\circ} (\boldsymbol{\varepsilon} - \xi_M \boldsymbol{\alpha}) \\ &\quad + \bar{a}_1 - \frac{1}{2} \boldsymbol{\alpha} \overset{2}{\circ} \mathsf{B} \overset{2}{\circ} \boldsymbol{\alpha} - \vartheta \widetilde{\Lambda}^{\xi_M}, \\ \boldsymbol{\Pi}^{\mathbf{M}} &= \xi_M [\mathsf{C} \overset{2}{\circ} (\boldsymbol{\varepsilon} - \xi_M \boldsymbol{\alpha}) - \mathsf{B} \overset{2}{\circ} \boldsymbol{\alpha}] + a_1 \mathbf{I} \\ &\quad + [\mathbf{I} \otimes \mathbf{I} - (\mathbf{I} \otimes \mathbf{I})^{\wedge\sim\wedge}] \overset{2}{\circ} (\mathbf{F}^{\mathbf{M}} \overset{1}{\circ} \mathbf{c}_1)^T, \end{aligned}$$

where the functions $\widetilde{\Lambda}^{\xi_A}$ and $\widetilde{\Lambda}^{\xi_M}$ characterize the specific expression of $\widetilde{\Gamma}_{PT}$ and can depend on the present value of the state as well as of the values attained by the phase fractions at the switching instants (Bernardini and Pence, 2002b). For constant $\rho$ consider the case of vanishing source fields $\mathbf{b}$, $b^{\xi_A}$, $b^{\xi_M}$, $\mathbf{b}^{\mathbf{M}}$ in conjunction with inactive inequality constraints. Then the governing equations reduce to the equality constraints

$$\xi_A + \xi_M = 1, \tag{76}$$

$$\boldsymbol{\alpha} \cdot \mathbf{I} = 0, \tag{77}$$

the stress equations of equilibrium

$$\mathrm{Div}[\mathsf{C} \overset{2}{\circ} (\boldsymbol{\varepsilon} - \xi_M \boldsymbol{\alpha})] = \mathbf{0}, \tag{78}$$

the heat equation

$$\kappa \nabla^2 \vartheta + r + \vartheta(\widetilde{\Lambda}^{\xi_A}\dot{\xi}_A + \widetilde{\Lambda}^{\xi_M}\dot{\xi}_M) - \vartheta \Delta\eta_0 \dot{\xi}_M = c\dot{\vartheta}, \tag{79}$$

the reorientation balance

$$\begin{aligned} &- \mathrm{Div}(\mathbf{I} \otimes \mathbf{c}_1) + \xi_M[\mathsf{C} \overset{2}{\circ} (\boldsymbol{\varepsilon} - \xi_M \boldsymbol{\alpha}) - \mathsf{B} \overset{2}{\circ} \boldsymbol{\alpha}] \\ &\quad + a_1 \mathbf{I} + [\mathbf{I} \otimes \mathbf{I} - (\mathbf{I} \otimes \mathbf{I})^{\tilde{\ }\tilde{\ }}] \overset{2}{\circ} (\mathbf{F}^{\mathbf{M}} \overset{1}{\circ} \mathbf{c}_1)^T = \mathbf{0}, \end{aligned} \tag{80}$$

and the phase transformation balance

$$\begin{aligned} &\Omega(\xi_M - \xi_A) + \Delta\eta_0(\vartheta - \vartheta_0) \\ &\quad + \boldsymbol{\alpha} \overset{2}{\circ} \mathsf{C} \overset{2}{\circ} (\boldsymbol{\varepsilon} - \xi_M \boldsymbol{\alpha}) - \frac{1}{2} \boldsymbol{\alpha} \overset{2}{\circ} \mathsf{B} \overset{2}{\circ} \boldsymbol{\alpha} = -\vartheta(\widetilde{\Lambda}^{\xi_A} - \widetilde{\Lambda}^{\xi_M}). \end{aligned} \tag{81}$$

If the fields $(\boldsymbol{\varepsilon}, \vartheta)$ are now taken as spatially homogeneous (independent of $\mathbf{X}$) then these equations support a spatially homogenous solution for $(\xi_A, \xi_M, a_1, \boldsymbol{\alpha})$ with $\mathbf{F}^{\mathbf{M}} = \mathbf{0}$, and arbitrary spatially homogeneous $\mathbf{c}_1$. Specifically, the stress equations of equilibrium (78) are satisfied identically and the heat equation (79) gives the requisite process heat supply $r$. The time evolution of the microstructural variables $\boldsymbol{\alpha}$, $\xi_A$, $\xi_M$ and the remaining constraint variable $a_1$ are then determined by the constraints (76), (77) in conjunction with (80), (81). If, in this context, stress and temperature are regarded as prescribed, then $(\mathbf{S}(t), \vartheta(t))$ are both intensive in the usual thermodynamic sense. This provides a standard Gibbs framework for considering the response of the remaining variables $\boldsymbol{\alpha}$, $\xi_A$, $\xi_M$. Substituting from (75) into (80), (81) gives

$$\xi_M(\mathbf{S} - \rho \mathsf{B} \overset{2}{\circ} \boldsymbol{\alpha}) + \rho a_1 \mathbf{I} = \mathbf{0}, \tag{82}$$

$$\begin{aligned} &\rho\Omega(\xi_M - \xi_A) + \rho\Delta\eta_0(\vartheta - \vartheta_0) \\ &\quad + \boldsymbol{\alpha} \overset{2}{\circ} \mathbf{S} - \frac{\rho}{2} \boldsymbol{\alpha} \overset{2}{\circ} \mathsf{B} \overset{2}{\circ} \boldsymbol{\alpha} = -\rho\vartheta(\widetilde{\Lambda}^{\xi_A} - \widetilde{\Lambda}^{\xi_M}). \end{aligned} \tag{83}$$

Equations (82) and (77) can be solved for $\boldsymbol{\alpha}$ and $a_1$ giving

$$\boldsymbol{\alpha} = \frac{1}{\rho} \widehat{\mathbf{A}}(\mathbf{S}), \tag{84}$$

with

$$\widehat{\mathbf{A}}(\mathbf{S}) := \mathsf{A} \overset{2}{\circ} \mathbf{S} - \frac{\mathbf{I} \overset{2}{\circ} \mathsf{A} \overset{2}{\circ} \mathbf{S}}{\mathbf{I} \overset{2}{\circ} \mathsf{A} \overset{2}{\circ} \mathbf{I}} \mathsf{A} \overset{2}{\circ} \mathbf{I}, \tag{85}$$

where $\mathsf{A}$, as the inverse to $\mathsf{B}$, inherits symmetries from (74). Note that if $\mathsf{B}$ is isotropic, then (84) renders $\boldsymbol{\alpha}$ as a scalar multiple of the stress deviator.

With $\boldsymbol{\alpha}$ thus determined from (84), the remaining equations (76) and (83) determine the response of $\xi_A$, $\xi_M$ as for example discussed in Bernardini and Pence (2002b). This response is generally hysteretic due to the effect of the right side of (83). If, however, the right side of (83) vanishes, then $\xi_A$, $\xi_M$ can be given explicitly in terms of the current values of $\mathbf{S}$ and $\vartheta$ as

$$\xi_A = \widehat{\xi}_A(\mathbf{S}, \vartheta), \qquad \xi_M = \widehat{\xi}_M(\mathbf{S}, \vartheta),$$

with

$$\widehat{\xi}_A(\mathbf{S},\vartheta) := \frac{1}{2} + \frac{1}{2\Omega}\Delta\eta_0(\vartheta-\vartheta_0) + \frac{1}{4\Omega\rho^2}\mathbf{S} \overset{2}{\circ} \widehat{\mathbf{A}}(\mathbf{S}),$$
$$\widehat{\xi}_M(\mathbf{S},\vartheta) := \frac{1}{2} - \frac{1}{2\Omega}\Delta\eta_0(\vartheta-\vartheta_0) - \frac{1}{4\Omega\rho^2}\mathbf{S} \overset{2}{\circ} \widehat{\mathbf{A}}(\mathbf{S}).$$

More generally, the inequality constraints associated with (3) and (7) give

$$\boldsymbol{\alpha} = \begin{cases} \frac{1}{\rho}\widehat{\mathbf{A}}(\mathbf{S}) & \text{if } \widehat{\mathbf{A}}(\mathbf{S})\cdot\widehat{\mathbf{A}}(\mathbf{S}) \le \frac{1}{4}(\rho g^*)^2, \\ \frac{g^*}{2\sqrt{\widehat{\mathbf{A}}(\mathbf{S})\cdot\widehat{\mathbf{A}}(\mathbf{S})}}\widehat{\mathbf{A}}(\mathbf{S}) & \text{if } \widehat{\mathbf{A}}(\mathbf{S})\cdot\widehat{\mathbf{A}}(\mathbf{S}) \ge \frac{1}{4}(\rho g^*)^2, \end{cases}$$
$$\xi_A = \begin{cases} 0 & \text{if } \widehat{\xi}_A(\mathbf{S},\vartheta) \le 0, \\ \widehat{\xi}_A(\mathbf{S},\vartheta) & \text{if } 0 \le \widehat{\xi}_A(\mathbf{S},\vartheta) \le 1, \\ 1 & \text{if } \widehat{\xi}_A(\mathbf{S},\vartheta) \ge 1, \end{cases} \tag{86}$$
$$\xi_M = \begin{cases} 0 & \text{if } \widehat{\xi}_M(\mathbf{S},\vartheta) \le 0, \\ \widehat{\xi}_M(\mathbf{S},\vartheta) & \text{if } 0 \le \widehat{\xi}_M(\mathbf{S},\vartheta) \le 1, \\ 1 & \text{if } \widehat{\xi}_M(\mathbf{S},\vartheta) \ge 1, \end{cases}$$

thus providing stops to the phase fraction evolution and a limitation on the magnitude (but not the direction) of $\boldsymbol{\alpha}$ as otherwise given by (84), (85). The explicit determination of $\boldsymbol{\alpha}$, $\xi_A$, and $\xi_M$ in terms of the current $\mathbf{S}$ and $\vartheta$ given by (86) is consistent with standard Austenite–Martensite equilibrium phase diagrams for hysteresis-free phase mixing. Consideration of a nonzero right side to (83) then describes hysteretic phase fraction evolution about this Gibbs equilibrium.

## 9.11. Conclusions

The framework provided by multifield theory combined with a suitable constitutive prescription for the entropy production is well suited for modeling the macroscopic themomechanical behavior of shape memory materials. An appropriate multifield framework involves two scalar microstructural fields describing the phase partitioning and a second-order tensor microstructural field describing Martensite orientation. Together this generates a description for pseudoelasticity, shape memory effect, and reorientation of Martensitic variants at both low and high temperature. The present work has focused mainly on the discussion of general aspects of the theory paying special attention to the treatment of balance equations and to the exploitation of the constitutive structure. A first specific example involving an explicit prescription for the constitutive functions has been presented with reference to linearized kinematics. Further models obtained from richer expressions of the constitutive functions, as well as numerical simulations aimed so as to reproduce the experimentally observed spatial localization of the phase transformations, will be presented elsewhere.

## References


R. Abeyaratne and J. K. Knowles, Kinetic relations and the propagation of phase boundaries in solids, *Arch. Rational Mech. Anal.*, **114** (1991), 119–154.

J. G. Ball and R. D. James, Fine phase mixtures as minimizers of energy, *Arch. Rational Mech. Anal.*, **100** (1987), 13–52.

D. Bernardini, On the macroscopic free energy functions for shape memory alloys, *J. Mech. Phys. Solids*, **49** (2001), 813–837.

D. Bernardini and T. J. Pence, Shape memory materials: Modeling, in M. Schwartz, ed., *Encyclopedia of Smart Materials*, Vol. 2 John Wiley, New York, 2002a, 964–980.

D. Bernardini and T. J. Pence, Models for one-variant shape memory materials based on dissipation functions, *Internat. J. Nonlinear Mech.*, **37**-8 (2002b), 1299–1317.

K. Bhattacharya and R. V. Kohn, Elastic energy minimization and the recoverable strains of polycrystalline shape-memory materials, *Arch. Rational Mech. Anal.*, **139** (1997), 99–180.

E. N. Bondaryev and C. M. Wayman, Some stress-strain-temperature relationships for shape memory alloys, *Metall. Trans.* A, **19** (1988), 2407–2413.

J. G. Boyd and D. C. Lagoudas, A thermodynamical constitutive model for shape memory materials, Part I: The monolithic shape memory alloy, *Internat. J. Plasticity*, **12** (1996), 805–842.

L. C. Brinson, One-dimensional constitutive behavior of shape memory alloys: Thermomechanical derivation with non-constant material functions and redefined Martensite internal variable, *J. Intell. Material Systems Struct.*, **4** (1993), 229–242.

M. Brokate and J. Sprekels, *Hysteresis and Phase Transitions*, Springer-Verlag, Berlin, 1995.

G. Capriz, *Continua with Microstructure*, Springer-Verlag, Berlin, 1989.

G. Capriz and E. G. Virga, Interactions in general continua with microstructure, *Arch. Rational Mech. Anal.*, **109** (1990), 323–342.

G. Capriz, Continua with substructure, Parts I and II, *Phys. Mesomech.*, **3** (2000), 5–14 and 37–50.

B. D. Coleman, Thermodynamics of materials with memory, *Arch. Rational Mech. Anal.*, **17** (1964), 1–46.

B. D. Coleman and M. E. Gurtin, Thermodynamics with internal state variables, *J. Chemical Phys.*, **47** (1967), 597–613.

G. Del Piero and D. R. Owen, Structured deformations of continua, *Arch. Rational Mech. Anal.*, **124** (1993), 99–155.

J. E. Dunn and J. Serrin, On the thermodynamics of interstitial working, *Arch. Rational Mech. Anal.*, **88** (1985), 95–133.

J. L. Ericksen, Equilibrium of bars, *J. Elasticity*, **3–4** (1975), 191–201.

A. C. Eringen, *Microcontinuum Field Theories* I: *Foundations and Solids*, Springer-Verlag, Berlin, 1999.

F. Falk, Model free energy, mechanics and thermodynamics of shape memory alloys, *Acta Metall.*, **28** (1980), 1773–1780.

F. D. Fischer, M. Berveiller, K. Tanaka, and E. R. Oberaigner, Continuum mechanical aspects of phase transformations in solids, *Arch. Appl. Mech.*, **64** (1994), 54–85.

F. D. Fischer, Q. P. Sun, and K. Tanaka, Transformation-induced plasticity (TRIP), *Appl. Mech. Rev.*, **49** (1996), 317–364.

M. Frémond and S. Miyazaki, *Shape Memory Alloys*, CISM Courses and Lectures 351, Springer-Verlag, Berlin, 1996.

E. Fried and M. E. Gurtin, Dynamic solid-solid transitions with phase characterized by an order parameter, *Phys.* D, **72** (1994), 287–308.

K. Gall and H. Sehitoglu, The role of texture in tension-compression asymmetry in polycrystalline NiTi, *Internat. J. Plasticity*, **15** (1999), 69–92.

K. Gall, H. Sehitoglu, Y. I. Chumlyakov, I. V. Kireeva, Tension-compression asymmetry of the stress-strain response in aged single crystal and polycrystalline NiTi, *Acta Materials*, **47** (1999), 1203–1217.

X. Gao, M. Huang, and L. C. Brinson, A multivariant micromechanical model for SMAs, Part 1: Crystallographic issues for single crystal model, *Internat. J. Plasticity*, **16** (2000), 1345–1369.

A. E. Green and P. M. Nagdhi, On the thermodynamics and the nature of the second law, *Proc. Roy. Soc. London* A, **357** (1977), 253–270.

M. Huang, X. Gao, and L. C. Brinson, A multivariant micromechanical model for SMAs, Part 2: Polycrystal model, *Internat. J. Plasticity*, **16** (2000), 1371–1390.

Y. Huo and I. Müller, Non equilibrium thermodynamics of pseudoelasticity, *Cont. Mech. Thermodynam.*, **5** (1993), 163–204.

Y. Ivshin and T. J. Pence, A constitutive model for hysteretic phase transitions, *Internat. J. Engng. Sci.*, **32** (1994a), 681–704.

Y. Ivshin and T. J. Pence, A thermomechanical model for a one-variant shape memory material, *J. Intell. Material Systems Struct.*, **5** (1994b), 455–473.

V. I. Levitas, Thermomechanical theory of martensitic transformations in inelastic materials, *Internat. J. Solids Struct.*, **35** (1998), 889–940.

C. Liang and C. A. Rogers, One-dimensional thermomechanical constitutive relations for shape memory alloys, *J. Intell. Material Systems Struct.*, **1** (1990), 207–234.

J. Lubliner and F. Auricchio, Generalized plasticity and shape memory alloys, *Internat. J. Solids Struct.*, **33** (1996), 991–1003.

P. M. Mariano, Multifield theories in mechanics of solids, *Adv. Appl. Mech.*, **38** (2001), 1–93.

K. R. Melton, General applications of shape memory alloys and smart materials, in K. Otsuka and C. M. Wayman, eds., *Shape Memory Materials*, Cambridge University Press, New York, 1999.

I. Müller, On the entropy inequality, *Arch. Rational Mech. Anal.*, **26** (1967), 118–141.

I. Müller, *Thermodynamics*, Pitman, London, 1985.

I. Müller, On the size of the hysteresis in pseudoelasticity, *Cont. Mech. Thermodynam.*, **1** (1989), 125–142.

A. I. Murdoch, *Elastic Materials of Second Grade*, Research Report ES 78-132, University of Cincinnati, Cincinnati, 1978.

A. I. Murdoch, On material frame-indifference, *Proc. Roy. Soc. London* A, **380** (1982), 417–426.

A. I. Murdoch, On objectivity and material symmetry for simple elastic solids, *J. Elasticity*, **60** (2000), 233–242.

W. Noll, Lectures on the foundations of continuum mechanics and thermodynamics, *Arch. Rational Mech. Anal.*, **52** (1973), 62–92.

W. Noll and E. Virga, Fit regions and functions of bounded variation, *Arch. Rational Mech. Anal.*, **102** (1988), 1–21.

R. W. Ogden, *Non-Linear Elastic Deformations*, Dover Publications, Mineola, NY, 1997.

K. Otsuka and K. Shimizu, Pseudoelasticity and shape memory effects in alloys, *Internat. Metals Rev.*, **31** (1986), 93–114.

K. Otsuka and C. M. Wayman, *Shape Memory Materials*, Cambridge University Press, New York, 1998.

D. R. Owen and W. O. Williams, On the time derivatives of equilibrated response functions, *Arch. Rational Mech. Anal.*, **33** (1969), 288–306.

E. Patoor, A. Eberhardt, and M. Berveiller, Thermomechanical behavior of shape memory alloys, *Arch. Mech.*, **40** (1988), 755–794.

T. J. Pence, On the mechanical dissipation of solutions to the Riemann problem for impact involving a two-phase elastic material, *Arch. Rational Mech. Anal.*, **117** (1992), 1–52.

K. R. Rajagopal and A. R. Srinivasa, On the thermomechanics of shape memory wires, *Z. Angew. Math. Phys.*, **50** (1999), 459–496.

B. Raniecki, C. Lexcellent, and K. Tanaka, Thermodynamic models for pseudoelastic behavior of shape memory alloys, *Arch. Mech.*, **44** (1992), 261–288.

B. Raniecki and C. Lexcellent, RL models of pseudoelasticity and their specification for some shape memory solids, *Europ. J. Mech.* A *Solids*, **13** (1994), 21–50.

J. R. Rice, Inelastic constitutive relations for solids: An internal variable theory and its application to metal plasticity, *J. Mech. Phys. Solids*, **19** (1971), 433–455.

P. Rosakis and H. Tsai, Dynamic twinning processes in crystals, *Internat. J. Solids Stuct.*, **32** (1995), 2711–2723.

N. Siredey, E. Patoor, M. Berveiller, A. Eberhardt, Constitutive equations for polycrystalline thermoelastic shape memory alloys, Part I: Intragranular interactions and behavior of the grain, *Internat. J. Solids Stuct.*, **36** (1999), 4289–4315.

T. W. Shield, Orientation dependence of the pseudoelastic behavior of single crystals of CuAlNi in tension, *J. Mech. Phys. Solids*, **43** (1995), 869–895.

M. Šilhavý, *The Mechanics and Thermodynamics of Continuous Media*, Springer-Verlag, Berlin, 1997.

Q. P. Sun, K. C. Hwang, Micromechanics constitutive description of thermoelastic martensitic transformation, in J. Hutchinson and T. W. Wu, eds., *Advances in Applied Mechanics*, Vol. 31, Academic Press, New York, 1994, 249–298.

K. Tanaka, S. Kobayashi, and Y. Sato, Thermomechanics of transformation pseudoelasticity and shape memory effect in alloys, *Internat. J. Plasticity*, **2** (1986), 59–72.

L. Truskinovsky, Nucleation and growth in classical elastodynamics, in P. M. Duxbury and T. J. Pence, eds., *Dynamics of Crystal Surfaces and Interfaces*, Fundamental Materials Science Series, Plenum, New York, 1997, 185–197.

L. Truskinovsky and G. Zanzotto, Elastic crystals with a triple point, *J. Mech. Phys. Solids*, **50** (2002), 189–215.

X. Wu and T. J. Pence, Two variant modeling of shape memory materials: Unfolding a phase diagram triple point, *J. Intell. Material Systems Struct.*, **9** (1998), 335–354.

# 10. Balance at a Junction among Coherent Interfaces in Materials with Substructure

Gianfranco Capriz* and Paolo Maria Mariano†

**Abstract.** The junction among coherent interfaces separating phases in continua exhibiting material substructure is modeled as a virtual bar endowed by its own measures of interaction (macro- and microstresses and self-forces); balance equations at the junction are derived within a multifield theory. To involve bodies with substructure is not an idle formal exercise; strong anchoring conditions at the boundary rarely allow the existence of smooth solutions. Thus interfaces are more frequent and junctions inevitable if overall substructural order need be preserved. Besides, our development frames together old remarks (e.g., a remark by Gibbs) and new suggestions. Also we have the chance, here, to return to some nontrivial open issues of the general theory.

## 10.1. Introduction

We derive here balance equations at a junction among coherent interfaces in bodies exhibiting material substructure. The junction is a line in three dimensions along which different phases meet; it is endowed by an excess energy (line free energy) which may be evaluated by experiments (see Gibbs, 1948).

The substructure within the material is treated here as in (Capriz, 1985, 1989): We associate with each material patch **P** the position **x** of its center of mass and some information about its texture, the latter through an order parameter chosen as an element $\nu$ of an appropriate finite-dimensional differentiable manifold $\mathcal{M}$. The geometric properties of $\mathcal{M}$ determine the nature of each specific substructure. Substructural interactions provide extra power, are balanced, and are typically associated with $\nu$ and, perhaps, its covariant gradient; their rendering depends on the geometrical properties of $\mathcal{M}$: If $\mathcal{M}$ admits a physically significant connection, then it is possible to resort to appropriate distinct quantities called conventionally *microstress* and *self-force* to represent substructural interactions.

Where discontinuity surfaces are present, even across one another at a junction, the picture becomes more complex because such surfaces and lines may possess their

*Dipartimento di Matematica, Università di Pisa, via Buonarroti 2, 56127 Pisa, Italy, capriz@dm.unipi.it.

†Dipartimento di Ingegneria Strutturale e Geotecnica, Università di Roma "La Sapienza," via Eudossiana 18, 00184 Roma, Italy, paolo.mariano@uniroma1.it.

own energy. Thus a deep ambiguity between form and substance arises. On the one hand, the surfaces model thin transition layers involving a crowd of atoms; the layers are not "intrinsic" in that they can "reform" so as to adjust to given side conditions, the layers comprising other atoms. In mathematical terms the discontinuity surfaces must be interpreted as "free internal boundaries"; in particular, balance laws must be of the local type and pointwise balances of interactions at the interfaces and at the possible junction must be considered (or, better, deduced).

It is for us to conceive a consistent and adequate model of the physical circumstances, hopefully summoning only easily available tools. Experimental evidence seems to show that the membrane model is satisfactory for the interfaces; would then the classical theory of strings offer adequate hints to describe the behavior of the junction? In the standard case we need a rod to hold firm the margin of a real membrane, but here, of course, the junction is the margin of a number of "membranes" and it could move and adjust its shape to respond to particular circumstances. Thus to answer the question one need refer back to experimental evidence. Besides, the involvement of substructures begs the question of what is then a "membrane," a "string," or a "rod"; thus we must provide first an answer still to another question.

## 10.2. The standard instance

### 10.2.1. Notation

To focus on some ideas essential in the study of a junction, let us start by considering first *simple materials*; then the body $B$ is identified with its reference placement $\mathcal{B}$ which may be assumed to be a *fit region*.

Given a frame of reference $\{\mathbf{X}\}$ for $\mathcal{B}$, a sufficiently smooth bijection $\mathbf{x}(\mathbf{X})$ shows the present place of $\mathbf{P}$ at $\mathbf{X}$ in $\mathcal{B}$; $\mathbf{F}$ denotes, as usual, the placement gradient $\nabla\mathbf{x}$ (assumed to be such that $\det\mathbf{F} > \mathbf{0}$).

Although we always have in mind the present setting of $\mathcal{B}$ and seek balances there, we write equations in terms of $\mathbf{X}$ as it belongs to $\mathcal{B}$ and consider images of surfaces presently occurring, as they are bijected within $\mathcal{B}$. We use standard notation, briefly recalled below for convenience.

Within $\mathcal{B}$ let $N$ oriented intersecting surfaces $\Sigma_k$ $(k = 1, \ldots, N)$ be selected by the assignment of $N$ smooth functions $f_k(\mathbf{X})$ so that

$$\Sigma_k \equiv \{\mathbf{X} \in cl\mathcal{B},\ f_k(\mathbf{X}) = 0\}, \quad k = 1, \ldots, N; \tag{1}$$

the orientation of each $\Sigma_k$ is given through the normal vector field $\mathbf{n}_k = \nabla f_k/|\nabla f_k|$. Notice that, later, we do not exclude the influence of time $t$, but for simplicity, we do not mark that dependence here.

Given any smooth vector field $a$ on $\Sigma_k$, we denote its surface gradient by $\nabla_\Sigma a$. The *curvature tensor* $\mathsf{L}$ is the opposite of the surface gradient of the normal vector field $\mathsf{L} = -\nabla_\Sigma\mathbf{n}$, while the *overall curvature* $\mathcal{K}$ is the opposite of the trace of $\mathsf{L}$: $\mathcal{K} = \mathrm{tr}(\nabla_\Sigma\mathbf{n}) = \mathrm{Div}_\Sigma\,\mathbf{n}$.

Given a field $e$ that is continuous on $\mathcal{B}\backslash\Sigma_k$ and admits the limits ($\varepsilon > 0$)

$$e_k^{\pm} = \lim_{\varepsilon\to 0} e(\mathbf{X} \pm \varepsilon \mathbf{n}_k), \quad \mathbf{X} \in \Sigma_k, \tag{2}$$

we use the notation $[e]_k$ for the *jump* $e_k^+ - e_k^-$ of $e$ at $\Sigma_k$ and $\langle e\rangle_k$ for the average $\frac{1}{2}(e_k + e_k)$, with $e$ taking values in a linear space. Of course, given two fields, say, $e_1$ and $e_2$, then, with some significance assigned to the product,

$$[e_1e_2]_k = [e_1]_k\langle e_2\rangle_k + \langle e_1\rangle_k[e_2]_k. \tag{3}$$

If the field $\mathbf{F}$ has the properties just mentioned for $e$, then the surface $\Sigma_k$ is *coherent* if

$$[\mathbf{F}]_k(\mathbf{I} - \mathbf{n}_k \otimes \mathbf{n}_k) = 0, \quad \mathbf{X} \in \Sigma_k, \tag{4}$$

where $\mathbf{I}$ is the unit tensor while $(\mathbf{I} - \mathbf{n}_k \otimes \mathbf{n}_k)$ is the projector on $\Sigma_k$.

The condition of coherence is presumed satisfied from now on, and we find it convenient to have a notation for the surface gradient

$$\mathsf{F}_k = \nabla_\Sigma \mathbf{x} = \langle\mathbf{F}\rangle_k(\mathbf{I} - \mathbf{n}_k \otimes \mathbf{n}_k), \quad \mathbf{X} \in \Sigma_k. \tag{5}$$

Our interest here centers on the case when the surfaces $\Sigma_k$ all "fan out" of a single "junction" $\mathfrak{J}$. We assume that $\mathfrak{J}$ be a *simple connected curve* and that, during the deformative process, *it remains a simple curve*; i.e., we impose the requirement of *stability* of the junction. We assume also that $\mathfrak{J}$ be smooth so that it can be represented by a smooth function $\mathbf{r}(\mathsf{s})$ such that $|\mathbf{r}_{,\mathsf{s}}| = 1$; $\mathsf{s} \in [0, l]$: $l$ the length of $\mathfrak{J}$. If $\mathbf{r}_{,\mathsf{s}}$ and $\mathbf{r}_{\mathsf{ss}}$ are linearly independent, then in each point of $\mathfrak{J}$ a unique triple is defined: the tangent vector $\mathsf{t} = \mathbf{r}_{,\mathsf{s}}$, the principal normal $\mathfrak{n}$, and the binormal. The curvature vector $\mathfrak{h}$ of $\mathfrak{J}$ and the scalar curvature $\mathfrak{K}$ can be obtained as $\mathfrak{h} = \mathsf{t}_{,\mathsf{s}} = \mathbf{r}_{,\mathsf{ss}}$, $\mathfrak{K} = \mathfrak{h} \cdot \mathfrak{n}$. In any case, the plane normal to $\mathsf{t}$ is clearly available.

We render explicit the suggestion that the surfaces $\Sigma_k$ "fan out" of $\mathfrak{J}$ by requiring that (i) $\mathfrak{J}$ is the margin of all $\Sigma_k$ and (ii) the $k$ limits $\lim_{\mathbf{X}\to\mathbf{r}(\mathsf{s})} \mathbf{n}_k(\mathbf{X})$ exist at all $\mathbf{r}(\mathsf{s})$ and their values, say, $\mathbf{n}_k(\mathbf{r}(\mathsf{s})) \equiv \mathbf{n}_k(\mathsf{s})$, at each $\mathbf{r}(\mathsf{s})$ are all different.

We also require the existence for all fields $e$ of interest of $\lim_{\mathbf{X}\to\mathfrak{J}} e_k^{\pm}(\mathbf{X}, t)$, $\mathbf{X} \in \Sigma_k$. In particular, if $\mathbf{F}_{k,k+1}$ is the gradient of deformation in the region $\mathcal{B}_{k,k+1}$ of $\mathcal{B}$ between the surfaces $\Sigma_k$ and $\Sigma_{k+1}$, we presume the existence of the limit $\mathbf{F}_{k,k+1}^{\mathfrak{J}} = \lim_{\mathbf{X}\to\mathfrak{J}} \mathbf{F}_{k,k+1}(\mathbf{X})$, $\mathbf{X} \in \mathcal{B}_{k,k+1}$.

The coherence condition (4) implies that $(\mathbf{F}_{k,k+1}^{\mathfrak{J}} - \mathbf{F}_{k+l,k+l+1}^{\mathfrak{J}})\mathsf{t} = 0$ for any $l = 2, \ldots, N-2$ and $k + l \leq N - 1$.

### 10.2.2. Balances at the junction

In the present subsection the following measures of interaction are involved: the Piola–Kirchhoff *stress tensor* $\mathbf{T}$ in the bulk, the *surface stress* $\mathsf{T}$ on the discontinuity surfaces, the *line force* $\mathbf{h}$, and the *line moment* $\mathsf{m}$ along the junction.

The line force $\mathbf{h}$ is a vector at each point of $\mathfrak{J}$ and can be decomposed into the sum of the *line tension* $\sigma\mathsf{t}$, with $\sigma = \mathbf{h} \cdot \mathsf{t}$, and the *line shear* $\mathbf{g} = (\mathbf{I} - \mathsf{t} \otimes \mathsf{t})\mathbf{h}$.

$\mathbf{T}$ is assumed to be smooth outside the surfaces and to suffer jumps along the surfaces $\Sigma_k$. For each $\Sigma_k$, $\mathsf{T}_k$ is such that $\mathsf{T}_k\mathbf{n}_k = 0$. On each $\Sigma_k$ the corresponding $\mathsf{T}_k$ is supposed to be *continuous* up to the junction.

Cauchy's equations of balance apply in $\mathcal{B} \setminus (\bigcup_{k=1}^N \Sigma_k)$:

$$\rho\mathbf{f} + \operatorname{Div}\mathbf{T} = 0, \tag{6}$$

$$\mathbf{T}\mathbf{F}^T = \mathbf{F}\mathbf{T}^T, \tag{7}$$

where, as in standard notation, $\rho$ is the density of mass and $\mathbf{f}$ represents body force per unit mass in the reference placement; (7) is the symmetry requirement of the Cauchy stress $(\det \mathbf{F})^{-1}\mathbf{T}\mathbf{F}^T$. A standard "pillbox" reasoning leads to pointwise balances on each $\Sigma_k$,

$$[\mathbf{T}]_k\mathbf{n}_k + \operatorname{Div}_\Sigma \mathsf{T}_k = 0, \tag{8}$$

$$\mathsf{T}_k\mathsf{F}_k^T = \mathsf{F}_k\mathsf{T}_k^T \tag{9}$$

(see Angenent and Gurtin, 1989), where (9) is again a symmetry requirement but for the current surface stress $\mathsf{T}_k\mathsf{F}_k^T$.

To obtain the balance conditions along $\mathfrak{J}$ consider the tube $\mathfrak{b}_R \backslash \mathfrak{b}_r$ wrapped around $\mathfrak{J}$. The tube is the difference between two coaxial "curved" cylinders $\mathfrak{b}_R$ and $\mathfrak{b}_r$ scanned by discs $D_R$ and $D_r$ of radii $R$ and $r$, respectively, $(r < R)$ everywhere normal to $\mathfrak{J}$ and with center on that piece of $\mathfrak{J}$, where $\mathsf{s} \in [\mathsf{s}_1, \mathsf{s}_2]$. $R$ is supposed to be small enough so that the disc $D_R$ for any chosen value $\bar{\mathsf{s}}$ of $\mathsf{s}$ does not intersect any one of the discs corresponding to other values of $\mathsf{s}$ different from $\bar{\mathsf{s}}$.

For the balance of forces on $\mathfrak{b}_R$ it seems reasonable to require that

$$\int_{\mathfrak{b}_R} \rho\mathbf{f} + \int_* \mathbf{T}\mathbf{m} + \sum_{k=1}^N \int_{\partial\mathfrak{b}_R\cap\Sigma_k} \mathsf{T}_k\bar{\mathbf{n}}_k + \mathbf{h}|_{\mathbf{r}(\mathsf{s}_2)} - \mathbf{h}|_{\mathbf{r}(\mathsf{s}_1)} = 0 \tag{10}$$

if we interpret $\mathbf{h}$ as the traction on a cross-section of the junction and $\int_*$ as the limit for $r \to 0$ of the surface integral over $\partial(\mathfrak{b}_R/\mathfrak{b}_r)$ less the two circles at $\mathsf{s} = \mathsf{s}_1$ and $\mathsf{s} = \mathsf{s}_2$, where $\bar{\mathbf{n}}_k$ is the normal to the boundary of $\partial\mathfrak{b}_R \cap \Sigma_k$ within the tangent plane to $\Sigma_k$. Applying Gauss's theorem to the surface integral, we obtain

$$\int_* \mathbf{T}\mathbf{m} = \int_{\mathfrak{b}_R} \operatorname{Div}\mathbf{T} + \sum_{k=1}^N \int_{\mathfrak{b}_R\cap\Sigma_k} [\mathbf{T}]_k\mathbf{n}_k + \lim_{r\to 0} \int_{\partial\mathfrak{b}_r} \mathbf{T}\mathbf{m}. \tag{11}$$

We *assume* that

$$\int_{\partial\mathfrak{b}_r} |\mathbf{T}\mathbf{m}| \tag{12}$$

is *bounded* as $r \to 0$ so that we have $\lim_{r\to 0} \int_{\partial\mathfrak{b}_r} \mathbf{T}\mathbf{m} = \int_{\mathsf{s}_1}^{\mathsf{s}_2} \lim_{r\to 0} \int_{\partial D_r} \mathbf{T}\mathbf{m} = 0$. Taking into account also the assumption of continuity of $\mathsf{T}_k$ up to the junction, Gauss theorem allows us to change (10) into

$$\int_{\mathfrak{b}_R} (\rho\mathbf{f} + \operatorname{Div}\mathbf{T}) + \sum_{k=1}^{N} \int_{\mathfrak{b}_R \cap \Sigma_k} ([\mathbf{T}]_k \mathbf{n}_k + \operatorname{Div}_\Sigma \mathsf{T}_k) + \int_{\mathsf{s}_1}^{\mathsf{s}_2} \left( \mathbf{h}_\mathsf{s} + \sum_{k=1}^{N} \mathsf{T}_k \bar{\mathbf{n}}_k \right) = 0. \tag{13}$$

Taking into account (6) and (8) and exploiting the arbitrariness of the interval $[\mathsf{s}_1, \mathsf{s}_2]$, we get the pointwise balance

$$\mathbf{h}_\mathsf{s} + \sum_{k=1}^{N} \mathsf{T}_k \bar{\mathbf{n}}_k = 0 \quad \text{along } \mathfrak{J}. \tag{14}$$

Circumstantial evidence similar to that subsumed in stating the balance equation (10) suggests the following condition of balance of moments:

$$\begin{aligned} \int_{\mathfrak{b}_R} (\mathbf{x} - \mathbf{x}_0) \times \rho\mathbf{f} + \int_{*} (\mathbf{x} - \mathbf{x}_0) \times \mathbf{Tm} + \sum_{k=1}^{N} \int_{\partial\mathfrak{b}_R \cap \Sigma_k} (\mathbf{x} - \mathbf{x}_0) \\ \times \mathsf{T}_k \bar{\mathbf{n}}_k + (\mathbf{x}(\mathbf{r}(\mathsf{s}_2)) - \mathbf{x}_0) \times \mathbf{h}|_{\mathbf{r}(\mathsf{s}_2)} - (\mathbf{x}(\mathbf{r}(\mathsf{s}_1)) - \mathbf{x}_0) \\ \times \mathbf{h}|_{\mathbf{r}(\mathsf{s}_1)} + \mathsf{m}(\mathsf{s}_2) - \mathsf{m}(\mathsf{s}_1) = 0. \end{aligned} \tag{15}$$

By applying the Gauss theorem and by taking into account (6)–(9) and (13), we obtain

$$\mathsf{t} \times \mathbf{h} + \mathsf{m}_\mathsf{s} = 0 \quad \text{along } \mathfrak{J}. \tag{16}$$

again thanks to the arbitrariness of the interval $[\mathsf{s}_1, \mathsf{s}_2]$.

Balance equations at junctions in three-dimensional Cauchy continua have been derived in (Simha and Batthacharya, 2000). The differences between their final results and ours must be sorted out by experiments. They do not exclude a singular behavior of $\mathbf{T}$ at the junction leading to a nonnull limit for the integral $\int_{\partial D_r} \mathbf{Tm}$. Their omission of line moments along the junction may be justified by the type of dependence of the energy on the shape of the junction. Contrarily and naively one may follow this line of thought: When one endows the interfaces with surface stress, one imagines them to behave as membranes; when one stretches a membrane by fixing it at a prescribed boundary (as for some strange drum) one needs the boundary to behave like a beam, which may be the case of junctions that are margins of interfaces. Leaving the matter open, we adopted a prudent stance and admitted the possibility that the junction have the capacity of a rod.

## 10.3. Considering substructures

When the material patches exhibit a substructure, its descriptors need be introduced and relevant interactions balanced; new relations at the junction are then obtained.

### 10.3.1. Placements

Here we model the substructure through an order parameter $\nu$ chosen as element of an appropriate finite-dimensional differentiable manifold $\mathcal{M}$ whose *tangent space* at $\nu$ is indicated with $T_\nu\mathcal{M}$. Two fields are defined on $\mathcal{B}$, namely, the smooth bijection

$$\mathbf{x} : \mathcal{B} \to \mathcal{E}^3 \tag{17}$$

representing the *classical placement field* mentioned in the previous section and the *order parameter field*

$$\nu : \mathcal{B} \to \mathcal{M}. \tag{18}$$

Examples can be quoted that are apparently very different from one another but in fact share the basic framework. Nematic liquid crystals are characterized by directions and a realization of $\mathcal{M}$ for them could be the projective plane or the unit sphere in $\mathbb{R}^3$ with identification of antipodes. A single parameter $\nu$ taking values in $[0, 1]$ may describe porous solids or two-phase materials. Elements of $\mathbb{R}^3$ are chosen to declare the kinematic perturbation induced on the displacement field by diffused microcracks. Symmetric second-order tensors may figure out the macroscopic effects of interacting macromolecules or partially ordered nematics. Skew-symmetric second-order tensors occur in theories of composites with diffused stiff fibers. Different phases of a body could be characterized by values of $\nu$ belonging to specific disjoint subsets of $\mathcal{M}$.

The geometrical properties of $\mathcal{M}$ determine the characteristic features of each special model. The substructural kinetic energy (if any) induces a metric on $\mathcal{M}$; though the metric may be used to induce a connection, such a connection may be physically significant or not. Only the existence on $\mathcal{M}$ of a connection with physical meaning allows one to represent substructural interactions through appropriate distinct quantities termed conventionally *microstress* and *self-force*. The alternative must be spelled out for the correct representation of the interactions. (See Capriz, 1985, 1989, 2001 for a general framework for multifield theories; see also Mariano, 2001 for additional results and a detailed list of references.) Here we restrict developments to cases where a physically significant connection on $\mathcal{M}$ exists and $\nabla\nu$ has covariant meaning. In what follows $\nu$ is assumed to be smooth on $\mathcal{B}/\cup\Sigma_k$.

Motions are represented through time-parametrized families of fields $\mathbf{x}_t$ and $\nu_t$ such that the position of the point $\mathbf{X} \in \mathcal{B}$ at the instant $t \in [0, d] \subset \mathbb{R}$ and the relevant value of the order parameter are given by

$$\mathbf{x}_t(\mathbf{X}) = \mathbf{x}(\mathbf{X}, t), \qquad \nu_t(\mathbf{X}) = \nu(\mathbf{X}, t), \tag{19}$$

where $\mathbf{x}_t$ and $\nu_t$ are considered to be at least twice differentiable with respect to time. Thus velocities at $\mathbf{X}$ and $t$ are given by

$$\dot{\mathbf{x}}(\mathbf{X}, t), \qquad \dot{\nu}(\mathbf{X}, t), \tag{20}$$

and for convenience we will indicate in the following with Vel the set of all possible pairs of fields $(\dot{\mathbf{x}}, \dot{\nu})$ at $(\mathbf{x}, \nu)$.

Let us consider a time-parametrized family of proper orthogonal tensors $\mathbf{Q}(t)$, with $\mathbf{Q} \in SO(3)$ at each $t$, and the associated family of corresponding vectors $\mathbf{q}(t)$ such that $\mathbf{Q}(t) = \exp(\mathsf{e}\mathbf{q}(t))$, $\mathsf{e}$ being Ricci's alternator.

If we take $\mathbf{Q}(t)$ as the descriptor of a rigid body motion, the rigid velocity field $\dot{\mathbf{x}}_R$ associated with the placement field is given by

$$\dot{\mathbf{x}}_R = \mathbf{c}(t) + \dot{\mathbf{q}} \times (\mathbf{x} - \mathbf{x}_0), \tag{21}$$

where $\mathbf{c}(t)$ is the *translation* velocity, $\mathbf{x}_0$ is some fixed point, and $\dot{\mathbf{q}}(t)$ is the rotational velocity.

$\nu$ may sense a change of observer. After a rigid change of its frame described by $\mathbf{Q}(t)$ an external observer measures a new value $\nu_{\mathbf{q}}$ of the order parameter related to the value $\nu$ evaluated before the change of frame by

$$\nu_{\mathbf{q}} = \nu + \mathcal{A}\mathbf{q} + o(|\mathbf{q}|), \qquad \mathcal{A} = \frac{d\nu_{\mathbf{q}}}{d\mathbf{q}}|_{\mathbf{q}=0}, \tag{22}$$

where $\mathcal{A}(\nu)$, an operator mapping vectors of $\mathbb{R}^3$ into elements of the tangent space of $T_\nu\mathcal{M}$, is the *infinitesimal generator of the action of* SO(3) on $\mathcal{M}$ generated by the element $\dot{\mathbf{q}}\,\times$ of the Lie algebra $\mathfrak{so}(3)$:

$$\dot{\nu}_{\mathbf{q}} = \mathcal{A}\dot{\mathbf{q}}. \tag{23}$$

Two different observers moving independently read two different pairs of velocity fields, $(\dot{\mathbf{x}}, \dot{\nu})$ and $(\dot{\mathbf{x}}^*, \dot{\nu}^*)$, respectively,

$$\dot{\mathbf{x}}^* = \dot{\mathbf{x}} + \mathbf{c}(t) + \dot{\mathbf{q}} \times (\mathbf{x} - \mathbf{x}_0), \tag{24}$$

$$\dot{\nu}^* = \dot{\nu} + \mathcal{A}\dot{\mathbf{q}} \tag{25}$$

if $\mathbf{c}$ and $\dot{\mathbf{q}}$ characterize the relative motion.

### 10.3.2. Power and balance of standard and substructural interactions in the bulk

Balance equations for substructural interactions are proposed and justified variously. Whereas in standard continua we know how the requirements of balance of forces and moments may be written directly and our imagination is sufficiently far-reaching to cover, without qualms, even rather unusual situations, when nonstandard interactions intervene, as in the presence of substructures, the path to the goal is much stonier. Then a smoother bypass begins by associating a measure of interaction to the time rate of each descriptor of the placement of the body and thus by coming to an educated guess of the virtual power of all external actions on any subbody $\mathcal{P}_{\mathfrak{b}}^{\text{ext}}$ and then proceeds by requiring its invariance under changes of observer (not only those due to the changes of frame in Euclidean space but also those, physically permissible, due to "change of frame" on $\mathcal{M}$).

Stretching analogies (fully conscious of limits of validity of our proposal and of its possible falsifiability in border cases), we take (omitting from now on the dependence on $\mathbf{X}$)

$$\mathcal{P}^{\text{ext}}_{\mathfrak{b}} = \int_{\mathfrak{b}} (\rho \mathbf{f} \cdot \dot{\mathbf{x}} + \rho \beta \cdot \dot{\nu}) + \int_{\partial \mathfrak{b}} (\mathbf{Tm} \cdot \dot{\mathbf{x}} + \mathcal{S}\mathbf{m} \cdot \dot{\nu}), \tag{26}$$

where $\beta$ represents *external bulk interactions* acting on the substructure per unit mass and $\mathcal{S}$ is the *microstress*, and we set down the following fundamental axiom of invariance of power.

**Axiom.** *$\mathcal{P}^{\text{ext}}_{\mathfrak{b}}$ is invariant under any change of observer, in primis under the changes ruled by* (24) *and* (25).

To satisfy this foremost condition it is necessary and sufficient that, for all $\mathbf{c}$ and $\dot{\mathbf{q}}$ and for all choices of the subbody $\mathfrak{b}$,

$$\begin{aligned} &\mathbf{c} \cdot \left( \int_{\mathfrak{b}} \rho \mathbf{f} + \int_{\partial \mathfrak{b}} \mathbf{Tm} \right) \\ &\quad + \dot{\mathbf{q}} \cdot \left( \int_{\mathfrak{b}} ((\mathbf{x} - \mathbf{x}_0) \times \rho \mathbf{f} + \mathcal{A}^T \rho \beta) + \int_{\partial \mathfrak{b}} ((\mathbf{x} - \mathbf{x}_0) \times \mathbf{Tm} + \mathcal{A}^T \mathcal{S}\mathbf{m}) \right) \\ &\quad = 0. \end{aligned} \tag{27}$$

The arbitrariness of $\mathbf{c}$ and $\dot{\mathbf{q}}$ implies the integral balances of forces and moments

$$\int_{\mathfrak{b}} \rho \mathbf{f} + \int_{\partial \mathfrak{b}} \mathbf{Tm} = 0, \tag{28}$$

$$\int_{\mathfrak{b}} ((\mathbf{x} - \mathbf{x}_0) \times \rho \mathbf{f} + \mathcal{A}^T \rho \beta) + \int_{\partial \mathfrak{b}} ((\mathbf{x} - \mathbf{x}_0) \times \mathbf{Tm} + \mathcal{A}^T \mathcal{S}\mathbf{m}) = 0. \tag{29}$$

From (28), since $\mathfrak{b}$ is arbitrary, the standard Cauchy equation (6) follows, whereas (7) is substituted by the specification of a contribution of the substructural actions to the skew part of Cauchy's stress

$$\mathcal{A}^T \rho \beta = \mathsf{e}\mathbf{T}\mathbf{F}^T - \operatorname{Div}(\mathcal{A}^T \mathcal{S}), \tag{30}$$

where $\mathsf{e}$ is Ricci's permutation tensor.

A deeper interpretation of (30) leads to the introduction of the self-force. By developing the divergence of $\mathcal{A}^T \mathcal{S}$, we see that (30) requires $(\nabla \mathcal{A}^T)\mathcal{S} - \mathsf{e}\mathbf{T}\mathbf{F}^T$ to belong to the range of $\mathcal{A}^T$: There exists an element $\zeta$ of the cotangent space $T^*_\nu \mathcal{M}$ to $\mathcal{M}$ at $\nu$ such that

$$\mathcal{A}^T \zeta = \mathsf{e}\mathbf{T}\mathbf{F}^T - (\nabla \mathcal{A}^T)\mathcal{S} \tag{31}$$

and the following differential inclusion applies:

$$\rho \beta - \zeta + \operatorname{Div} \mathcal{S} \in \text{null space of } \mathcal{A}^T. \tag{32}$$

Thus $\zeta$ is determined apart from a component along the null space of $\mathcal{A}^T$. Actually, there is a class of bodies for which the indetermination is obviously avoided. In fact, since

$$(\text{range of } \mathcal{A})^\perp = \text{null space of } \mathcal{A}^T \tag{33}$$

(when a notion of orthogonality is available and (range of $\mathcal{A})^{\perp}$ is the set of "vectors" of $T_{\nu}^{*}\mathcal{M}$ orthogonal to all vectors in (range of $\mathcal{A}$)), then whenever the range of $\mathcal{A}$ coincides with the whole tangent space of $\mathcal{M}$, the null space of $\mathcal{A}^{T}$ is reduced to the singleton $\{0\}$ and thus the inclusion (32) implies the equation

$$\rho\beta - \zeta + \operatorname{Div}\mathcal{S} = 0. \tag{34}$$

That conditions are not so trivially simple, in general, is shown by the following example. Take as order parameter a stretchable vector $\mathsf{d}$ so that $\mathcal{A}$ is the second-order tensor $\mathsf{ed}$; in such a case the tangent space comprises also vectors parallel to $\mathsf{d}$ (in pure stretching), whereas the range of $\mathcal{A}$ comprises only vectors normal to $\mathsf{d}$.

A way out of the impasse is suggested by the following remarks. Given coordinate systems in $\mathcal{E}^3$ and on $\mathcal{M}$, $\mathcal{A}$ is expressed by a $(3\times\dim\mathcal{M})$-matrix of characteristic $k$, say $(k \le 3, k \le \dim\mathcal{M})$ and the null space of $\mathcal{A}^{T}$ is a subspace of $T_{\nu}^{*}\mathcal{M}$ with dimension $\dim\mathcal{M} - k$.

Actually one can choose the coordinates on $\mathcal{M}$ so that any vector of $T_{\nu}^{*}\mathcal{M}$ with the last $\dim\mathcal{M} - k$ components null belongs to the null space of $\mathcal{A}^{T}$. Hence a vector $\dot{\nu}^{\alpha}$ of $T_{\nu}\mathcal{M}$ with the last $\dim\mathcal{M} - k$ components null gives rise to a virtual power that is observer independent, whichever choice is made for its first $k$ components. Now, requiring that the virtual power (26) be invariant under changes of $\dot{\nu}$ allowed by that arbitrariness, still large when $k > 0$, one eliminates the residual indetermination inherent in (32) and that inclusion collapses into the balance equations (34).

An illuminating example is provided by stretchable vectors mentioned above: Instead of the components of the vector $\mathsf{d}$, we suggest taking as coordinates two independent quantities: the unit vector $\mathsf{d}|\mathsf{d}|^{-1}$ and $|\mathsf{d}|$. For $\mathsf{d}|\mathsf{d}|^{-1}$ one follows developments as for nematic liquid crystals, where $\zeta$ is known only up to a vector parallel to $\mathsf{d}$. For the length $|\mathsf{d}|$ (or any strictly increasing function of that length, which is not affected by rotations), the dual variable is the stretching action otherwise undetermined.

### 10.3.3. Rods with substructure

As a preliminary, we find it expedient to analyze the equilibrium of rods with substructure because this seems to be an adequate model for our picture of the junction. To this end, consider a rod as the region scanned by translating a regular plane region $A$ along a regular simple curve $\mathbf{r}(\ell)$, with $\ell$ the arc-length parametrization $\ell \in [0, \bar{\ell}]$, while maintaining the center of $A$ on the curve and $A$ always orthogonal to the tangent $\mathsf{t}$. We indicate with $\mathbf{n}_B$ the normal at the bases $A(0)$ and $A(\ell)$ and with $\mathbf{n}_L$ the normal along the lateral boundary $\partial A \times [0, \bar{\ell}]$.

The theory of rods, in which a peculiarly thin three-dimensional body is the object, necessarily involves totals of actions over the body and on separate portions of the boundary; in particular, totals over cross-sections and first moments over those sections, too. Thus in the classical case, one is led to tractions $\mathbf{h}$ and bending and twisting moments (adding up to a vector $\mathsf{m}$) and to balance equations

$$\mathbf{h}_\ell + \bar{\mathbf{b}} = 0, \tag{35}$$

$$\mathsf{m}_\ell + \mathsf{t} \times \mathbf{h} + \mu = 0, \tag{36}$$

where $\bar{\mathbf{b}}$ is the external force per unit length (bulk action plus traction on the side boundary) and $\mu$ is the external moment per unit length.

When rods with substructure are object of study, the peculiarity of the representation of microactions through covariant vectors in the cotangent spaces is an obstacle to the evaluation of totals. One must face, first of all, the need of a deeper look at the properties of the manifold $\mathcal{M}$ embodying all values of the substructure. We are mainly concerned here with problems in statics; however, to justify the assumption that $\mathcal{M}$ be Riemannian, it is expedient to take heed of the kinetic energy associated with rates of change of the substructure, a quadratic form with coefficients that provide a metric for $\mathcal{M}$. Then a celebrated theorem of Nash (1954) assures us that a $C^1$-isometric embedding of the Riemannian manifold $\mathcal{M}$ into an Euclidean space $\mathfrak{L}$ of dimension $2\dim\mathcal{M}+1$ exists; some delicate points arise in connection with that embedding, in particular, the question of its possible insufficient regularity for all developments that follow below. Using Euclidean spaces of higher dimension, Nash has proved that $C^k$-smoothness ($k > 1$) can be achieved (Nash, 1956). However, optimistically we brush aside these preoccupations and proceed as though no further obstacles intervene. The widening of the dimension, the nonuniqueness of the embedding, the physical meaning of the possible lack of smoothness, all must be placed against experimental evidence case by case, and for each, hopefully, a justification must be found. Convenience and simplicity may push one in accepting less regular embeddings in lower-dimensional Euclidean space; one must be aware, however, that one may thus forecast transitions less smooth than occur in reality.

We will work below using the same notation for variables and fields as already introduced, as though we were still on $\mathcal{M}$ (now a hypersurface in a convenient $\mathfrak{L}$) rather than in $\mathfrak{L}$, though, as a modest reminder, we use bold letters to denote the extended versions of the same quantities (exception is made for $\mathcal{S}$ with a slight abuse of notation). Perhaps within the domain of each separate phase the values fall only on $\mathcal{M}$, but to discuss transitions and give values to phase differences, we must allow ourselves to exit from $\mathcal{M}$ and move freely in $\mathfrak{L}$. Such is the case, for instance, of the so-called chevron transition in nematic liquid crystals: Even if we consider such transitions sudden (i.e., first order), the jump is represented by an isotropic state (not on $\mathcal{M}$, now the manifold of directions). If we take the transition to be through a thin layer (and smooth within it, i.e., second order), it can be described only in the embedding space (of symmetric, traceless tensors).

Another totally different reason for widening the dimension of the spaces where states and interactions are placed is the possible wrinkling of the interface. Then, a convenient mathematical model may consist in the smooth surface obtained through an "ironing out" the wrinkles, on which added variables (e.g., a roughness tensor) account for the wrinkles; the latter, even in a classical elastic solid for instance, allow the intrusion of distributed torques otherwise obviously absent. However, a general study of such a class of interfaces seems to be missing as yet.

We denote by $\bar{\beta}$ the interactions per unit length on the substructure which are the sum of the volume actions due to $\rho\beta$ per unit volume and the surface actions per unit area on the lateral skin measured by $\mathcal{S}\mathbf{n}_L$ and by $\mathfrak{p}$ the internal force per unit length. We also set

$$\mathfrak{s} = \int_A \mathcal{S}\mathbf{n}_B, \qquad \bar{\mathsf{m}} = \int_A \mathcal{A}^T \mathcal{S}\mathbf{n}_B. \tag{37}$$

The second-order tensor $\mathcal{A}^T \mathcal{S}$ applied to $\mathbf{n}_B$ gives rise to a standard couple. Using the same notation as before, we return to the integral balances (28) and (29) applied to a generic cutting of the rod ($\ell \in [\ell_1, \ell_2] \subseteq [0, \bar{\ell}]$) and accordingly write the integral balances of forces and couples, respectively, as

$$\mathbf{h}_2 - \mathbf{h}_1 + \int_{\ell_1}^{\ell_2} \bar{\mathbf{b}} = 0, \tag{38}$$

$$(\mathbf{x}(\ell_2) - \mathbf{x}_0) \times \mathbf{h}_2 - (\mathbf{x}(\ell_1) - \mathbf{x}_0) \times \mathbf{h}_1 + \mathsf{m}_2 - \mathsf{m}_1$$
$$+ \int_{\ell_1}^{\ell_2} ((\mathbf{x} - \mathbf{x}_0) \times \bar{\mathbf{b}} + \mu) + \mathcal{A}^T \mathfrak{s}_2 - \mathcal{A}^T \mathfrak{s}_1 + \bar{\mathsf{m}}_2 - \bar{\mathsf{m}}_1 + \int_{\ell_1}^{\ell_2} \mathcal{A}^T \bar{\beta} = 0. \tag{39}$$

The arbitrariness of $[\ell_1, \ell_2]$ implies both (35) and

$$\mathsf{t} \times \mathbf{h} + \mu + (\mathsf{m} + \bar{\mathsf{m}})_{,\ell} + \mathcal{A}^T_{,\ell}\mathfrak{s} + \mathcal{A}^T (\mathfrak{s}_{,\ell} + \bar{\beta}) = 0. \tag{40}$$

We then see the need of the existence of a self-force $\mathfrak{p}'$ such that

$$\mathcal{A}^T \mathfrak{p}' = -\mathsf{t} \times \xi - \mu - \mathcal{A}^T_{,\ell}\mathfrak{s} - (\mathsf{m} + \bar{\mathsf{m}})_{,\ell}. \tag{41}$$

This allows us to write (40) as

$$\mathfrak{s}_{,\ell} + \bar{\rho}\bar{\beta} - \mathfrak{p} = 0 \tag{42}$$

under the condition that

$$\mathfrak{p} = \mathfrak{p}' + \mathfrak{p}'' \quad \text{and} \quad \mathcal{A}^T \mathfrak{p}'' = 0. \tag{43}$$

The indetermination in (43) may be eliminated by reasoning akin to that suggested at the end of Section 10.3.2. However, if the matter had been already investigated in three dimensions, then the integration over the cylindrical body of (34) would settle the question. Actually, complications may arise if wrinkling were present, as mentioned above; then $\mathfrak{p}$ may not be simply linked to $\zeta$. In conclusion, on a rod, the material substructure generates basically an increment $\bar{\mathsf{m}}$ of the standard torque. The standard equation (36) is evidently a special case that one reaches in the absence of substructure.

Analogies and differences with the theory of filiform bodies with substructure (Capriz and Cohen, 1988) are evident: The basic difference is the presence here of the term $(\mathsf{m}+\bar{\mathsf{m}})_{,\ell}$ collecting the standard moment $\mathsf{m}$ and the extra standard moment $\bar{\mathsf{m}}$ generated by the microstress (together with the applied couples $\mu$).

### 10.3.4. Membranes with substructure

Let $\bar{\Sigma}$ be a compact regular surface embedded in the three-dimensional Euclidean space $\mathcal{E}^3$ given by

$$\bar{\Sigma} = \{\mathbf{X} \in \mathcal{E}^3 \text{ s.t. } \bar{f}(\mathbf{X}) = 0\} \tag{44}$$

with $\bar{f}(\cdot)$ a smooth function in $\mathcal{E}^3$. $\bar{\Sigma}$ is oriented by the normal vector field $\mathbf{n} = \nabla \bar{f}/|\nabla \bar{f}|$ and has a regular boundary $\partial\bar{\Sigma}$ endowed with normal $\bar{\mathbf{n}}$ belonging to the tangent plane of $\bar{\Sigma}$ at $\mathbf{X}$.

The surface $\bar{\Sigma}$ is a natural geometric model of a material membrane, one presumes the existence of a surface stress $\mathsf{T}$ (say, a Piola–Kirchhoff surface stress) such that $\mathsf{T}\mathbf{n} = 0$ and thus associates tractions tangential to $\bar{\Sigma}$ with tangent vectors on $\bar{\Sigma}$.

If $\tilde{\mathbf{b}}$ is the *body forces per unit area* and $\tilde{\rho}$ the *density of mass per unit area*, the balance of standard forces becomes

$$\mathrm{Div}_{\bar{\Sigma}}\, \mathsf{T} + \tilde{\rho}\tilde{\mathbf{b}} = 0 \quad \text{on } \bar{\Sigma}, \tag{45}$$

while the balance of torques implies

$$\mathsf{T}\mathsf{F}^T = \mathsf{F}\mathsf{T}^T, \tag{46}$$

where $\mathsf{F}(\mathbf{X})$ is the surface deformation gradient $\nabla_{\bar{\Sigma}}\mathbf{x}(\mathbf{X})$ as in Section 10.2.

To consider membranes with substructure, we first associate an element of $\mathcal{M}$ with each point of $\bar{\Sigma}$, then consider in addition to $\mathsf{T}$ a surface microstress $\mathbb{S}$ such that $\mathbb{S}\mathbf{n} = 0$. $\mathbb{S}$ associates elements of the cotangent space of $\mathcal{M}$ with tangent vectors on $\bar{\Sigma}$.

We indicate with $\tilde{\beta}$ the *body interactions per unit area* on the substructure. We consider also an arbitrary subsurface $\bar{\Sigma}^*$ of $\bar{\Sigma}$ and write the balances of forces and torques on $\bar{\Sigma}^*$ as follows:

$$\int_{\bar{\Sigma}^*} \tilde{\rho}\tilde{\mathbf{b}} + \int_{\partial\bar{\Sigma}^*} \mathsf{T}\bar{\mathbf{n}} = 0, \tag{47}$$

$$\int_{\bar{\Sigma}^*} ((\mathbf{x} - \mathbf{x}_0) \times \tilde{\rho}\tilde{\mathbf{b}} + \tilde{\rho}\mathcal{A}^T\tilde{\beta}) + \int_{\partial\bar{\Sigma}^*} ((\mathbf{x} - \mathbf{x}_0) \times \mathsf{T}\bar{\mathbf{n}} + \mathcal{A}^T\mathbb{S}\bar{\mathbf{n}}) = 0. \tag{48}$$

By (47), the arbitrariness of $\bar{\Sigma}^*$ again implies the pointwise balance (45), while by (47) we obtain

$$\mathsf{e}\mathsf{T}\mathsf{F}^T - (\nabla_{\bar{\Sigma}}\mathcal{A}^T)\mathbb{S} - \mathcal{A}^T(\mathrm{Div}_{\bar{\Sigma}}\, \mathbb{S} + \tilde{\beta}) = 0, \tag{49}$$

which suggests the existence of a self-force $\mathfrak{z}'$ such that

$$\mathcal{A}^T\mathfrak{z}' = \mathsf{e}\mathsf{T}\mathsf{F}^T - (\nabla_{\bar{\Sigma}}\mathcal{A}^T)\mathbb{S}. \tag{50}$$

Consequently, (49) becomes

$$\mathrm{Div}_{\bar{\Sigma}}\, \mathbb{S} + \tilde{\rho}\tilde{\beta} - \mathfrak{z} = 0 \tag{51}$$

with

$$\mathfrak{z} = \mathfrak{z}' + \mathfrak{z}'' \quad \text{and} \quad \mathcal{A}^T\mathfrak{z}'' = 0. \tag{52}$$

Remarks at the end of Section 10.3.2 apply and allow us to recognize the special nature of $\mathfrak{z}''$ in each relevant case. Of course, had we already decided how to avoid the indetermination in three dimensional bodies, we would only need to obtain again the final balance equation by integration of (34).

### 10.3.5. Balance of standard and substructural interactions at a discontinuity surface

We turn now to the topics discussed in Section 10.3.2 and deduce the balance equations at each discontinuity surface taking into account the results in Section 10.3.4. We do not consider also surface body interactions like $\tilde{\mathbf{b}}$ and $\tilde{\beta}$, but assume the order parameter field continuous across the interface.

Let $\mathfrak{b}_\Sigma$ be an arbitrary subbody crossing one interface $\Sigma$ and far from the others. The subbody $\mathfrak{b}_\Sigma$ is considered as a "pillbox" around $\Sigma$: It is such that the intersection between the boundary $\partial\mathfrak{b}_\Sigma$ of $\mathfrak{b}_\Sigma$ and $\Sigma$ is a curve $\partial\mathfrak{b}_\Sigma \cap \Sigma$.

Taking into account the developments in Section 10.3.4, the integral balances of forces (28) and moments (29) become on $\mathfrak{b}_\Sigma$, respectively,

$$\int_{\mathfrak{b}_\Sigma} \rho\mathbf{f} + \int_{\partial\mathfrak{b}_\Sigma} \mathbf{Tm} + \int_{\partial\mathfrak{b}_\Sigma\cap\Sigma} \mathsf{T}\bar{\mathbf{n}} + \int_{\mathfrak{b}_\Sigma\cap\Sigma} \tilde{\mathbf{b}} = 0, \quad (53)$$

$$\int_{\mathfrak{b}_\Sigma} ((\mathbf{x}-\mathbf{x}_0)\times\rho\mathbf{f} + \mathcal{A}^T\rho\beta) + \int_{\partial\mathfrak{b}_\Sigma} ((\mathbf{x}-\mathbf{x}_0)\times\mathbf{Tm} + \mathcal{A}^T\mathcal{S}\mathbf{m})$$
$$+ \int_{\partial\mathfrak{b}_\Sigma\cap\Sigma} ((\mathbf{x}-\mathbf{x}_0)\times\mathsf{T}\bar{\mathbf{n}} + \mathcal{A}^T\mathbb{S}\bar{\mathbf{n}}) + \int_{\mathfrak{b}_\Sigma\cap\Sigma} ((\mathbf{x}-\mathbf{x}_0)\times\tilde{\mathbf{b}} + \mathcal{A}^T\tilde{\beta}) = 0. \quad (54)$$

By shrinking $\mathfrak{b}_\Sigma$ at the interface, we obtain

$$\int_{\mathfrak{b}_\Sigma} \rho\mathbf{f} \to 0, \qquad \int_{\mathfrak{b}_\Sigma} ((\mathbf{x}-\mathbf{x}_0)\times\rho\mathbf{f} + \mathcal{A}^T\rho\beta) \to 0. \quad (55)$$

as $\mathfrak{b}_\Sigma \to \mathfrak{b}_\Sigma \cap \Sigma$, because the integrands are continuous on $\mathcal{B}$, and

$$\int_{\partial\mathfrak{b}_\Sigma} \mathbf{Tm} \to \int_{\mathfrak{b}_\Sigma\cap\Sigma} [\mathbf{T}]\mathbf{n}, \quad (56)$$

$$\int_{\partial\mathfrak{b}_\Sigma} ((\mathbf{x}-\mathbf{x}_0)\times\mathbf{Tm} + \mathcal{A}^T\mathcal{S}\mathbf{m}) \to \int_{\mathfrak{b}_\Sigma\cap\Sigma} (\mathbf{x}-\mathbf{x}_0)\times[\mathbf{T}]\mathbf{n} + \mathcal{A}^T[\mathcal{S}]\mathbf{n} \quad (57)$$

as $\mathfrak{b}_\Sigma \to \mathfrak{b}_\Sigma \cap \Sigma$. Thanks to the arbitrariness of $\mathfrak{b}_\Sigma$ and the use of the Gauss theorem, we obtain

$$\tilde{\mathbf{b}} + [\mathbf{T}]\mathbf{n} + \mathrm{Div}_\Sigma\,\mathsf{T} = 0 \quad \text{on } \Sigma \quad (58)$$

from (53) and

$$\mathcal{A}^T\tilde{\beta} + \mathcal{A}^T[\mathcal{S}]\mathbf{n} = \mathsf{eTF}^T - \mathrm{Div}_\Sigma(\mathcal{A}^T\mathbb{S}) \quad \text{on } \Sigma \quad (59)$$

from (54). By developing the divergence of $\mathcal{A}^T\mathbb{S}$, we note that (58) requires the term $(\nabla_\Sigma\mathcal{A}^T)\mathbb{S} - \mathsf{eTF}^T$ to belong to the null space of $\mathcal{A}^T$; i.e., there exists an element $\mathfrak{z}$ of $T_\nu^*\mathcal{M}$ at each $\nu$ such that

$$\mathcal{A}^T\mathfrak{z}' = \mathsf{eTF}^T - (\nabla_\Sigma\mathcal{A}^T)\mathbb{S} \quad \text{on } \Sigma. \quad (60)$$

Note that equation (60) indicates that the presence of material substructure renders nonsymmetric the surface Cauchy stress tensor as well as it renders nonsymmetric the standard Cauchy stress $\mathbf{TF}^T$, as stated by equation (31).

Previous conditions imply the following differential inclusion

$$\tilde{\beta} + [\mathcal{S}]\mathbf{n} - \mathfrak{z} + \mathrm{Div}_\Sigma \mathbb{S} \in \text{null space of } \mathcal{A}^T \quad \text{on } \Sigma, \tag{61}$$

which reduces to

$$\tilde{\beta} + [\mathcal{S}]\mathbf{n} - \mathfrak{z}' + \mathrm{Div}_\Sigma \mathbb{S} = -\mathfrak{z}'' \quad \text{on } \Sigma, \tag{62}$$

with

$$\mathfrak{z} = \mathfrak{z}' + \mathfrak{z}'' \quad \text{and} \quad \mathcal{A}^T \mathfrak{z}'' = 0, \tag{63}$$

an indetermination that can be avoided by a pillbox argument based directly on (33), with the usual proviso of the absence of wrinkling.

Even in this case, remarks of Section 10.3.2 apply and allow us to recognize in each special case the nature of $\mathfrak{z}''$. (The measures of surface substructural interactions, i.e., surface microstresses and surface self-forces, have been introduced in Mariano (2000, 2001), and the related balances have been deduced there.)

### 10.3.6. Balance of standard and substructural interactions at the junction

To obtain the balances of standard and substructural interactions at the junction, we consider the subbody $\mathfrak{b}_R$ described in Section 10.2.2, coinciding with a cylinder wrapped around the junction. As in Section 10.2.2, if we take an arbitrary point of the piece of the junction in $\mathfrak{b}_R$ and consider the section of $\mathfrak{b}_R$, obtained cutting $\mathfrak{b}_R$ with a plane orthogonal to the tangent at the junction at the same point, we find a disc $D_R$ of diameter $R$. (For geometrical details, see Section 10.2.2.) The piece of $\mathfrak{J}$ included in $\mathfrak{b}_R$ is a curve with arc-length parametrization $\mathsf{s} \in [\mathsf{s}_1, \mathsf{s}_2]$: the points $\mathbf{r}(\mathsf{s}_1)$ and $\mathbf{r}(\mathsf{s}_2)$ are the intersections of $\mathfrak{J}$ with the boundary $\partial\mathfrak{b}_R$ of $\mathfrak{b}_R$.

We do not consider here line body interactions $\bar{\beta}$ on the substructure at the junction and write for $\mathfrak{b}_R$ the integral balances of forces and torques as

$$\int_{\mathfrak{b}_R} \rho \mathbf{f} + \int_{\partial \mathfrak{b}_R} \mathbf{Tm} + \sum_{k=1}^{N} \int_{\partial \mathfrak{b}_R \cap \Sigma_k} \mathsf{T}_k \bar{\mathbf{n}}_k + \int_{\mathsf{s}_1}^{\mathsf{s}_2} \mathbf{h}_{\mathsf{s}} + \int_{\mathfrak{b}_\Sigma \cap \Sigma} \tilde{\mathbf{b}} = 0, \tag{64}$$

$$\begin{aligned}
&\int_{\mathfrak{b}_R} ((\mathbf{x} - \mathbf{x}_0) \times \rho \mathbf{f} + \mathcal{A}^T \rho \beta) + \int_{\partial \mathfrak{b}_R} ((\mathbf{x} - \mathbf{x}_0) \times \mathbf{Tm} + \mathcal{A}^T \mathcal{S}\mathbf{m}) \\
&+ \sum_{k=1}^{N} \int_{\partial \mathfrak{b}_R \cap \Sigma_k} ((\mathbf{x} - \mathbf{x}_0) \times \mathsf{T}_k \bar{\mathbf{n}}_k + \mathcal{A}^T \mathbb{S}_k \bar{\mathbf{n}}_k) \\
&+ (\mathsf{m} + \bar{\mathsf{m}})(\mathsf{s}_2) - (\mathsf{m} + \bar{\mathsf{m}})(\mathsf{s}_1) + \sum_{k=1}^{N} \int_{\mathfrak{b}_\Sigma \cap \Sigma} ((\mathbf{x} - \mathbf{x}_0) \times \tilde{b} + \mathcal{A}^T \tilde{\beta}) \\
&+ (\mathbf{x} - \mathbf{x}_0) \times \mathbf{h}(\mathsf{s}_2) - (\mathbf{x} - \mathbf{x}_0) \times \mathbf{h}(\mathsf{s}_1) + \mathcal{A}^T \mathfrak{s}|_{\mathbf{r}(\mathsf{s}_2)} - \mathcal{A}^T \mathfrak{s}|_{\mathbf{r}(\mathsf{s}_1)} = 0,
\end{aligned} \tag{65}$$

which are the counterparts of (28) and (29) and account for the results in Sections 10.3.3 and 10.3.4.

To obtain the integral balances of standard and substructural interactions at the junction, we take the limit $R \to 0$ in (64) and (65) by assuming not only the boundedness of (12) but also that

$$\int_{\partial \mathfrak{b}_R} |\mathcal{S}\mathbf{m}| \tag{66}$$

is bounded as $R \to 0$. This implies that

$$\lim_{R\to 0} \int_{\partial \mathfrak{b}_R} \mathcal{S}\mathbf{m} = \int_{\mathsf{s}_1}^{\mathsf{s}_2} \lim_{R\to 0} \int_{\partial D_R} \mathcal{S}\mathbf{m} = 0. \tag{67}$$

As a consequence of (12), we find (as in Section 10.2) that (64) reduces to

$$\int_{\mathsf{s}_1}^{\mathsf{s}_2} \left( \mathbf{h}_{\mathsf{s}} + \sum_{k=1}^{N} \mathsf{T}_k \bar{\mathbf{n}}_k \right) = 0 \tag{68}$$

as $R \to 0$, which is the *integral balance of forces at the junction*. The arbitrariness of the interval $[\mathsf{s}_1, \mathsf{s}_2]$ leads to the *pointwise balance* (14), which we rewrite here as

$$\mathbf{h}_{,\mathsf{s}} + \sum_{k=1}^{N} \mathsf{T}_k \bar{\mathbf{n}}_k = 0 \quad \text{along } \mathfrak{J}. \tag{69}$$

We apply the same procedure to (65) and, thanks to (68), obtain the integral balance

$$\int_{\mathsf{s}_1}^{\mathsf{s}_2} \left( \mathsf{t} \times \mathbf{h} + (\mathsf{m} + \bar{\mathsf{m}})_{,\mathsf{s}} + (\mathcal{A}^T \mathfrak{s})_{,\mathsf{s}} + \mathcal{A}^T \sum_{k=1}^{N} \mathbb{S}_k \bar{\mathbf{n}}_k \right) = 0. \tag{70}$$

The arbitrariness of the interval $[\mathsf{s}_1, \mathsf{s}_2]$ implies

$$\mathsf{t} \times \mathbf{h} + (\mathsf{m} + \bar{\mathsf{m}})_{,\mathsf{s}} + \mathcal{A}^T_{,\mathsf{s}} \mathfrak{s} + \mathcal{A}^T \mathfrak{s}_{,\mathsf{s}} + \mathcal{A}^T \sum_{k=1}^{N} \mathbb{S}_k \bar{\mathbf{n}}_k = 0, \tag{71}$$

which suggests the existence of a self-force $\mathfrak{p}$ such that

$$-\mathcal{A}^T \mathfrak{p} = \mathsf{t} \times \mathbf{h} + (\mathsf{m} + \bar{\mathsf{m}})_{,\mathsf{s}} + \mathcal{A}^T_{,\mathsf{s}} \mathfrak{s}. \tag{72}$$

This allows one to write (72) as

$$\mathcal{A}^T \left( \mathfrak{s}_{,\mathsf{s}} - \mathfrak{p} + \sum_{k=1}^{N} \mathbb{S}_k \bar{\mathbf{n}}_k \right) = 0, \tag{73}$$

which implies the differential inclusion

$$\mathfrak{s}_{,\mathsf{s}} - \mathfrak{p} + \sum_{k=1}^{N} \mathbb{S}_k \bar{\mathbf{n}}_k \in \text{null space of } \mathcal{A}^T. \tag{74}$$

With the help of the developments in Sections 10.3.2 and 10.3.5, we see that *in the case in which $\mathcal{A}$ covers at each $\nu$ the whole tangent space of $\mathcal{M}$*, the differential inclusion (74) reduces to

$$\mathfrak{s}_{,\mathrm{s}} - \mathfrak{p} + \sum_{k=1}^{N} \mathbb{S}_k \bar{\mathbf{n}}_k = 0 \quad \text{along } \mathfrak{J}. \tag{75}$$

When $\mathcal{A}(\nu)$ does not cover the whole tangent space of $\mathcal{M}$ at $\nu$, (74) implies that

$$\mathfrak{s}_{,\mathrm{s}} - \mathfrak{p} + \sum_{k=1}^{N} \mathbb{S}_k \bar{\mathbf{n}}_k = 0 \quad \text{along } \mathfrak{J}, \tag{76}$$

where

$$\mathfrak{p} = \mathfrak{p}' + \mathfrak{p}'' \quad \text{and} \quad \mathcal{A}^T \mathfrak{p}'' = 0. \tag{77}$$

Had we argued directly on the basis of (62), the indetermination would not have appeared.

## 10.4. Constitutive restrictions

All measures of interaction introduced so far are state functions; for an ample class of bodies, they may be related to the expression of their free energy. It has become fashionable for this purpose to exploit a mechanical version of the second law of thermodynamics. We are aware of the disdain accruing from the overwhelming resort to developments à la Coleman and Noll (1963) and of the disputes attending thereof (to the point of accusing those authors of having revived a long-discredited phlogiston conjecture). We realize also that a satisfactory and general thermodynamics of continua with substructure is not yet available. However, the corollaries we draw are simple and elegant and may even find alternative justifications.

The version of the second law that we exploit enjoins that for a subbody $\mathfrak{b}_R$,

$$\frac{d}{dt}\{\text{free energy of } \mathfrak{b}_R\} - \mathcal{P}^{\text{ext}}_{\mathfrak{b}_R} \leq 0. \tag{78}$$

We consider *bulk* ($\psi$), *surface* ($\phi$), and *junction* ($\omega$) free energy densities such in a way that the free energy of a subbody $\mathfrak{b}_R$ around the junction is given by

$$\{\text{free energy of } \mathfrak{b}_R\} = \int_{\mathfrak{b}_R} \psi + \sum_{k=1}^{N} \int_{\mathfrak{b}_R \cap \Sigma_k} \phi_k + \int_{\mathrm{s}_1}^{\mathrm{s}_2} \omega. \tag{79}$$

Once explicit expressions of the free energy densities have been selected, we can derive restrictions on possible constitutive expressions for the measures of interaction.

First we consider a subbody $\mathfrak{b}$ far from the interfaces. With reference to $\mathfrak{b}$, we write the mechanical dissipation inequality as

$$\frac{d}{dt}\int_{\mathfrak{b}} \psi - \int_{\mathfrak{b}} (\rho \mathbf{f} \cdot \dot{x} + \rho \beta \cdot \dot{\nu}) - \int_{\partial \mathfrak{b}} (\mathbf{Tm} \cdot \dot{x} + \mathcal{S}\mathbf{m} \cdot \dot{\nu}) \leq 0. \tag{80}$$

By using the Gauss theorem on the last integral in (80) and taking into account the balances (6) and (34), we may rewrite (80) as

$$\frac{d}{dt}\int_{\mathfrak{b}}\psi - \int_{\mathfrak{b}}(\mathbf{T}\cdot\dot{F} + \zeta\cdot\dot{\nu} + \mathcal{S}\cdot\nabla\dot{\nu}) \leq 0. \tag{81}$$

We consider a constitutive structure of the bulk free energy $\psi$ of the form

$$\psi = \hat{\psi}(\mathbf{F}, \nu, \nabla\nu), \tag{82}$$

and presume that the measures of interaction depend on the same objects entering the constitutive list of $\psi$ only. By developing the time derivative in (81), we obtain

$$\int_{\mathfrak{b}}((\partial_{\mathbf{F}}\psi - \mathbf{T})\cdot\dot{\mathbf{F}} + (\partial_{\nu}\psi - \zeta)\cdot\dot{\nu} + (\partial_{\nabla\nu}\psi - \mathcal{S})\cdot\nabla\dot{\nu}) \leq 0. \tag{83}$$

In principle, given any state $(\mathbf{F}, \nu, \nabla\nu)$ of the material elements in the bulk, we may choose arbitrarily velocity fields $(\dot{\mathbf{F}}, \dot{\nu}, \nabla\dot{\nu})$ from $(\mathbf{F}, \nu, \nabla\nu)$. Consequently, as $\mathfrak{b}$ is arbitrary (but taken away the discontinuity surfaces and the junctions) and (83) is linear in the velocity fields, to satisfy the inequality it is necessary that

$$\mathbf{T} = \partial_{\mathbf{F}}\psi(\mathbf{F}, \nu, \nabla\nu), \tag{84}$$

$$\zeta = \partial_{\nu}\psi(\mathbf{F}, \nu, \nabla\nu), \tag{85}$$

$$\mathcal{S} = \partial_{\nabla\nu}\psi(\mathbf{F}, \nu, \nabla\nu). \tag{86}$$

For subbodies $\mathfrak{b}_\Sigma$ around the $k$th discontinuity surface, like the "pillbox" adopted in Section 10.3.5, we write the mechanical dissipation inequality as

$$\begin{aligned}\frac{d}{dt}\left(\int_{\mathfrak{b}_\Sigma}\psi + \int_{\mathfrak{b}_\Sigma\cap\Sigma}\phi\right) - \int_{\mathfrak{b}_\Sigma}(\rho\mathbf{f}\cdot\dot{\mathbf{x}} + \rho\beta\cdot\dot{\nu}) - \int_{\partial\mathfrak{b}_\Sigma}(\mathbf{Tm}\cdot\dot{\mathbf{x}} + \mathcal{S}\mathbf{m}\cdot\dot{\nu})\\ - \int_{\partial\mathfrak{b}_\Sigma\cap\Sigma}(\mathbb{T}\bar{\mathbf{n}}\cdot\dot{x}^{\pm} + \mathbb{S}\bar{\mathbf{n}}\cdot\dot{\nu}^{\pm}) - \int_{\mathfrak{b}_\Sigma\cap\Sigma}(\tilde{\mathbf{b}}\cdot\dot{x}^{\pm} + \tilde{\beta}\cdot\dot{\nu}^{\pm}) \leq 0.\end{aligned} \tag{87}$$

If we adopt the constitutive relation (82) and a constitutive structure for the surface free energy of the form

$$\phi = \hat{\phi}(\mathsf{F}, \nu^{\pm}, \mathsf{N}), \tag{88}$$

where $\mathsf{N} = \nabla_\Sigma\nu = \langle\nabla\nu\rangle(\mathbf{I} - \mathbf{n}\otimes\mathbf{n})$, and presume that the surface measures of interactions depend only on the entries of $\phi$, by developing the time derivatives of $\psi$ and $\phi$ and using (84)–(86), we reduce the inequality (87) to

$$\begin{aligned}\int_{\mathfrak{b}_\Sigma\cap\Sigma}((\partial_{\mathsf{F}}\phi - \mathbb{T})\cdot\dot{\mathsf{F}} + (\partial_{\nu^{\pm}}\phi - \mathfrak{z})\cdot\dot{\nu}^{\pm} + (\partial_{\mathsf{N}}\phi - \mathbb{S})\cdot\dot{\mathsf{N}})\\ - \int_{\mathfrak{b}_\Sigma\cap\Sigma}(\langle\mathbf{T}\rangle\mathbf{n}\cdot[\dot{\mathbf{x}}] + \langle\mathcal{S}\rangle\mathbf{n}\cdot[\dot{\nu}]) \leq 0.\end{aligned} \tag{89}$$

Even in this case, for any state $(\mathsf{F}, \nu^{\pm}, \mathsf{N})$ of any material element of the interface $\Sigma$, we may choose arbitrarily velocity fields $(\dot{\mathsf{F}}, \dot{\nu}^{\pm}, \dot{\mathsf{N}})$. Consequently, since (89) is linear in the rates $(\dot{\mathsf{F}}, \dot{\nu}^{\pm}, \dot{\mathsf{N}})$, thanks to the arbitrariness of $\mathfrak{b}_\Sigma$, it is necessary that for each $k$th discontinuity surface

$$\mathsf{T}_k = \partial_{\mathsf{F}_k} \phi_k(\mathsf{F}_k, \nu_k^{\pm}, \mathsf{N}_k), \tag{90}$$

$$\mathfrak{z}_k = \partial_{\nu} \phi_k(\mathsf{F}_k, \nu_k^{\pm}, \mathsf{N}_k), \tag{91}$$

$$\mathbb{S}_k = \partial_{\mathsf{N}_k} \phi_k(\mathsf{F}_k, \nu_k^{\pm}, \mathsf{N}_k). \tag{92}$$

Moreover, the arbitrariness of $\mathfrak{b}_\Sigma$ leads to the *local surface dissipation inequality*

$$\langle \mathbf{T} \rangle \mathbf{n} \cdot [\dot{\mathbf{x}}] + \langle \mathcal{S} \rangle \mathbf{n} \cdot [\dot{\nu}] \geq 0. \tag{93}$$

Note that, as $\nu^{\pm}$ is the average $\langle \nu \rangle$ around $\Sigma$, for $\mathcal{M}$ a nonlinear manifold, $\langle \nu \rangle$ would lose significance had we not proceeded to the embedding in a linear space. If $\nu$ is continuous across $\Sigma_k$, as assumed in previous sections, then $\nu^{\pm} = \nu$.

To obtain constitutive restrictions at the junction, we consider a subbody $\mathfrak{b}_R$ around $\mathfrak{J}$ (as we have described in Sections 10.2.2 and 10.3.6) and we write the mechanical dissipation inequality as

$$\begin{aligned} &\frac{d}{dt}\left(\int_{\mathfrak{b}_R} \psi + \sum_{k=1}^{N} \int_{\mathfrak{b}_R \cap \Sigma_k} \phi_k + \int_{\mathsf{s}_1}^{\mathsf{s}_2} \omega\right) - \int_{\mathfrak{b}_R} (\rho \mathbf{f} \cdot \dot{\mathbf{x}} + \rho \beta \cdot \dot{\nu}) \\ &- \int_{\partial \mathfrak{b}_R} (\mathbf{T}\mathbf{m} \cdot \dot{\mathbf{x}} + \mathcal{S}\mathbf{m} \cdot \dot{\nu}) - \sum_{k=1}^{N} \int_{\partial \mathfrak{b}_R \cap \Sigma_k} (\mathsf{T}_k \bar{\mathbf{n}}_k \cdot \dot{\mathbf{x}}_k^{\pm} + \mathbb{S}_k \bar{\mathbf{n}}_k \cdot \dot{\nu}_k^{\pm}) \\ &- (\mathbf{h} \cdot \dot{\mathbf{x}}(\mathsf{s}_2) - \mathbf{h} \cdot \dot{\mathbf{x}}(\mathsf{s}_1)) - ((\mathsf{m} + \bar{\mathsf{m}})(\mathsf{s}_2) \cdot \dot{\theta}(\mathsf{s}_2) - (\mathsf{m} + \bar{\mathsf{m}})(\mathsf{s}_1) \cdot \dot{\theta}(\mathsf{s}_1)) \\ &- \sum_{k=1}^{N} \int_{\mathfrak{b}_\Sigma \cap \Sigma} (\tilde{\mathbf{b}}_k \cdot \dot{\mathbf{x}}_k^{\pm} + \tilde{\beta}_k \cdot \dot{\nu}_k^{\pm}) - (\mathfrak{s} \cdot \dot{\nu}|_{\mathbf{r}(\mathsf{s}_2)} - \mathfrak{s} \cdot \dot{\nu}|_{\mathbf{r}(\mathsf{s}_1)}) \leq 0, \end{aligned} \tag{94}$$

where $\theta(\mathsf{s})$ is the rotation of the plane coincident with the plane orthogonal to $\mathfrak{J}$ at $\mathbf{r}(\mathsf{s})$ in the reference configuration.

If we use the balances in the bulk and at each discontinuity surface, together with the balance of standard and substructural interactions at the junction, namely equations (69) and (76), we may rewrite (94) as

$$\begin{aligned} &\frac{d}{dt}\left(\int_{\mathfrak{b}_R} \psi + \sum_{k=1}^{N} \int_{\mathfrak{b}_R \cap \Sigma_k} \phi_k + \int_{\mathsf{s}_1}^{\mathsf{s}_2} \omega\right) - \int_{\mathfrak{b}_\Sigma} (\mathbf{T} \cdot \dot{\mathbf{F}} + \zeta \cdot \dot{\nu} + \mathcal{S} \cdot \nabla \dot{\nu}) \\ &- \int_{\mathfrak{b}_\Sigma \cap \Sigma} (\langle \mathbf{T} \rangle \mathbf{n} \cdot [\dot{\mathbf{x}}] + \langle \mathcal{S} \rangle \mathbf{n} \cdot [\dot{\nu}]) - \int_{\mathfrak{b}_\Sigma \cap \Sigma} (\mathsf{T} \cdot \dot{\mathsf{F}}_k + \mathfrak{z} \cdot \dot{\nu}^{\pm} + \mathbb{S} \cdot \dot{\mathsf{N}}) \\ &- \int_{\mathsf{s}_1}^{\mathsf{s}_2} (\xi \cdot \dot{\lambda} + (\mathsf{m} + \bar{\mathsf{m}}) \cdot \dot{\kappa} + \mathfrak{p} \cdot \dot{\nu} + \mathfrak{s} \cdot \overline{(\nabla \nu)\mathsf{t}}^{\,\cdot}) \leq 0, \end{aligned} \tag{95}$$

where $\lambda = \mathbf{F}\mathsf{t}$ and $\dot{\kappa} = \theta_{,\mathsf{s}}$. We use the constitutive relations (82) and (88) together with

$$\omega = \hat{\omega}(\lambda, \kappa, \nu, (\nabla \nu)\mathsf{t}) \tag{96}$$

(presuming that all the line measures of interaction depend only on the entries of $\omega$), develop the time derivatives, and insert them in (95). The results (84)–(86) and (90)–(92) allow us to reduce the mechanical dissipation inequality (95) to

$$\int_{\mathsf{s}_1}^{\mathsf{s}_2} \Big( (\partial_\lambda \omega - \mathbf{h}) \cdot \dot{\lambda} + (\partial_\kappa \omega - (\mathsf{m} + \bar{\mathsf{m}})) \cdot \dot{\kappa} \\ + (\partial_\nu \omega - \mathfrak{p}) \cdot \dot{\nu} + (\partial_{(\nabla\nu)\mathsf{t}} \omega - \mathfrak{s}) \cdot \overline{(\nabla\nu)\mathsf{t}}^{\boldsymbol{\cdot}} \Big) \geq 0. \tag{97}$$

For any state $(\lambda, \kappa, \nu, (\nabla\nu)\mathsf{t})$ of a material element of the junction, we can choose rates $(\dot{\lambda}, \dot{\kappa}, \dot{\nu}, \overline{(\nabla\nu)\mathsf{t}}^{\boldsymbol{\cdot}})$ arbitrarily from $(\lambda, \kappa, \nu, (\nabla\nu)\mathsf{t})$. However, as (97) is linear in the rates involved and $[\mathsf{s}_1, \mathsf{s}_2]$ is arbitrary, the following constitutive restrictions follow:

$$\xi = \partial_\lambda \omega(\lambda, \kappa, \nu, (\nabla\nu)\mathsf{t}), \tag{98}$$

$$\mathfrak{m} = \partial_\kappa \omega(\lambda, \kappa, \nu, (\nabla\nu)\mathsf{t}), \tag{99}$$

$$\mathfrak{p} = \partial_\varphi \omega(\lambda, \kappa, \nu, (\nabla\nu)\mathsf{t}), \tag{100}$$

$$\mathfrak{s} = \partial_{(\nabla\nu)\mathsf{t}} \omega(\lambda, \kappa, \nu, (\nabla\nu)\mathsf{t}). \tag{101}$$

The relations (98)–(101) are peculiar results of the present paper, together with the balance of substructural interactions at the junction.

**Remark 1** (independence of the free energy from the rate of the state variables)**.** In principle, we can choose for the free energies expressions different from (82), (88), and (96). However, well-known developments would exclude such a dependence.

**Remark 2** (parabolic evolution due to dissipative phenomena)**.** The constitutive restrictions (84)–(86), (90)–(92), and (98)–(101) are peculiar of the thermodynamic equilibrium. Nonequilibrium parts of the measures of interaction, depending constitutively on the velocity fields, could be considered along thermodynamical processes of nonequilibrium. This happens, e.g., when viscosity phenomena occur and the stress $\mathbf{T}$ has the form

$$\mathbf{T} = \partial_{\mathbf{F}} \psi(\mathbf{F}, \nu, \nabla\nu) + \mathbf{T}^{ne}(\mathbf{F}, \nu, \nabla\nu; \dot{\mathbf{F}}). \tag{102}$$

It may even involve the rate of the order parameter and possibly of its gradient.

An interesting case occurs when one considers an expression analogous to (102) for the self-force $\zeta$, namely,

$$\zeta = \partial_\nu \psi(\mathbf{F}, \nu, \nabla\nu) + \mathbf{z}^{ne}(\mathbf{F}, \nu, \nabla\nu; \dot{\nu}), \tag{103}$$

and $\zeta^{ne}$ is such that

$$\zeta^{ne} \cdot \dot{\nu} \leq 0. \tag{104}$$

A solution of (104) is

$$\zeta^{ne} = \mathbf{A}\dot{\nu} \tag{105}$$

with **A** an appropriate definite negative tensor (possibly scalar in some special model) such that

$$\mathbf{A} = \hat{\mathbf{A}}(\mathbf{F}, \nu, \nabla\nu; \dot{\nu}). \tag{106}$$

In common special cases, decomposed free energy densities of the form

$$\psi = \hat{\psi}_1(\mathbf{F}, \nu) + \hat{\psi}_2(\nu, \nabla\nu) \tag{107}$$

(with $\hat{\psi}_1(\mathbf{I}, 0) = 0$, **I** being the unit tensor) may be selected and $\hat{\psi}_2(\nu, \nabla\nu)$ chosen as

$$\psi_2 = \frac{1}{2} b\nabla\nu \cdot \nabla\nu + \iota(\nu) \tag{108}$$

with $b$ some appropriate constant and $\iota(\nu)$ a coarse grained potential, possibly with double wells as in the case of solidification or solid-solid phase transitions (see, e.g., Penrose and Fife, 1990; Anderson et al., 2000; Nestler et al., 2000). When this happens, from (85) and (86) the bulk balance of substructural interactions becomes

$$\mathbf{A}\dot{\varphi} = b\Delta\nu - \partial_\nu\iota(\nu) - \partial_\nu\hat{\psi}_1(\mathbf{F}, \nu), \tag{109}$$

which is a generalized form of Ginzburg–Landau equation. When, in fact, both **A** and $\nu$ are scalars and the body does not undergo deformations, equation (109) reduces to

$$A\dot{\varphi} = b\Delta\varphi - \partial_\nu\iota(\nu), \tag{110}$$

which is the standard Ginzburg–Landau equation.

**Acknowledgments**

The support of the Italian National Group of Mathematical Physics (GNFM) and CNR is gratefully acknowledged.

## References

D. M. Anderson, G. B. McFadden, and A. A. Wheeler, A phase-field model of solidification with convection, *Phys.* D, **135** (2000), 175–194.

S. Angenent and M. E. Gurtin, Multiphase thermomechanics with interfacial structure II: Evolution of an isothermal interface, *Arch. Rational Mech. Anal.*, **108** (1989), 323–391.

G. Capriz, Continua with latent microstructure, *Arch. Rational Mech. Anal.*, **90** (1985), 43–56.

G. Capriz, *Continua with Microstructure*, Springer-Verlag, Berlin, 1989.

G. Capriz, Continua with substructure I and II, *Phys. Mesomech.*, (2001), 5–14 and 37–50.

G. Capriz and H. Cohen, Mechanics of filiform bodies, *Mech. Res. Comm.*, **15** (1988), 315–325.

B. D. Coleman and W. Noll, The thermodynamics of elastic materials with heat conduction and viscosity, *Arch. Rational Mech. Anal.*, **13** (1963), 245–261.

J. L. Ericksen, Liquid crystals with variable degree of orientation, *Arch. Rational Mech. Anal.*, **113** (1991), 97–120.

J. W. Gibbs, *Collected Works* I, Yale University Press, New Haven, CT, 1948, 289 (reprinted from *Trans. Connect. Acad. Arts Sci.*, 1874).

P. M. Mariano, Configurational forces in continua with microstructure, *Z. Angew. Math. Phys.*, **51** (2000), 752–791.

P. M. Mariano, Multifield theories in mechanics of solids, *Adv. Appl. Mech.*, **38** (2001), 1–93.

J. F. Nash, $C^1$ isometric imbeddings, *Ann. Math.*, **60** (1954), 383–396.

J. F. Nash, The imbedding problem for Riemannian manifold, *Ann. Math.*, **63** (1956), 20–63.

B. Nestler, A. A. Wheeler, L. Ratke, and C. Stöcker, Phase-field model for solidification of a monotectic alloy with convection, *Phys.* D, **141** (2000), 133–154.

O. Penrose and P. C. Fife, Thermodinamically consistent models of phase-field type for the kinetics of phase transitions, *Phys.* D, **43** (1990), 44–62.

N. K. Simha and K. Battacharya, Kinetics of multiphase boundaries with edges and junctions in a three-dimensional multiphase body, *J. Mech. Phys. Solids*, **48** (2000), 2619–2641.